Land, water and development

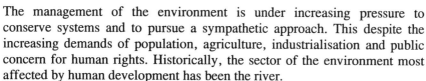

D1494998

The management of the environment is under increasing pressure to conserve systems and to pursue a sympathetic approach. This despite the increasing demands of population, agriculture, industrialisation and public concern for human rights. Historically, the sector of the environment most affected by human development has been the river.

In the second edition of *Land, Water and Development*, Malcolm Newson explores the meaning of sustainability in river basin development; the ecosystem model is made more explicit than in the first edition. The rapid evolution of practical concepts of sustainable river management since the 'Earth Summit' in 1992 is charted for a number of countries. New national case studies include those of Australia, South Africa, Israel and Iran. The links between land and water management are strengthened and it is concluded that catchment management planning, incorporating wide public consultation, is the best device by which river managers can influence sustainable development.

The book contains sections on water and sediment transfer systems, their management in the developed and developing worlds, technical issues (now including irrigation and river restoration), and river basin institutions and the way in which they use knowledge.

The book provides an integrating holistic treatment of a complex field. The first edition has been widely used as a textbook and professional guide and has helped underline the case for catchment management planning in several countries.

Malcolm Newson is Professor of Physical Geography at the University of Newcastle upon Tyne.

Frontispiece Flooding in the Thames valley near Oxford. Shows the interplay of natural and human functions of the floodplain.

Land, water and development

Sustainable management of river basin systems

Second edition

Malcolm Newson

London and New York

To Willow

First published 1992
by Routledge
11 New Fetter Lane, London EC4P 4EE

Reprinted in 1995
Second edition 1997

Simultaneously published in the USA and Canada
by Routledge
29 West 35th Street, New York, NY 10001

©1997 Malcolm Newson

Typeset in Times by
Mathematical Composition Setters Ltd, Salisbury Wiltshire
Printed and bound in Great Britain by
Biddles Ltd, Guildford and King's Lynn

British Library Cataloguing in Publication Data
A catalogue record for this book is available from the British Library

Library of Congress Cataloguing in Publication Data
Newson, Malcolm David.
    Land, water, and development: sustainable management of river
basin systems / Malcolm Newson. – [2nd ed.]
    Includes bibliographical references and index.
    1. Watershed management.   2. Water resources development.
3. Sustainable development.   I. Title.
TC409.N48   1997
333.91' 15–dc21   97–7592

ISBN 0-415-15506-1 (hbk)
ISBN 0-415-15507-X (pbk)

Contents

Plates

Figures

Tables

Preface to the first edition

'It stands to reason', said the farmer, 'we've only had these quick, high floods since the foresters ploughed those hills up there.'

This man's knowledge of, dependence on, and reaction to his local river made his reasoning easy. Yet to a government hydrologist, as was the author at the time of the conversation, proof of a link between preparing upland soils for successful afforestation and a change in the unit hydrograph for the basin would take a decade of expensive research. After its completion the logical outcome of the proven link, between land there and water here, i.e. modifying forestry practice, compensating the farmer or afforesting a less sensitive hillside, would not translate into public policy. There were simply no river-related land planning policies in many countries; the UK was no exception.

The outcome was that local authority engineers built the farmer a bridge over the newly flood-prone stream. Perhaps it is the heroic talent of the civil engineer to solve in this way point problems where they arise which has discouraged the 'look upstream' mentality of the local, the peasant, the river enthusiast. Societies have built dams, canals, flood walls, bridges and other structures to 'stabilise' river systems without questioning the cause of the instability. Rather like early technical medicine, we have used the equivalents of drugs and pain-killers to cure 'now' problems whereas some claim the true human talent lies with holism and the longer term.

This book tries to assemble a body of knowledge which supports a very broad approach to river problems; physically destabilised river systems will be a major theme but polluted and biologically sterilised systems are all amenable to 'the treatment'. At the other extreme, there is increasing demand to conserve, by management intervention, those relatively few pristine wilderness river systems which remain.

'The treatment' as a concept comes relatively easily to the geographer. Fluvial geomorphology, a major feature of geographical research in the second half of this century, has provided chapter and verse on the natural dynamics of the river basin, which function to transfer water and sediments to the ocean, leaving a characteristic morphology – river channel, floodplain, valley side – to form the basis of river habitat. While the engineer has applied knowledge of precise physical laws to the 'now' problems of river basins for millennia, the geomorphological view puts these laws into the context of their boundary conditions in a variety of global environments, at a variety of scales, and, most critically for enlightened management, over a range of timescales.

Seen as a total transport system in the longer term it becomes axiomatic that Mankind's use of land settlement, agriculture and forestry will have an eventual impact on water. In some cases governments appreciate this link quickly, particularly if impacts are rapid, spectacular and very damaging. This book will draw out the particular case of flooding and erosion following deforestation; in countries such as New Zealand public policies were reactive but science-driven. Proactive river policies are scarcer!

No scientific research finds its way directly into public policy; the democrat reader would not wish it. However, scientists become considerably frustrated by the political filters through which their results are put and so this book also pays attention to those ideas, attitudes, policies and laws which relate land practice to river management and which derive from the people. It is perhaps variability across the globe in this cultural, ethical approach to river basins which sustains another element of the geographer's interest. This internationalism finds particular contrasts between the developing and developed world. The developed world consultant can do well to remember that, while the laws of physics are universal, relativity has a new perspective when their application lies, for example, in the hands of an Ethiopian peasant woman. 'Where is the river?', asked the consultant, 'I have designed a weir for it which will help you to grow your crops.' 'It has not been here for thirty years', said the African, 'and we grow excellent crops because it no longer floods us.' Seldom is the misunderstanding this great but on many river basin development schemes it might as well be so.

This book is idealistic but idealism is tempered by practicalities. Upon what kind of knowledge base should we mount a river basin management programme and how is such knowledge derived? If we are to forsake direct and heroic interventions in favour of a greater willingness to make indirect approaches, through land planning and management, what are the risks? What sorts of institutions provide successful regulation of river basins? The clear message must be that people should identify with both the basin as a unit and the river itself. Furthermore they must participate, as in Ontario, Canada, and not merely be moved around to fit the right river scheme, as in Indonesia. Scale, another geographical tool, can help us to understand the problem of identity; the circulation of information around units of different sizes is critical, be they drainage basins or institutions managing them.

Fortunately, the international movement to conserve ecosystems and those concerned with human amenity have recently adopted river systems much more actively than in the past. The International Rivers Network is part of this 'new wave' and it may well recruit support in the developed world from the host of 'yuppie' riverside dwellers, the millions enjoying water sports and those who regularly walk river banks or fish for recreation.

This book is undertaken some twenty years after the author drew great inspiration from *Water, Earth and Man* (Chorley, 1969). Geographers have, since then, become widespread in research, education and administration in connection with rivers; at the very least this volume will record their progress. Since 1969 our knowledge of the rate of environmental change has exploded; among those ecosystems whose response to extrinsic change is critical for human occupation of the planet, river basin systems take a high priority, having been the first to be exploited by settled human societies. Can science help?

A note to readers from other disciplinary backgrounds

If you have had no experience of the workings of the hydrological cycle it may be necessary to begin your reading of this text at Chapter 3, reversing through to Chapter 1! Years of experience of teaching this material to final year students with a background in hydrology (though a growing minority join us from social science!) make me feel that the chapter order as presented sets up the questions to be answered by the book (see Prologue and Chapter 1) and that the placing of transfer systems ahead of hydrology forces attention to the morphological condition of the river basin.

If you have had no background in river systems please consult the many simple texts available at school or college level (e.g. Newson, 1979 and Newson, 1994).

Preface to the second edition

The manuscript for the first edition of this book was sent to the publishers without a settled title and I only acquiesced to 'Land, Water and Development' after conceding that three words balanced those of the influential *Water, Earth and Man* (Chorley, 1969), while being more politically correct and advancing the agenda for geography to a more practical position (the subtitle referred to 'management').

This second edition has become inevitable because of the weasel word which qualified 'management' in that subtitle – 'sustainable'. Another irony of 1992 was that 'sustainable' became the war-cry of environmental politics all over the world, thanks to the UN Conference on Environment and Development (the 'Earth Summit') held in Rio de Janeiro that summer. Like a war-cry, each nation, tribe and community has a different version, or meaning of the word. However, the Earth Summit's 'Agenda 21', a UN Commission and a host of community and NGO projects have kept the debate running; the Summit focused the world's attention on the huge discrepancies between environmental problems and their perception by nation states at different levels of development. Perhaps because of the disappointments many felt about attitudes adopted by state leaders, Agenda 21 spends much of its time on other groupings of humans and other spatial units around which the environment might be managed sustainably by popular agreement and sound administration; river basins were prominent among these Environmental Management Units (EMUs) or 'bioregions'. Clearly, a second edition needs to update the context of the debate and some of the outcomes since 1992. It must also address another discrepancy raised in Rio – that between the power of science to explore and illuminate complex systems and the power of people in their environment to judge development pathways 'from the gut'.

Thus, within a consensus that 'development' is perhaps the most important word in the environmental dictionary – and that it should be sustainable – we have a physical science agenda which is seeking *inter alia* for the critical capital and limits to change in natural systems and a social science agenda which is considering indigenous knowledge systems (or 'vernacular science'), participation, conciliation, economic instruments and subsidiarity in decision-making. There is a clear need to overlap these two contributary perspectives – a traditional role for geography.

In detail there are, therefore, several major tasks for a second edition to face. The first is, in my view, to strengthen the position in the book of the *ecosystem approach to the river basin*. The little-known publication by Marchand and Toornstra (1986) first inspired me to this vision of the river basin and proved that any basin was amenable to modelling as an eco-system, followed by analysis of its key characteristics, critical properties and resistance/resilience to change, whether natural or imposed by development. The fact that many of the 'spontaneous regulation functions' of the Marchand and Toornstra model are synonymous with the geomorphological (sediment transfer) and hydrological (water transfer) systems – given a chapter each in the first edition – makes the ecosystem approach even more attractive. They also, at least partly, equate with the 'critical natural capital' approach to defining sustainability. Perhaps the one big danger of this new river basin image (see Prologue, first edition) is that it seldom shows the *groundwater* elements of the ecosystem, yet there is a looming crisis in many parts of the world in the way we plunder or pollute groundwater; I have amended the ecosystem diagram (Figure vi) to this end.

To the physical scientist, systems analysis, suitably directed, can provide both understanding and our much needed predictive capability. The second edition therefore gives more attention to what we understand about the *sensitivity of the river basin system* under different conditions of climate and physiography – to both intrinsic and extrinsic influences. It is a prime fault of environmental politicians to see every natural system as sensitive to every stress we put on it; we need also to look at resilience in relation to both artificial *and natural* stress.

As we know more, we also have more information; knowledge and information are different commodities and their combination requires judgement. The second edition, therefore, explores in more detail the deluge of *decision-making* tools such as knowledge-based systems which have been deployed at the interface of river basin management and those professionals, individuals and communities affected by it. Despite the information explosion, 'data', 'ground truth' and other 'hard' facts which reflect the often unique conditions of a locality remain in short supply, especially in the developing world – the paradox is that, where data are needed most they are most expensive and difficult to collect.

To the social scientist the key questions at the interface with science include the nature and availability of *indigenous knowledge*; on this depends its power as an alternative (at all stages of development) to the profession-al's ambitious data-collection programmes and greedy models. Gathering indigenous knowledge has the added advantage of beginning the process of *community involvement*, basis of the other 'S' word of the 1990s –

'subsidiarity'. Social scientists have developed a number of programmes, such as 'rapid rural appraisal', which attempt to render to the unique a place in the face of the general, to incorporate the identities of those to whom sustainability will have direct meaning.

Such separation of agendas is hard to defend outside the walls of academic institutions; Agenda 21 allows development to set its own priorities for knowledge and those in the field of river management helped create this tone at the Dublin conference which preceded the Earth Summit and reported to it. The Agenda has a chapter devoted to protection of freshwater but also highly relevant treatments of atmosphere, deforestation, desertification and drought, and mountain development.

Before the reader enters some transcendental world of immaculate plans and comfortingly elegant solutions, the 'Devil's advocate' of *free-market economics* should be heard. There are those who consider river basin planning and management to be a long-winded luxury indulged in by academics! The answer, it is claimed, is a proper pricing system for water and water-related benefits (James Winpenny is very scathing of complex planning structures); however, it is of course impossible to allow the economics of water to be truly 'free market' and in tuning the economic system, an activity beloved of capitalists and governments, some form of model of the resource system is needed. I see economics as one potential delivery system for the ideas conveyed in this book, not a *de novo* solution to all ills!

The geographical focus of attention in river basin development has spread to drylands, which dominate the world's fastest growing population centres; their nearby mountains are equally important. Many of the rivers near the centre of this new focus are *international basins*, adding an extra dimension of challenge (and often tension threatening 'water wars'). The rapidity of dryland development lures nations into all the troubled areas of basin development – the financial project cycle, the technocratic consultancy industry, inadequate environmental assessment, infringement of rights. Thanks to the kindness of a number of individuals and organisations I have been able to gather information for the second edition from Israel, Iran and South Africa, nations in which these problems have to some extent been countered.

Worldwide, the efficiency of water use in agriculture is less than 40 per cent; if drylands are the geographical focus of modern basin development, *irrigation* is the key technical issue within sustainability arguments. Current practices are unsustainable – there is perhaps no better demonstration of the meaning of that word than the fact that as much agricultural land goes out of agricultural production each year (because of system failures) as comes in through new schemes. Malin Falkenmark's continual message that water allocation in biomass production and water's role in environmental degradation should be the basis of development programmes, rather than water technology (supply and removal engineering) finds an accord here. I attempt to overlay her allocation per capita model across the ecosystem model; we may speak of the 'hydraulic density of population' based on this simple notion of carrying capacity but should not be duped by its simplicity. The rights of people (and non-human biota) must be considered in any such allocation procedure. The black population of South Africa have demonstrated the impact of rights in predicating water development programmes which have at their core river basin planning, even though South Africa is renowned for its highly technical water transfer systems.

What is the role of a book such as this? Clearly, from the comments of reviewers the first edition has been used to teach students of geography and related disciplines but it has also made some encouraging breakthroughs into policy, for example in the National Rivers Authority's catchment planning initiative in the UK and in the European Community's 'Project Fluvius'. The hopes are that more graduates and more professional converts will ensure that the idealism of the academic 'angle' which must be the core of such a text becomes translated into practical actions (or at least provides support for those who have taken such actions off their own bat). I am always encouraged by Bruce Mitchell's separation of river management principles into strategic (the holistic vision characteristic of the ecosystem concept) and operational (the integrated deployment of the necessary specialisms and techniques, both physical and socio-economic). If this book helps convey ideas between these two modes of work and if it additionally throws some light on 'sustainable development' as it applies to river basins it will have succeeded. I increasingly feel that those involved with river basins have the special responsibility of being in the vanguard of innovation in environmental management, thanks to the critical position of the resources with which we deal; if we cannot 'get it right' there is less room for optimism about sustainable management of other ecosystems and their resources.

Malcolm Newson
Ipswich, April 1996

Acknowledgements

The author wishes to acknowledge the many contributors over the years to the set of ideas presented here; as a work of synthesis it necessarily contains much imported material – too much to acknowledge by individual citation. As an avid bibliophile the author is indebted to those who have produced the books which are stacked in piles all round every room he works in! There is a particular debt to those who, in their river writings, dared to break fresh ground – all the way from Richard Chorley's *Water, Earth and Man* in 1969 to Stan Schumm's *The Fluvial System* in 1977 to John Gardiner's *River Projects and Conservation: A Manual for Holistic Appraisal* in 1991.

The author also acknowledges a life-long debate with engineers; levels of mutual appreciation have grown as the contribution of geographers has become clearer and less self-effacing, but the engineers known to me have also been faced with environmental and social problems with which they have found it hard to cope – I believe they have responded with great flair.

Acknowledgement is due, by name, to Lynne Martindale for producing the typescript and Ann Rooke for the artwork, both to punishing schedules – one marvels at the accuracy. The gatekeeper for those 'gremlins' that got through was Ros Ramage, who also worked at great speed and with stunning accuracy.

Finally I acknowledge the tolerance of all those whom I have failed for about two years, while wandering around with what Laurens van der Post once described as 'the river look' on my face!

ACKNOWLEDGEMENTS FOR THE SECOND EDITION

In addition to the inspirational and practical inputs acknowledged in the first edition of *Land, Water and Development* the author would like to thank all those who have reviewed (or actually used!) the book and therefore helped with praise or criticism. If nothing else 'LWD1' has become a useful reference for those wishing to summarise (more concisely) the whole 'bag' of ideas it contains. I am once again extremely grateful for Ann Rooke's hard work on the illustrations in the second edition.

I am grateful to those who have since invited me to see their land, water or development: Iran Ghazi, Betti Tiberini, Kate Rowntree, Bill Rowlston, Lea Wittenberg, Moshe Inbar, John Pitlick, Joe Chappell, Mark Major, among many. To have been 'an outsider, inside' the National Rivers Authority (and now the Environment Agency) as both research and development contractor and public representative has kept me up to date with the ever-changing and increasingly influential UK water sector.

Nearer home still, I thank my Head of Department, Tony Stevenson, for allowing me to double up my teaching in autumn 1995 in order to have sabbatical leave in early 1996. The death of both my parents, to whom I acknowledge my enthusiasm, was a severe blow at the start of that period but I have been more than restored by the company of Cath and her cats. Her remarkable paintings of water are deserving of Ruskin's praise for Turner: 'more ... than any philosophy of reflection which will not admit of our whys and hows'. They have made me realise that to love rivers is to know rivers and vice versa, and that such a balance is a guide to life itself.

Malcolm Newson, Corbridge, Northumberland

Prologue

RIVER BASIN IMAGES

Our perceptions and concepts are increasingly structured by visual experiences. Figures i–viii illustrate a range of published images of river basins. Any hydrologist asked to draw one will produce a prototype hot air balloon, slightly crumpled and with a very simple stream network, possibly second or third order according to Strahler's (1957) classification (Figure i). This is the small catchment beloved of water supply engineers and one over which land use control is relatively easily practised; it is also the largest scale used by experiments in hydrology – because of the unsuitability of larger units (say > 100 km^2) for controls and experimental treatments.

If we consider, however, that at another extreme the Amazon river basin may be fourteenth order on the Strahler scale, has a basin area of 5 Mkm^2 and is in desperate need of efficient management under changing political and environmental circumstances, the hydrologist's simplicity of concept may be dangerously illusory. The river manager, even on a small island like

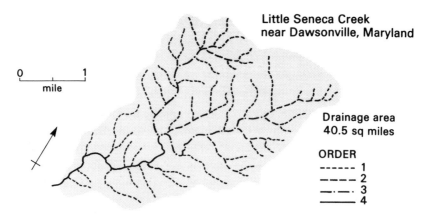

Figure i The geomorphologist's view of the river basin: ordered stream network within a defined boundary (Leopold, 1974)

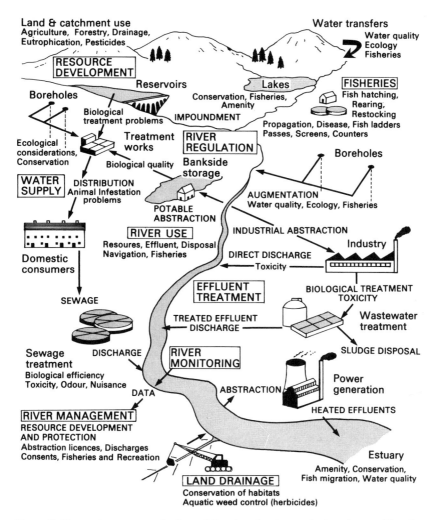

Figure ii   The water manager's view of the river basin: control systems without natural boundaries

Great Britain, may turn up the alternative image shown in Figure ii. The reader should note, however, that the dangerous illusion in this image is that the basin's boundary is not shown. Certainly 'land and catchment use' is noted as a label but, along with the other labels, this connotes the manager as something of a hero, coping with problems thrown at the channel network by land use on the banks. Engineers, particularly, tend to use 'responsibility', 'dirty' and other anthropocentric and value-laden terms when dealing with river basin management. Figure ii admits that water management itself imposes certain further 'duties' of planning and engineering, although a river system as completely controlled as the Tennessee example (Figure iii) would, to the tidy mind of the traditional engineer, perhaps need no further intervention. Such pictorial 'overkill' would not nowadays be a feature of the PR for a river scheme but the tidy-minded, rather simplistic, positivism of the river engineer is still represented in Figure iv. Here management problems are presented as amenable to mathematical or physical solution. Once again, any suggestion of the catchment boundary, or of indirect influences on the river, is missing.

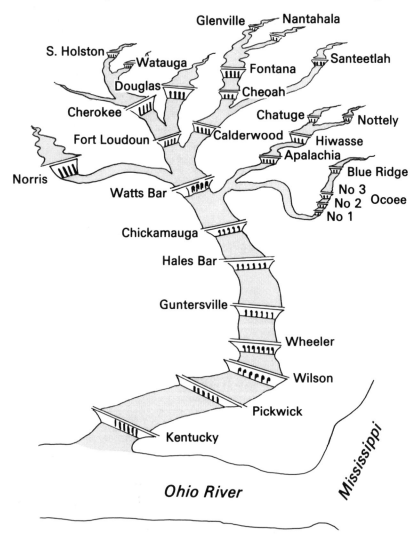

Figure iii The Tennessee Valley Authority view of the river basin: total control by dams (Huxley, 1943)

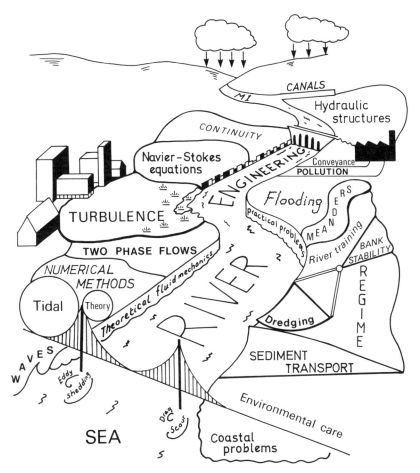

Figure iv The engineer's view of the river basin: a series of hydraulic problems (Knight, 1987)

Figure v, a recent offering from geomorphology, is more useful because it is not scale-dependent and because it emphasises sources, transfers and transformations in the drainage basin. Schumm (1977) applied the original version of this image to the basin sediment system but it is relevant, too, to water flows and pollution. Headwater areas are largely sources for the types of river management problems which this book addresses. In the transfer zone, riparian land use is no less critical but efforts in planning and management can be largely directed to specific sites, for example on floodplains and valley floors. At some stage in the system water, sediments or a pollutant will be deposited, or used, or will enter the food chain. The beauty of Schumm's image is that it is educational but very simple. The

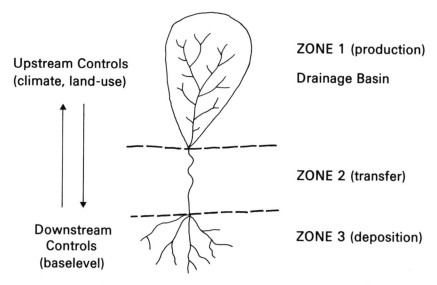

Figure v S. A. Schumm's river basin as a sediment transfer system (Schumm, 1977)

connectivity and interdependence which it draws out for the sediment transfer system is typical of the full basin ecosystem.

In recent years ecologists have been successful in translating the principles and properties of ecosystems into simplified structures which stress two key elements of environmental protection: the need for diversity in the objects (e.g. species) within the system, and the importance of flows of energy and materials between them. 'Healthy' ecosystems are relatively stable, offering few 'surprises' to participating organisms. Organisms exploiting the resources of ecosystems face the danger of reducing diversity or appropriating too much/too many of the system's resources.

The ecosystem model of river basins has been readily accepted by physical geographers, whose use of natural systems theory has been beneficial to the subject as a whole (Gregory, 1985). The two images presented from geomorphology in Figures i and v are, in fact, demonstrations of the physical basis of the river ecosystem. However, the most complete presentation of the basin as an ecosystem in relation to human exploitation of its resources is that by Marchand and Toornstra (1986). Their model (modified and simplified in Figure vi) sets up tensions between the 'spontaneous regulation functions' of the basin in nature and the artificial regulation imposed by human exploitation of the basin. When human exploitation is extensive (for example, the system is used for fishing, livestock, small-scale irrigation, hunting or transport) – essentially the picture under low levels of development – the spontaneous regulation functions of the basin aid and support our resource use. When, however, exploitation becomes intensive, with damming, channel manipulation for navigation/flood control, power generation and large-scale irrigation, the necessary artificial substitutes for natural regulatory functions are inherently unsustainable and may, in fact, lead to environmental degradation.

This book treats the basin in terms of the detail of the spontaneous regulation functions in the Marchand and Toornstra model (with the exception of the function of supplying genetic diversity). We deal with the following spontaneous regulation functions:

● Regulation of water regime (terrestrial and climatic)

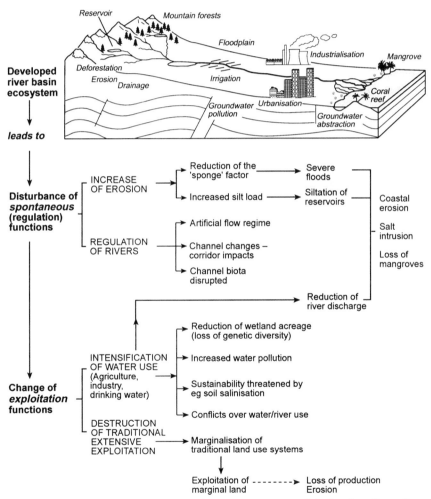

Figure vi The catchment ecosystem, its spontaneous exploitation functions: the basis for sustainable development (after Marchand and Toornstra, 1986)

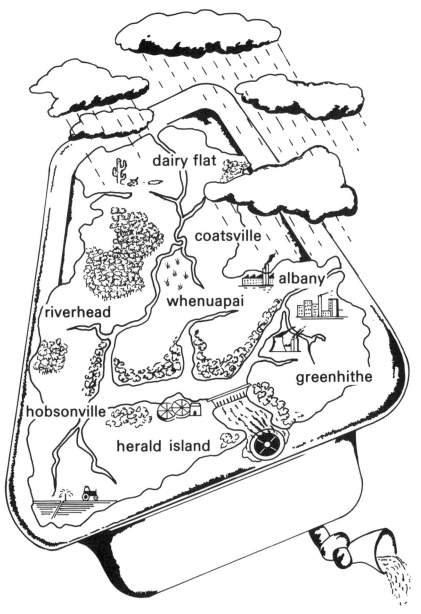

Figure vii Public relations view: your river basin (modified from *The Upper Waitemata Harbour Catchment Study*, National Water and Soil Conservation Authority, Wellington, New Zealand, undated)

- Regulation of erosion and sedimentation
- Water purification.

Marchand and Toornstra's description of these functions is:

> The drainage basin of a river fulfils a series of functions for man that require no human intervention, i.e. with no need of investments or regulatory systems.
>
> (p. 11)

The history of river basin development (Chapter 1) illustrates the way in which humankind has consistently sought to intensify exploitation, impose artificial regulation and then regret either the investment needed or the environmental degradation caused; environmental degradation inevitably bespeaks more artificial regulation. We leave the last word in this Prologue based on images of the basin with Black (1970):

> The watershed is an intricate natural resource which demands varied practices, complex management decisions, and manifold research efforts in order to ensure its efficient utilisation.
>
> (Black, 1970, p. 161)

Readers may like to note the convention used in terminology in this book. The large management units over which we shall try to specify outcomes (such as sustainable development, ecosystem equilibrium and human welfare) are described as river basins. 'Watershed' or 'catchment' are taken as much more specific terms referring to small-scale, identifiable, sources of water and less desirable products, amenable to treatment (in this case positive land-use allocation or accommodation).

Possibly the humorous image deriving from New Zealand (Figure vii) most closely summarises the unit for which we have here most to say. Furthermore, it is intended to deliver the management problem, at a basin scale, in an understandable form to those who live in it.

Chapter 1

History of river basin management

The purpose of this chapter is to explore briefly the nature of Man's occupation of river basins; the adoption of a conscious modern attempt at holistic management will almost certainly involve cultural attitudes to the problems, with their roots in history. Too often scientists ignore the importance of such elements in the translation of research results into policy and practice. For example, religious attitudes to the significance of water and its uses date back to the dawn of recorded history, as recently portrayed for the British Isles by Bord and Bord (1986).

The first hominids of 6–8 million years ago emerged as a savanna species and therefore into a seasonal climate; elements of our species as fundamental as bipedalism and communication are attributed to this environmental context. The savanna forced adaptation to finding, harvesting and storing food and water. Settlement, when it developed, inevitably produced advantages for the levelling out of supplies; in the case of water, however, considerable technological intervention was required. Societal repercussions of the need for efficiency and some equity in the distribution of water included the highly structured 'hydraulic' civilisations of the Indus, China, Egypt and the first of all: 'the Fertile Crescent'.

1.1 HYDRAULIC CULTURES AND RELIGIOUS CODES: MANAGEMENT IN ADVANCE OF SCIENCE

The closest and probably the most widespread association of past human activity with the hydrological balance, relief, slopes and stream networks of the drainage basin has been achieved through the operation of irrigation systems.

(Smith, 1969, p. 107)

Irrigation began to form a strong bond between humans and river basins in the sixth millennium BC; two important river basin civilisations, Mesopotamia and then Egypt, manipulated water to sustain settled agriculture. Both irrigation and elementary flood control were practised. The food surpluses which were generated by the success of these elementary management strategies were the basis for excess labour to be put into creating the architecture and other artefacts from which we have come to know so much about the Tigris–Euphrates and Nile valleys between 5000 and 3000 BC (Hawkes, 1976). The Sumerians built temples to the gods whom they considered responsible for the success of agriculture, while the Egyptians built memorials to the kings who were paramount in the strongly structured societies essential to primitive water management. Toynbee (1976) describes the Sumerian achievement as the source from which Egypt and later the Indus civilisation drew their basic water technologies; he also stresses the importance of *social structures*:

> Before this alluvium (of the Tigris and Euphrates) was drained and irrigated for human occupation and cultivation it was inhospitable to Man and to his domesticated plants and animals. It was a maze of waters threading their way through reed-beds – the marshland to which the district around the lower course of the Euphrates has now reverted.
>
> The mastering of the jungle swamp was a social, far more than a technological achievement. The human conquest of alluvium must have been planned by leaders who had the imagination, foresight and self-control to work for returns that would be lucrative ultimately but not immediately. The one indispensable new tool was a script. The leaders needed this instrument for organising people and water and soil in quantities and magnitudes that were too vast to be handled efficiently by the unrecorded memorising of oral arrangements and instructions.
>
> (Toynbee, 1976, pp. 45, 51)

Toynbee sees in this dependence on social structures and *flows of information* the reason for the eventual demise of the Sumerian culture: 'The Sumerian civilisation depended for its survival on an effective control and administration of the lower Tigris–Euphrates basin's waters; this control could not become fully effective unless and until it was brought under a unitary command' (p. 61). Instead Sumerian society became partitioned into a number of local city-states. Smith (1969) also stresses the tight structures responsible for any successful hydraulic culture; summarising Wittfogel (1957), he suggests that:

> the construction and maintenance of large-scale irrigation systems require the assembly of a considerable labour force which may be most efficiently created either by the institution of forced labour or the levy of tribute and taxation or both. A centralised administration is also needed for the maintenance of canals and to control water distribution. The administration in control of the distribution of water is, in effect, in complete control of agricultural activity, and is thus in a position to demand complete authority and complete submissiveness, subject only to

mass revolt and rebellion in the face of desperate conditions. Society becomes polarised, in fact, into an illiterate, dependent peasantry and an elite, as in the traditional bureaucratic governments of China.

(Smith, 1969, p. 108)

Biswas (1967) tabulates a chronology of hydrological engineering works by the Sumerians, the Egyptians and the Harappans who, by 2500 BC, had developed a very powerful (though less creative) civilisation in the Indus basin (Table 1.1). Among the most interesting artefacts remaining is the Sadd el-Kafara ('Dam of the Pagans') built *c.* 2800 BC just south of Cairo. It was apparently built without a spillway and with a capacity so small in relation to its catchment area that it failed early in its lifetime. Distribution of water was clearly more successful than collection; it requires, after all, much more organisation than understanding and it was to be 2,000 years before the study of nature began and 4,500 years before scientific hydrology! According to Vallentine (1967) the Egyptians had more success in their diversion of the Nile into a huge flood storage scheme (Lake Moeris), which was excavated, according to Herodotus, in the nineteenth century BC. Smith (1972), however, concludes that Lake Moeris is entirely natural and that Herodotus is an unreliable source for the record of Egypt's water engineering. Figure 1.1 illustrates the origins of river basin mapping; the reader may wish to compare it with Figure 5.6.

Irrigation practice along the Nile floodplain required the subdivision of land into embanked plots to be flooded during the annual flood cycle of the river between July and September. The Tigris and Euphrates were much less predictable and systems of canals and ditches, fed by diversion structures, took water directly to small plots. It is suggested by some writers that the need for efficient irrigation prompted the development of geometric ground survey techniques. A tablet in the British Museum illustrates algebraic

Table 1.1 Key dates in the development of hydraulic civilisations

Date (BC)	Event
3000	King Menes dammed the Nile and diverted its course.
3000	Nilometers were used to record the rise of the Nile.
2800	Failure of the Sadd el-Kafara dam.
2750	Origin of the Indus Valley water supply and drainage systems.
2200	Various waterworks of 'The Great Yu' in China.
1850	Lake Moeris and other works of Pharaoh Amenmhet III.
1750	Water codes of King Hammurabi.
1050	Water meters used at Oasis Gadames in North Africa.
714	Destruction of quanat systems at Ulhu (Armenia) by King Saragon II. Quanat system gradually spread to Persia, Egypt and India.
690	Construction of Sennacherib's Channel.

Source: After Biswas (1967)

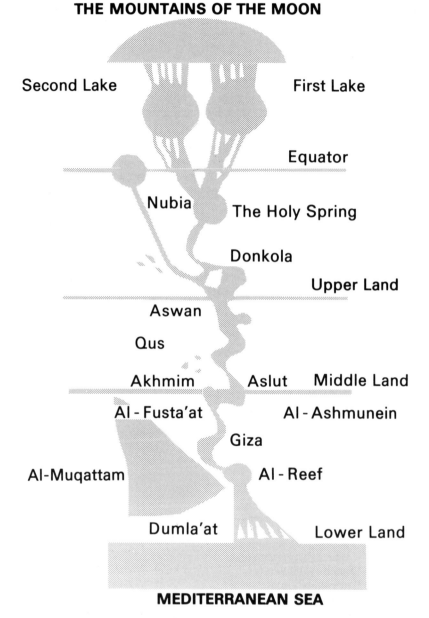

Figure 1.1 Ptolemy's map of the Nile basin (from T. E. Evans, 1990)

calculations for the design of dikes, dams and wells. The story of Noah almost certainly derives from the biggest flood to affect the Sumerian civilisation and therefore flood protection was highly developed. Rulers were extremely interested in legislating for efficient river management. The Code of Hammurabi (*c*. 1750 BC) is a codification of Sumerian and Babylonian water law. Flood damage due to negligence was particularly feared:

> Sec 53. If any one be too lazy to keep his dam in proper condition, and does not keep it so; if then the dam breaks and all the fields are flooded, then shall he in whose dam the break occurred be sold for money and the money shall replace the corn which he has caused to be ruined.
>
> Sec 55. If any one open his ditches to water his crop, but is careless, and the water flood the field of his neighbor, then he shall repay his neighbor with corn for his loss.
>
> Sec 56. If a man let out the water, and the water overflow the land of his neighbor, he shall pay 10 gur of corn for every 10 gan of land flooded.
>
> (Biswas, 1967, p. 128)

One must not neglect the military significance of water engineering at this time. Sennacherib the Assyrian destroyed Babylon in 689 BC by damming the Euphrates and then destroying the dam (Smith, 1972). Sennacherib became the agent of some extremely well surveyed and constructed dams and irrigation schemes.

Another important prehistoric achievement in the Middle East was the building of quanats, underground channels conveying water from springs to consumers (over a distance of more than 30 kilometres in some cases) at a constant gradient and free from pollution or losses by evaporation. Groundwater exploitation and underground storage of water are also recorded for both Egypt and Mesopotamia. Clearly, therefore, potable water supplies to urban centres were also a feature of these hydraulic civilisations. It is, however, from the archaeological record of the Indus civilisation that houses served by both water supply and drainage systems have been found, dating from 2200 BC (Vallentine, 1967).

At about the same stage of prehistory the legendary Emperor Yu in China began to control rivers in the interests of land reclamation. As Biswas (1967, p. 120) recorded, 'He studied the rivers … he mastered the waters'; it is clear from the Chinese literature that mapping of river networks and, consequently, some concept of river basins, had been born.

At the dawn of Western civilisation the hydraulic cultures of the Middle and Far East were still strong; but rival military fortunes produced fluctuating prosperity and eventually Western influences became all-pervading. However, uniquely it seems, the Indus civilisation ended as a result of profound *environmental change* – climatic extremes of damaging floods and extensive droughts overcame the sophistication of the engineering. The evidence for wetter conditions at the height of the Indus culture is mainly faunal (chiefly rhinoceros) and it has been criticised by Raikes (1967) as giving a false impression; hydrological responses, he argues, are complex when climate changes and we may make reference to current arguments over desertification and salinisation (Chapter 6) in his support. The Indus civilisation, claims Raikes, 'could have survived and prospered on zero rainfall with or without irrigation for the Indus complex of rivers enjoys a well-marked seasonal flood that would have inundated vast areas of the flood-plain' (p. 113). Even with today's barrages and levees, he claims, uncontrolled flooding still occurs. Toynbee stresses the magnitude and scale of the Indus civilisation, with its two principal cities 400 miles apart; he does not consider the management problems of such a scale: possibly management failed to respond in a coordinated way to environmental change, however caused.

Water has been a central feature in the development of many, if not most, world religions; the birth of Islam, Judaism and Christianity in semi-arid environments has helped create this linkage − the fundamental shared resource, the basis of food and livelihoods, is bound to figure prominently in religious codes. However, water has an additional spirituality of its own nature, as an element, a vital force and an agent of destruction.

There has been an increasing recent interest in the way the natural environment as a whole figures in the major world religions (see references to Table 1.2); most of this interest has been conceptual because in most nations civil society has grown away from a religious code (which is why we focus here on Iran where the civil code has been appropriated and controlled by religion since the Islamic revolution of 1974).

Table 1.2 Salient elements of attitudes taken to freshwater by major world religions

Religion	Cultural attitudes to environment/ water/ rivers
Christianity	Stewardship: humans in God's image; work ethic; water purity, healing
Judaism	People and land; reclaiming wilderness; non-specific on water
Islam	Water a gift of the Almighty; holy laws on use; ethics of scarcity; protection
Buddhism	Harmony with nature; nature a moral concept: sacred sites include rivers
Hinduism	Polytheist − gods in Nature; caste system ecological; rivers sacred

Sources: highly compressed from Breuilly and Palmer, 1992; Rose, 1992; Khalid and O'Brien, 1992; Bagader et al., 1994; Batchelor and Brown, 1992; Gadgil and Guha, 1992; Prime, 1992

An additional spiritual dimension is needed to include the religious attitudes to water of 200 million indigenous peoples (Beauclerk *et al.*, 1988). Characteristics of indigenous peoples are said to include sustainable use of resources (as a result of the deification of the elements and the folklore of disasters resulting from abuse) and the sharing of land and wealth through kinship or codes. Warren and Rajasekaran (1995) have a fuller list of characteristics:

1   Adaptive skills communicated through oral traditions
2   Time-tested agricultural and natural resource management practices
3   Strategies to cope with socio-cultural and environmental changes
4   Constant informal experimentation and innovation
5   Trial-and-error problem-solving approaches to challenges of local environment
6   Highly developed decision-making skills.

1.2   THE RISE OF HYDRAULICS AND HYDROLOGY

After this tentative conclusion that water distribution can occur in advance of hydrological knowledge, how can empirical knowledge and theoretical understanding be put to work in support of engineering? The dichotomy between the reasoning science of the Greeks and the practical application of the Romans is traditionally drawn in deriving the origins of Western science.

Empirical records of river levels can be traced for the Nile back to 3000 BC; the famous Roda 'nilometer' (Figure 1.2) recorded the annual flood. A system of flood warning may have been developed, using watch towers and 'extremely good rowers' (Biswas, 1967, p. 125) who propelled their boats ahead of the flood wave. Biswas also records the 3,000-year history of simple water metering for irrigation supplies in North Africa. In *River God*, Wilbur Smith's imagination has the arrival of the Nile flood thus:

> we woke to find that during the night the river had swollen with the commencement of the annual flood. We had no warning of it until the joyous cries of the watchmen down at the port roused us. Both banks were already lined with the populace of the city. They greeted the waters with prayers and songs and waving palm fronds.
>
> (Smith, 1993, p. 305)

Greek philosophers were not able to advance our knowledge of hydrology, though Archimedes' observations led to the foundation of hydrostatics. The engineering skill of the Romans, however, led to great progress in urban water supply and drainage systems. Nevertheless, Xenophanes of Colophon (570–470 BC) stated that 'the sea is the source of the waters and the source of the winds without the great sea, not from the clouds could come the flowing rivers or the heaven's rain'. By contrast Nace (1974) dismisses the

Figure 1.2 The Roda nilometer, upon which the heights of the annual Nile flood
have been measured from antiquity (from Biswas, 1967)

Romans' contribution to hydrology: 'Despite their great hydraulic works, no
evidence has been found that Roman engineers as a group had any clear idea
of a hydrological cycle' (p. 44). This is an important message for modern
hydrologists!

Under the very different environmental conditions of these Western
civilisations irrigation was obviously less important; distribution systems
took water from constant, pure sources, such as large springs in the
countryside, to the streets and houses of Roman cities. One may speculate
that for the Roman empire the humid conditions of Europe encouraged
drainage, which led to sewerage and therefore to the use of remote sources
of water supply to avoid pollution.

The largest technical problems of achieving the most impressive Roman
feats, such as the Pont du Gard aqueduct (supplying Nîmes in southern

Plate 1.1 Roman water supply engineering, Corbridge, Northumberland (photo M. D. Newson)

France), would have been the design of capacity for flow and gradient. In an interesting review of the Pont du Gard's hydraulic design, Hauck and Novak (1987) stress the subtleties of conveying a steady flow of water down only 17 metres of fall in 50 kilometres. The Romans made a clear trade-off between the expense of a higher aqueduct (i.e. a longer span across the Gardon valley) and the need to maintain a steady gradient. In 19 BC the most precise level was a 6 metre-long bar, levelled by water in a groove or plumb bobs. Simple geometry and a knowledge of the flow rate would have provided the cross-sectional area of the aqueduct's channel. The authors report that the limestone spring water continually encrusted the channel and regular maintenance was clearly essential; in every other way the work was a masterpiece of applied hydraulics.

Bratt (1995) describes more recent masterpieces of mountain water transfers: the 376 *bisses* (total length 1,740 kilometres) which traverse the steep sides of the valleys in the canton of Valais, Switzerland, in the headwaters of the River Rhône to deliver irrigation water; they were constructed as early as the eleventh century, an indication that not all such skills were lost in the Dark Ages.

It is hard to document scientific progress during the Renaissance without reference to Leonardo da Vinci; of him Popham (1946) says, 'water played a

Plate 1.2 Leonardo da Vinci's perception of river turbulence: *Deluge* (Windsor Royal Library no. 12382)

very important part in his life. A great deal of his energies and his intellect were absorbed in directing and canalising rivers and in inventing or perfecting hydraulic machinery' (p. 70). He was obsessed with depicting water movement in his art (Plate 1.2) and careful observation aided his design of water wheels and pumps. However, his was not merely a brilliant combination of water engineering and art: he formalised the relationship between catchment and flow properties in his study of the Arno above Florence (Plate 1.3). The Arno catchment map (1502–3) shows very great care with both the stream network and the contributing slopes; mountains are not shown as isolated hills in the medieval tradition but by contour shading. To record so precisely the relationship between slopes and channels and between events over the river basin and those at a site (i.e. Florence) sets up the combination of hydrology and hydraulics which was eventually to guide modern river management. Levi (1995) considers Leonardo to be so important to the history of observational hydraulics that he devotes an entire chapter of his historical review to the man.

The next important step was the establishment of the *hydrological cycle.* Nace (1974) suggests that acceptable definitions of the hydrological cycle were published at a very early stage of recorded history, for example in the Bible (Ecclesiastes 1:7):

> All the rivers run into the sea; yet the sea is not full; unto the place from which the rivers come, thither they return again.

Plate 1.3 Leonardo da Vinci's map of northern Italy showing the watershed of the Arno (Windsor Royal Library no. 12277)

Palissy (1580) is acknowledged as having the first accurate insight into the general process of runoff (as credited by Ward, 1982). We have noted the antiquity of river level measurements on the Nile; rain gauging was not necessary in the dry climates of the river-fed hydraulic civilisations. It had developed in China by 200 BC; in India rain was measured as an index of tax liability for agricultural production. Not until John Dalton presented a paper to the Philosophical Society of Manchester in 1799 had rainfall and runoff been brought together quantitatively. It is interesting to note both the title and first paragraph of Dalton's paper:

> Experiments and Observations to determine whether a Quantity of Rain and Dew is equal to the Quantity of Water carried off by the Rivers and raised by Evaporation; with an Enquiry into the origin of Springs.

> Naturalists, however, are not unanimous in their opinions whether the rain that falls is sufficient to supply the demands of springs and rivers, and to afford the earth besides such a large portion for evaporation as it is well known is raised daily. To ascertain this point is an object of importance to the science of agriculture, and to every concern which the procuration and management of water makes a point, whether for domestic purposes or for the arts and manufacturers.

> (from Dooge, 1974)

Ward (1982) brings the development of hydrological concepts up to date in a review of the changing conclusions as to the sources and routes followed for runoff to streams. Most of the controversy of the twentieth century has concerned the balance between surface and subsurface runoff routes; most of the evidence has come from the fully instrumented catchment. We return to this source of scientific guidance to basin management in Chapter 8.

The rise of hydraulics and hydrology has clearly been slow and at each stage the simpler conceptual and theoretical advances were applied first (and almost universally). Hydrology, arguably, is more relevant to river basin management. The fact that it did not enter its main phase of numerical data collection until this century and become applied by river management agencies illustrates how meagre is the scientific guidance available for the management of complex systems. It is particularly unfortunate that our knowledge of runoff routes is so recent; ignorance has curtailed our appreciation of the importance of the catchment, its soils and land use.

1.3   MONKS, MILLS AND MINES: ORIGINS OF RIVER COORDINATION IN ENGLAND

C. T. Smith (1969) offers a variety of reasons why a river network can integrate the activity of societies within a drainage basin; irrigation is subordinate in most humid climates to direct water power, accessibility by water or by land along the valley floor, supplies of fish and game and supplies of water for humans and stock. The archaeological record in much of Europe demonstrates the attraction of early settlers to rivers and wetlands (Coles and Coles, 1989), with an increasing number of spectacular finds emerging through work on 'archaeology under alluvium'. During the Bronze Age the site at Flag Fen, near Peterborough, was clearly a prosperous settlement, benefiting from extensive exploitation of the interface resources between land and water; its occupants worshipped the waters and made ritual sacrifices of much of their wealth to them (Pryor, 1991).

Through the careful work of Rowland Parker (1976) it is possible to reconstruct a 2,000-year history of human settlement on the banks of a tributary of the River Rhee in Cambridgeshire. Up to the period of Anglo-Saxon conquest it is clear that navigation was an important function of even the smaller elements of the river network; indeed, invasion was by that route. However, Parker interprets the settlers as constructing a water mill which must have interrupted future navigation (since no mention is made of a mill leat). Later, with forest clearance on the interfluves, settlements moved from flood- and invasion-prone riversides to the terraces and low hills nearby. Those which did not were protected by the digging of moats and diversion channels.

The Domesday Book of 1086 enables us to gain insights into the distribution of mills (6,000 of them) and also of freshwater fisheries, a traditional water use by the very powerful monastic landowners of the time. After the Norman Conquest the rise of manorial estates and the use of

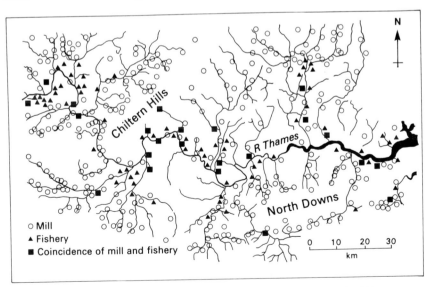

Figure 1.3 Domesday Book (i.e. AD 1086) mills and fisheries on the River Thames, England (compiled by Sheail, 1988)

milling tolls as part of a feudal structure ensured a rapid increase in obstructions to our rivers, but also in protections against selfish behaviour. The chaotic picture of mills and fisheries recorded by Domesday is shown in Figure 1.3.

The next step in Parker's record of the Brook, not surprisingly, involves some form of legislation from manorial courts to control private use by tenants of the channel to create ponds for stock watering, human bathing, and so on. Parker presents an almost continuous record of 'river offences' from 1318 until 1698. Gradually, with the obvious hiatus of the Black Death, pressure on the stream increases and the domination of ponding and diversion offences in the fourteenth century gives way to efficient land drainage ('cleaning and scouring') in the fifteenth and sixteenth and to pollution ('noysome sinkes and puggell water') in the sixteenth and seventeenth centuries. A selection of these offences is worthy of quotation to illustrate the ways in which English society was dealing with this two-mile stretch of minor stream.

In 1318 the following entries occur:

All the capital pledges of Foxton fined for not putting right the brook which was stopped up by Thomas Roys.
John Kersey fined 12*d*. for diverting the brook which flows through the middle of the manor, to a width of half a foot, and causing a nuisance.
Simon le Roo diverted the brook; fined 12*d*.
Roysia Kelle widened the stream by half a foot; fined 12*d*.
Ate Reeve did the same alongside his yard, widening the brook by letting the other bank fall in to a width of two feet; fined 3*d*.

1492   John Everard, butcher, allowed his dunghill to drain into the common stream of this village, to the serious detriment of the tenants and residents; fined 4*d*.; pain of 10*s*.

1541   Each tenant of this manor henceforth shall be ready and present at the cleaning and scouring of the watercourse called 'le Broke' whenever it shall be necessary, and they shall be warned to do so by the village officials or other inhabitants; pain of 12*d*.
No person shall wash linen called 'Clothes' in the common Broke; on pain of 20*d*. for everyone caught in default.

1590   Whoever has not appeared within one hour after the ringing of the bell to clean out the common brook shall forfeit 4*d*. to the Lord.
And whoever shall not clean out the brook fronting his land within the time appointed by the Heyward shall forfeit 12*d*.
No one shall wash any cloth in the brook before 8 of the clock at night on pain of forfeiting 12*d*. for every offence.

1594   No man shall lett out there sesterns or other noysome synkes untill eight of the clock at night uppon payne for every one doinge the contrary to forfeite unto the Lord for every tyme xii*d*.

1698   Item that any person who shall suffer any ducks to come into the Common Brooke shall fforfeit for every such offence 6*d.* to the Lord of the Manor.

Item that every inhabitant that shall not cleanse the Brooke or rivulet which runns through the towne of ffoxton soo far as is abutting upon their grounds att such time or times as shall bee appointed by the Churchwardens of the said towne shall fforfeit for every such offence 3*s.*4*d.* to the Lord of the Manner.

Item that any parson that shall lett out or suffer to runn any of their sinkes or puddles out of their yards into the Common Running Brooke of the town of ffoxton shall fforfeit for every such offence (if it bee att any time from ffoure of the clock in the morning until eight of the clock at night) 6*d.* to the Lord of the Manner.

(Parker, 1976, passim)

Steinberg (1991) has made a parallel exploration of the development of the Charles and Merrimack Rivers of New England, showing how common law and statutes evolved to cope with conflicts between the building of mills, navigation, fisheries and pollution. Guillerme (1988) reviews the period 300–1800 in France during which the focus of eighteen cities in the north passed from defence to milling to drainage, water supply and sanitation.

Whether manipulating streamflow for milling, stock watering, abstraction or fisheries, some form of obstruction to the natural regime was required and we have the origins of what we now call river regulation. Sheail (1988) reviews the history of river regulation in the UK; much of the available documentary evidence refers to drainage schemes in the Fens, Lincolnshire and Romney Marsh but the conclusion is universally applied:

Whatever the purpose of river regulation, no scheme could fulfil its potential without the cooperation of all the interests involved. A balance had to be struck between the protection of individual rights and the furtherance of the common good.

(Sheail, 1988, p. 222)

Central government became involved in drainage issues in 1427 with the establishment of a Commission of Sewers; a General Sewers Act followed in 1531. Interestingly, a complex hybrid of statute and common law was applied to water management, the presence of common law implying progress by precedents rather than a completely technocratic application of principles and hence standards (in the scientific sense).

The very extensive lowland drainage, under Dutch direction, of the seventeenth century has a literature of its own, particularly in the Fens (e.g. Darby, 1983). Less attention has been paid to the more hydrologically complex task of irrigation and drainage (of the same land) as was practised extensively in chalkland valleys from the seventeenth to the early twentieth century. The water-meadow farming system has much to commend it in an age of low-intensity production with plentiful labour while drainage of land in the Fens was considered to be of regional and national benefit, threatening only those who took the annual harvest from the flood (duck, reeds, fish). River regulation elsewhere quickly led to conflicts of interests between upstream and downstream users of water. While corn-millers frequently fought for effective use of low river flows, it was the arrival of industry, extractive then manufacturing, from the eighteenth century, with its demands on water for power, transport and processing, which put strain on the available structures for coordinating the use of the linear resource of the river. The common law of riparian rights, datable to the Chasemore v. Richards case of 1859, gives the following rights to those who own land adjacent to rivers.

> It has been now settled that the right to the enjoyment of a natural stream of water on the surface *ex jure naturae* belongs to the proprietor of the adjoining lands as a natural incident to the right to the soil itself; and that he is entitled to the benefit of it, as he is to all the other advantages belonging to the land of which he is the owner. He has the right to have it come to him in its natural state, in flow, quantity and quality, and to go from him without obstruction, upon the same principle that he is entitled to the support of his neighbours soil for his own in its natural state. His right in no way depends on prescription or the presumed grant of his neighbour.
>
> (Wisdom, 1979, p. 83)

Dams were built for a variety of purposes: for hydraulic mining (called 'hushing' in the northern Pennines), ore processing and separating, and to supply the canal network which expanded in the late eighteenth century. Binnie (1987) traces the origins of dams in Britain back to the Roman occupation; the occupants of fortifications along Hadrian's Wall clearly used reservoir storage on small streams. The Romans also used dams in connection with metal mining. The first modern dams were the mill and fishing weirs; in 1788 cotton milling alone accounted for 122 weirs on relatively large streams. More than 150 canal reservoir dams were built, the precursors of the modern water-supply reservoirs.

Industrial use of water power made mill operators a powerful voice in legal pressures for coordinated use of water. Finally, the growth of manufacturing and of large urban populations led to the construction of dams for domestic water supply. The conflict of interests between water storage and industrial water use led to the concept of 'compensation water', a minimum flow allowed out of reservoirs to maintain the rights of downstream users. Setting these flows was not easy in the absence of modern hydrological data. As Sheail comments:

> Parliamentary committees spent more time considering the issues of compensation flows than any other aspect of reservoir development. Much controversy stemmed from lack of data on, and understanding of, the hydrological processes involved in determining the reliable yield of catchment areas and the need to take account of the changing economic use of the river water.
>
> (Sheail, 1988, p. 228)

A detailed history of technical aspects of compensation flows is provided by Sheail in Gustard et al. (1987). Careful debate and evaluation was hardly necessary in the case of north Pennine lead mining. The 'hushing' technique, in which artificial flood waves were created (by the construction and subsequent breach of dams on small tributaries) to expose metal ores, produced sudden and often fatal flooding of the valley floors below.

Lead mining was also involved in early pollution litigation, an interesting example being that of Hodgkinson v. Ennor (1863) in which the owner of a paper-mill using pure water resurging in a limestone spring brought a successful case against a lead mine and works 7 kilometres away. This was, effectively, a case of groundwater pollution and the defendant attempted to escape the clutches of riparian rights by claiming that these pollutants percolated through the ground. The very dirty refuse from the mine washings was, however, its own 'tracer' material and the case was won.

1.4 THE RISE OF ENVIRONMENT

The early hydraulic civilisations perfected, with relatively simple science, but with highly structured management systems, the major distributional systems of irrigation and water supply. They carried out flood protection but very empirically, on the basis of common experience. While the Romans built sewer networks to collect waste and must therefore have understood the public health problems associated with river pollution, it is to the 'Workshop of the World', Britain during its Industrial Revolution, that we can look for an emerging approach to rivers as collection systems. Two elements of urbanisation and industrialisation prompted the need: the huge toll of life in epidemics of water-borne diseases (e.g. cholera in 1832), which forced attention to the classic source–pathway–target pollution system, and the need to establish 'gathering grounds' in the uplands to feed reservoirs of abundant pure water to be supplied under gravity to the developing lowland conurbations.

Binnie (1981) records the personal achievements of those Victorian water engineers who carried forward the heroic skills of Sennacherib or Vitruvius into an era of new design requirements. Aspects of coordinated management were quickly added to the agenda of social reformers of the time; we have already considered legislation related to compensation flows from the new upland reservoirs. Of much more profound importance in any historical review of river basin management was the rapid development of statutory law in relation to river pollution.

The early emphasis on both water supply and sewerage was municipal, that is, part of local government of towns and cities, and was a determined alternative to private enterprise because of the importance of both issues to public health. The technical shortfall in understanding river pollution, matched with a political desire not to halt development, led to a regulatory concept of the 'best practicable' solution.

Howarth (1988) provides an illuminating history of the development of water pollution law in England and Wales (Scottish law is a separate system). He cites the earliest enactment as that of 1388 'for Punishing Nuisances which Cause Corruption of the Air near Cities and Great Towns'; pre-industrial cities of England treated their streams badly:

> so much dung and other filth of the garbage and entrails as well of beasts killed, as of other corruptions, be cast and put in ditches, rivers and other waters, and also many other places, within, about, and nigh unto divers cities, boroughs, and towns of the realm, and the suburbs of them, that the air there is greatly corrupt and infect, and many maladies and other intolerable diseases do daily happen, as well to the inhabitants and those that are conversant in the said cities, boroughs, towns, and suburbs, as to others repairing and travelling thither, to the great annoyance, damage, and peril of the inhabitants, dwellers, repairers, and travellers aforesaid.
>
> (Howarth, 1988, p. 2)

The subsequent acts dealing with water pollution were apparently of little success in combating the public attitude to streams as dumps for all excrement and filth, not an unusual attitude in burgeoning developing-world cities of today.

> Sweepings from butchers' stalls, dung, guts and blood,
> Drowned puppies, stinking sprats, all drenched in mud,
> Dead cats and turnip tops, come tumbling down the flood.
> Swift, 'Description of a city shower'

By the arrival of the main phase of the Industrial Revolution, England and Wales had a plethora of legislation relating to the clearance of filth from towns using water-borne systems (e.g. Town Improvement Clauses Act 1847, Public Health Act 1848) and to the mitigation of the effect of these domestic and industrial wastes on rivers. One of the principal of the latter class of enactments was a Salmon Fisheries Act 1861; because the hard-working Victorian water engineers were busy bringing clean, fresh supplies from the upland streams, concern for urban lowland streams was directed mainly at the loss of livelihood from fisheries. In our major urban centres none of this legislation had a comprehensive effect; as is revealed by Engels' account of the Irk in Manchester during the 1840s:

> Above the bridge are tanneries, bone mills and gasworks, from which all drains and refuse find their way into the Irk, which receives further the contents of the neighbouring sewers and privies. It may be easily imagined, therefore, what sort of residue the stream deposits. Below the bridge you look upon the piles of debris, the refuse, filth, the offal from the courts on the steep left bank.

The principle of riparian rights, though set out again by Lord Wensleydale in 1859, did not bring the common law into any greater efficiency than statute in dealing with the relationship between towns on the same river. Clearly, the next stage was to change the geographical reference scale for administration of the water cycle, rather than to perfect new legislative principles, though the Royal Commission on Sewage Disposal provided chemical and physical principles for pollution control by 1912.

The Rivers Pollution Prevention Act 1876 was eventually implemented under the administration of the county councils, independent bodies neither operating sewerage systems nor organised deliberately around river basin units. Ironically such a spatial organisation came about through progress in flood protection, still a mainly rural preoccupation, in the 1930 Land Drainage Act. Local drainage boards were established on a catchment basis and the 1948 River Boards Act set up similar bodies to deal with water resources (see Chapter 7). From this point onward we see a steady move towards the addition of pollution control responsibilities to these authorities (called various names by the 1963, 1973 and 1989 water legislation; see Chapter 7).

Patterson (1987) sees in the last 100 years of river basin management in England and Wales a grave political symptom, the removal of the *democratic* element of municipal control in favour of, firstly, a *technocratic* element in the river basin authorities and, latterly, an increasing *commodification* of water. Regional state institutions in England and Wales, he claims, are particularly inaccessible to non-dominant groups such as consumers. As we have learned throughout the history of river basin management, such social issues are by no means irrelevant.

Since the 1960s the rise of environmentalism, aided by industrial stagnation and diversification, has led to a general desire to improve the purity of rivers. Toxicological studies reveal hidden dangers from impure water; fishing and boating have become major recreations and residential accommodation has returned to river waterfronts. It is the 600-year record of attempting to control the river environment which is now under most scrutiny in the 'New Environmental Age'. Not just water pollution is involved; flood protection causes environmental damage (Purseglove, 1988) and, elsewhere in the world, irrigation does so too. As a result of enormous resource pressures, Man's attitude to the water cycle and to river basin management is now as critical as it was in prehistoric Mesopotamia.

1.5   THE LESSONS OF HISTORY AND THE CHALLENGES OF THE FUTURE

This chapter has analysed the historical record of Man's intervention in the land phase of the hydrological cycle in order to draw out fundamental cultural attitudes to river management under a very wide variety of circumstances. A central aim has been to reveal that aspect of cultural attitude which may be labelled rational: what we would now call the scientific approach. This element is essential if we are to know the value of knowledge as a guide to management.

In order to learn lessons from historical sequences about the impacts of river basin development there needs to be a concerted academic approach to the subject. At the end of her entertaining sketchbook of river history Haslam (1991) makes the following plea:

> It is time the human ecology of rivers, a fascinating record of human ingenuity and endeavour, was rediscovered and re-appreciated. ...man's dependence on the river has changed, so that it is *no longer immediate, but distanced*. There is a gap, a distance, between man and the use of the river, so that the river has been set aside, indeed vandalised in the name of progress.

> (Haslam, 1991, p. 303; emphasis added)

Thus, in periods and cultures in which natural resources no longer have a divine power, to have them delivered in a denaturised way removes any corporate or individual conscience about how the source is treated; this is the broadest conclusion about the passage of the development process as far as rivers are concerned. The groundwater resource is even harder to contemplate because it is unseen; paradoxically, however, springs continue to evoke a protectionist attitude long after rivers are ruined.

Few authors have yet answered Haslam's call with detailed studies of the historical sequence in a river basin. An exception is the study by Decamps *et al.* (1989) of the Garonne which relates the date of changes to the spontaneous and exploitation functions of the river to their extent. The same volume of studies contains a yet more detailed study of the history of flood protection by 'river training' in Switzerland (Vischer, 1989) which yields great insights into the epic and heroic status of those charged, in the modern ecosystems view, with removing 'swamps' and controlling channels.

Retaining the ecosystem framework we get the patterns of impact and response shown in Table 1.3.

Table 1.3 An attempt to sequence the progress of exploitation of river basin ecosystems by human society during the archaeological record and recorded history (mainly based on the record of Britain)

Development activity	Ecosystem impact/human response
Navigation/fisheries/hunting	extensive – human dependency/faith
Floodplain settlement/small irrigation	removal of vegetation/wetlands
Large-scale irrigation	loss to rivers, rights/administration
Local offtake activities from river	start of system-wide impacts (local laws)
Local return flows (reduced/polluted)	dilution capacity retained
In-river activity (cattle, cleansing)	start of community concerns (local laws)
In-river activity (obstruction, drainage)	system-wide impacts (local laws)
Power generation (run-of-river)	system-wide impacts (rights issues)
River regulation by dams	non-natural flows – system redefined
Widespread point-source pollution	dilution capacity in national laws
Diffuse-source pollution	system redefined – land-use included
Recreation	all uses redefined – stress on values
Conservation	ecosystem re-evaluated – new laws

It is clearly impossible to structure a neat, linear, progressive impact analysis of historical exploitation – Table 4.1 shows how human needs for water progress during development processes and perhaps this is the most appropriate level at which to leave the analysis. However, with a few qualifications, Table 1.3 emphasises the following aspects:

(a) Truly extensive exploitation of the river ecosystem is a very restricted phase (though lasting long periods in low-density, less-developed populations); at this stage culture is set by spirituality and religious principles may later be the only 'carrier' of this culture through phases of development.

(b) The ability to move water around the system is a critical feature of exploitation, allowing human society to be free of water hazards yet to benefit from its essential resources (apart from the aesthetic – since this is essentially a period of abandoning the river). Engineering capability surges ahead of ecosystem information, rights issues and administrations (though the latter are well served by mega-projects).

(c) Societal responses to impacts on the ecosystem require, as development proceeds, information and administration; both are easier to assemble for small-scale systems unless national structures are strong (e.g. hydraulic civilisations) or ownership patterns appropriate (e.g. monastic/manorial systems in England). Highly important rights issues are settled early in the development process (e.g. riparian rights in Britain, prior appropriation in western USA) and highly important principles about ecosystem capacities (e.g. the 'best practicable option' approach in Britain) follow at a much later stage.

(d) Development pathways and societal responses chosen early in the development process normally compromise those that follow, undermining optimism that development can be truly sustainable unless it can be compressed, or the 'lessons of history' communicated. Sadly, it has been the engineering skill to move water and the administrative skill to move people which have so far dominated river basin development in all but those cultures where early on a strong religious attitude was taken to the river system and its values.

If river management is now to extend to the basin scale and is to take a broader role in global schemes of sustainable environmental intervention, what are the lessons of history? In summary, we have determined that:

(a) Social aspects of both coordination and control have been as important as, if not more important than, technical aspects.
(b) Distributional aspects of water management (e.g. early irrigation drainage and flood control) are capable of highly efficient development using relatively simple scientific foundations.
(c) Collection systems including flows into reservoirs but also flows of pollutants to streams are much more difficult to understand without a contribution from science. This contribution is only now becoming coherent.
(d) The fundamental legal principles upon which a society bases its approach to water management will powerfully determine and constrain the environmental outcome.
(e) Scale issues are critical because they control the distribution of information in the system, both technical and social.
(f) Response to environmental change is a key aspect of river basin management.

Chapter 2

Natural river basins
Transfer systems

In this chapter the aim is to set down those patterns and processes which lead us to the view of the drainage basin as a *systematic physical whole*. It is the key concept in the wider education of politicians, planners and the public that river systems are an *interconnected transport system,* albeit often working invisibly (as in the transfer of dissolved salts) or over extremely long timescales (as in the evolution of floodplains).

The ecosystem model of the river basin pays particular attention to the transfer system attributes, both sediment and water. Several of the spontaneous regulation functions of the basin rely for their operation on conservation of the natural dynamics of floodplains, wetlands, 'wild' channels and the slopes contributing water and sediment to those components. Climate is a major driving variable in the gross behaviour of the transfer system, controlling basin-scale inputs of water and (through plant cover) sediments, but the archaeological record and contemporary observations prove the progressive impacts of development. Humid zone basins are particularly vulnerable because of the influence of natural vegetation covers on evaporative loss and soil loss from slopes. By 'progressive impacts' we refer to the feedback mechanisms in the transfer system through which, for example, eroded slopes exhibit reduced infiltration capacity for rain and the resulting increase in runoff further increases slope erosion through gullying.

Professional perceptions are not without fault in their view of the basin system, particularly in terms of the timescales over which it comes to steady states and can therefore be managed by relatively simple, sustainable controls. Although remote sensing and interactive, real-time, mathematical modelling can now allow reactive as well as proactive control, the latter, as in river channel 'training' to improve land drainage, is still the least-cost solution in many cases to our exploitation of rivers or to protect ourselves from their extremes. There are, understandably, new professional viewpoints which see control of any kind as undesirable and river basin systems have become a focus for a variety of conservation and restoration approaches (see Chapter 6).

In formalising the importance of timescales in perceptions of the river basin system Hickin (1983) has demarcated the following groups of research interests in the field of river sediment dynamics:

Geologists   1,000,000 years
Geomorphologists   100–100,000 years
Engineers   100 years

Newson (1986) and Newson and Leeks (1987) take this a stage further and Figure 2.1 illustrates a corollary of the timescale criterion for river professionals – in this case a differentiation based upon the preferred spatial scale of research. These divergences lead to considerable problems of mutual understanding at conferences or, more expensively, in the design and carrying out of river development or control programmes. We return to this theme, therefore, in treating the importance of river basin management institutions in Chapter 7.

It is not surprising that much of the more holistic thinking about the practical management of rivers has come from geomorphologists. There is no superiority in the geomorphologist's viewpoint, we merely say that a conjunction of long-term and large-space scales has as many advantages for managing complex systems realistically and sustainably as it has disadvantages for conducting scientific experiments under controlled conditions. We return to comparing the geomorphological approach with the engineering approach to transfer systems at the close of this chapter.

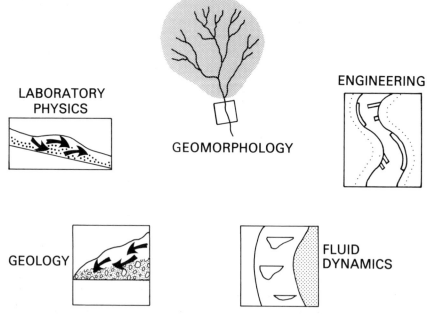

Figure 2.1 The scales of investigation adopted by river research specialisms

In the last analysis, however, a river basin which is *physically destabilised,* through natural or man-made perturbation of its morphological and sediment transfer elements, is extremely difficult to stabilise through management intervention; the ecosystem model views such a basin as having lost its major spontaneous regulation functions. The current scientific and political concensus over the desirability of longer-term aims has encouraged us, therefore, to understand and evaluate these natural functions before we run the risk of damaging them. There is a bias in this chapter to the sediment transfer system, although solutes are extremely important, not only for their contribution to geomorphological evolution but also as pollutants and nutrients. The transfer system for water itself and solutes is dealt with mainly in Chapter 3 where a hydrology bias is appropriate.

2.1   FLOW OF WATER AND TRANSPORT OF SEDIMENT

While hydrologists now have a variety of predictive models for runoff at the catchment scale, prediction of flows for river basin management is more difficult, requiring a general reduction of sophistication (or the incorporation of more sophisticated statistical components) as areal scales increase. If the basin manager then requires in addition to a successful prediction of river flow a precise estimate of the sediment transport, we simply do not have the robust techniques of the hydrologist.

Why should runoff prediction prove relatively easy but sediment transport more difficult? To answer this, one must attempt to unravel the complex interrelationships between the hydraulics of streamflow and the ability of flow to power transport. The most important aspect of sediment transport at the river basin scale is its supply from source areas, including headwater catchments, and the beds and banks of the channel network.

The concept of separated supply and transport components of the basin sediment system has only recently been found to be analytically and practically helpful, particularly in the formulation of geographical zonations of river basins. The American geologist W. M. Davis (1899) proposed a threefold subdivision of river basins into 'youth' (headwaters), 'maturity' and 'old age' (downstream reaches) – a classification helpful for the very long timescales over which the Davisian cycle of landform development was hypothesised to occur. However, for the last thirty years geomorphologists have become orientated towards active processes and the energy levels implied by terms such as 'youth' are misleading.

Schumm (1977) has proposed the subdivision of *supply, transfer* and *deposition* zones which is much more in tune with contemporary knowledge of processes (see Figure v). Further, classification schemes for river morphology within the transfer zone, also by Schumm (1963; see Table 2.1), emphasise the nature of the sediment load being transferred in the reach.

Table 2.1 Classification of alluvial channels

Mode of sediment transport	Channel sediment (M) %	Proportion of total load		Channel stability		
		Suspended load %	Bedload %	Stable (graded stream)	Depositing (excess load)	Eroding (deficiency of load)
Suspended load	30–100	85–100	0–15	Stable suspended load channel. Width-depth ratio >7; sinuosity >2.1; gradient relatively gentle.	Depositing suspended load channel. Major deposition of banks caused narrowing of channel; streambed deposition minor.	Eroding suspended load channel. Streambed erosion predominant; channel widening minor.
Mixed load	8–30	65–85	15–35	Stable mixed-load channel. Width-depth ratio >7 and <25; sinuosity <2.1 and >1.5; gradient moderate.	Depositing mixed-load channel. Initial major deposition on banks followed by streambed deposition.	Eroding mixed-load channel. Initial streambed erosion followed by channel widening.
Bedload	0–8	30–65	35–70	Stable bedload channel. Width-depth ratio >25; sinuosity <1.5; gradient relatively steep.	Depositing bedload channel. Streambed deposition and island formation.	Eroding bedload channel. Little streambed erosion; channel widening predominant.

Source: Schumm (1963)

2.1.1 Elementary hydraulics and sediment transport

In the headwater, sediment-supply zone of a river basin, particularly where this is mountainous, the coupling between slopes and channels controls the amount and timing of sediment removed from the system. Largely governed by natural conditions (but often exacerbated by development), slope transport of sediments occurs by the slow, progressive production of weathered rock and its gravitational movement, progressive or sudden, towards the nearest stream channel.

In the polarised case of a supply-zone river valley, slope inputs are almost direct to river channels: there is not the extensive floodplain and valley floor which intervenes to store sediments in the transfer zone. Clearly, therefore, there can be mismatches in time between the rate of slope-derived sediment supply and the rate of removal by the channel at its foot. Obviously, extreme events such as rare floods provide energy to both slope and channel environments but, even within the spectrum of extreme floods, there are also populations of *effectiveness*. Newson (1980) divides 'slope floods' from 'channel floods' on the basis of rainfall intensity and duration; later (Newson, 1989) he includes floods which are effective in both environments.

If we consider the stream channel transport process, we need to demarcate the different calibre or size of the grains supplied to the channel. Grain size determines, within broad limits, the process of transport within the flow. Figure 2.2 illustrates the components of *bedload* and *suspended load,* with an intermediate status of bouncing or saltating transport. This is a gross generalisation, particularly in the case of the saltation process, where the time effects of turbulent water flow are simply not known, other than at the statistical level or in experimental flume conditions. It is occasionally assumed that the suspended load, because of its fine grain size under most natural conditions, equates to the 'wash load' and derives from the surface washing of weathered material into the channel, for example from soil erosion. However, all forms of sediment transported in channels may derive from either the bed material of the channel itself or from bank erosion and inwash from slopes (or, in extreme events, from landslides or ephemeral gullies).

From whatever source, how are grains transported by river flows? While it is axiomatic that water flows downhill, the distribution of the energy so gained and its loss in overcoming friction, both internal and external, and in achieving work by transporting sediment are not yet fully described. Under most conditions of sediment transport during which there is an effective development of channel form, river flow will be turbulent. Turbulent flow is characterised by chaotic patterns of currents and vortices which, however, can be simplified within a treatment of the simple properties of flow; the most important of the properties for effective sediment transport is streamflow velocity and its variability with depth, across the channel and in distinct currents within the flow.

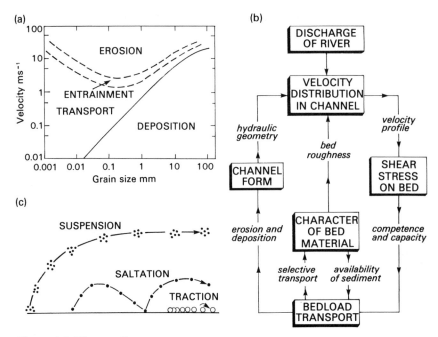

Figure 2.2 River sediment transport:
(a) The curves of Hjulstrom (1935)
(b) Feedback diagram between river flow, bed material transport and channel form (Ashworth and Ferguson, 1986)
(c) The major sediment transport processes in cartoon

'Velocity' (with little further specification) has been used as the predictive variable for many basin approaches to understanding – and thereby predicting – sediment transport. The best-known pictorial representation is that of Hjulstrom (1935), shown in Figure 2.2; these have educational value in that they:

(a) Differentiate conditions leading to entrainment by erosion of the particle, transport of the particle and deposition.
(b) Emphasise the importance of inter-particle relationships in the finer grain sizes where cohesion leads to a steep rise in the necessary velocity for erosion.

The curves do not, however:

(a) Extend sufficiently towards the coarser grain sizes which are important in sediment supply zones and during floods.
(b) Reveal the existence of inter-particle effects within coarser grades of sediment and in sediment mixes across the whole range of bed and bank materials and those delivered from slopes.

Alongside the effects of the normal mix of sediments available for transport by a known streamflow velocity we have also the quandary as to the location, within the turbulent mass of water, of that energy: average velocities may be plotted during a period of measurement with a current meter but these simple patterns demonstrate only the energy loss by friction at the channel boundaries and these patterns, too, are complex, as the measurements in mountain streams illustrate.

Much of the morphological development of river channels occurs as a result of the *sinuous planform* and uneven longitudinal profile of natural channels; in this connection, therefore, it is important to research:

(a) The existence, pattern and strength of *secondary currents* or flow cells which distribute the primary, downslope, energy of the flow into effective pathways.
(b) The distribution of velocity throughout a reach, as between high and low points (*riffles* and *pools* in the alluvial reaches) and between high and low flows.

These are major items which remain on the research agenda of those interested in refining our knowledge of fluvial systems. However, there is a clear ongoing need for predicting sediment transport rates, particularly of the coarser elements of bed material which are transported as bedload (by rolling or saltation), because this form of transport can produce the channel instability which threatens erosion of structures and flooding of property.

Bedload formulae (equations) have been developed and used for engineering applications for over a century.

They are of four main types:

(a) Those based on calculation of the shear stress produced by the flow on this stream bed.
(b) Those using stream discharge as an integrating predictor.
(c) Statistical/probabilistic approaches to the movement of grains.
(d) Stream power calculations – a generalised energy approach.

As a review by Gomez and Church (1989) illustrates, many transport formulae are vindicated by the use of data gathered in flume experiments or from field situations of unspecified relevance to the assumptions of the method in question. As the authors suggest,

> It remains a matter of some concern that there appear to be more bed load formulae than there are reliable data sets by which to test them. In consequence, no one formula, nor even a small group of formulae, has either been universally accepted or recognised as being especially appropriate for practical application.

> (Gomez and Church, 1989, p. 1161)

Much scientific effort, therefore, has gone into both the comparative testing of the available formulae on standardised data and the collection of large data sets from the field using innovative samplers (e.g. Helley and Smith, 1971). Gomez and Church conclude that for general river applications, where detailed hydraulic knowledge is normally impossible to gain, the stream power approach of the formula derived by Bagnold (1977) – using the analogy of an engine working at varying degrees of efficiency – is the most accurate, despite the fact that it lacks physical vigour and that actual operating efffiiciencies (commonly less than 10 per cent) are hard to predict.

2.1.2   Slopes, channel, storage zones

Because, in many regions of the world and under relatively frequent flood conditions, the sediment transport system can be shown to be *supply limited* (i.e. stream power to transport is not fully utilised), it becomes of critical importance to river basin management to understand the supply processes themselves.

Sediment supply occurs, in the Schumm model, mainly in the headwater zone where slopes and channels impinge closely and where gravitational energy is high and weathering processes active. This emphasises that there are at least two sources of sediment supply which may limit (or not) the transport of sediment out of the basin:

(a) Direct inputs from the slope weathering/transport system (which include man-made additional losses such as cultivation-induced soil erosion – see Chapter 6).

(b) Areas (and volumes) of stored sediments resulting from past phases of erosion under different climatic conditions or from an extreme flood in the recent past. The widespread Quaternary glaciations have meant that storage legacies are a common complication to the prediction of overall fluvial processes throughout large areas of the world.

At the statistical level of analysis there is clearly good adjustment, over long timescales, between sediment transport and sediment supply. Simplistically, any shortfall of supply is made up by increased erosion of bed and bank materials, leading to either incision of the transporting channel or its migration across the valley floor; the latter outcome produces the probability of undercutting an adjacent slope. The incision or migration of the channel therefore leads to conditions favouring supply. These links are often easiest to appreciate over the long term when, as Playfair wrote,

> Every river appears to consist of a main trunk, fed from a variety of branches, each running in a valley proportioned to its size, and all of them together forming a system of valleys, communicating with one another, and having such a nice adjustment of their declivities, that none

of them join the principal valley, either on too high or too low a level; a circumstance which would be infinitely improbable, if each of these valleys were not the work of the stream that flows in it.

(Playfair, 1802, p. 102)

These conditions of adjustment between slopes and channels occur over long timescales or during individual flood events when landslides, debris flows and other slope developments are shown to feed a wide range of sediment sizes into a channel system swollen by floodwaters and capable of transporting the resulting load considerable distances downstream.

This impression of perfect adjustment is at least partially illusory. Clearly, measurements of sediment transport loads in the field have shown that stream power, while it is a good indicator, is seldom fully utilised in transport and that the very existence of *storage zones* of fluvial material in the river basin indicates that the transport system is not a smooth one. As Ferguson (1981) puts it:

Rivers exist to carry water to the sea and they develop channels able to contain their normal flow. The form of the river channel affects the flow of water in it and, through erosion and deposition, the flow modifies the form. The channel (and if it migrates, the whole valley floor) acts as a jerky conveyor belt for alluvium moving intermittently seawards.

(Ferguson, 1981, p. 90)

The operation of the 'jerky conveyor belt' creates the characteristic diversity of basin fluvial geomorphology; each feature has a significance in the system and even those remote from the contemporary channel have a significance as stores of sediment, possibly nutrients and genetic information. A major characteristic of development is that the transfer processes of the basin ecosystem are accelerated and storage volumes/times reduced. Engineering in the channel network is then required to respond to the morphological impacts of this acceleration, further encouraging 'stability' by straightening and embanking rivers, minimising lateral transfers and drastically reducing the morphological diversity which is the fundamental basis of the channel and riparian biotic system. Figure 2.3 illustrates the major processes and morphological outcomes of the downstream transfer of sediment in a pristine basin. At any stage during the natural geomorphological development of the basin there may be only local adjustment of the morphology to the flows of water and sediment but overall there is a dynamic equilibrium.

Nevertheless, at the scales of measurement used in a morphometric approach to river basins (i.e. form indices largely from topographic maps), the 'nice adjustment' is demonstrated by close correlations, both positive and negative, which allowed Melton (1957) to set up the balanced system, inferring process adjustments from the interrelationships of forms such as channel slope and valley-side slope. There is a further, utilitarian, justification

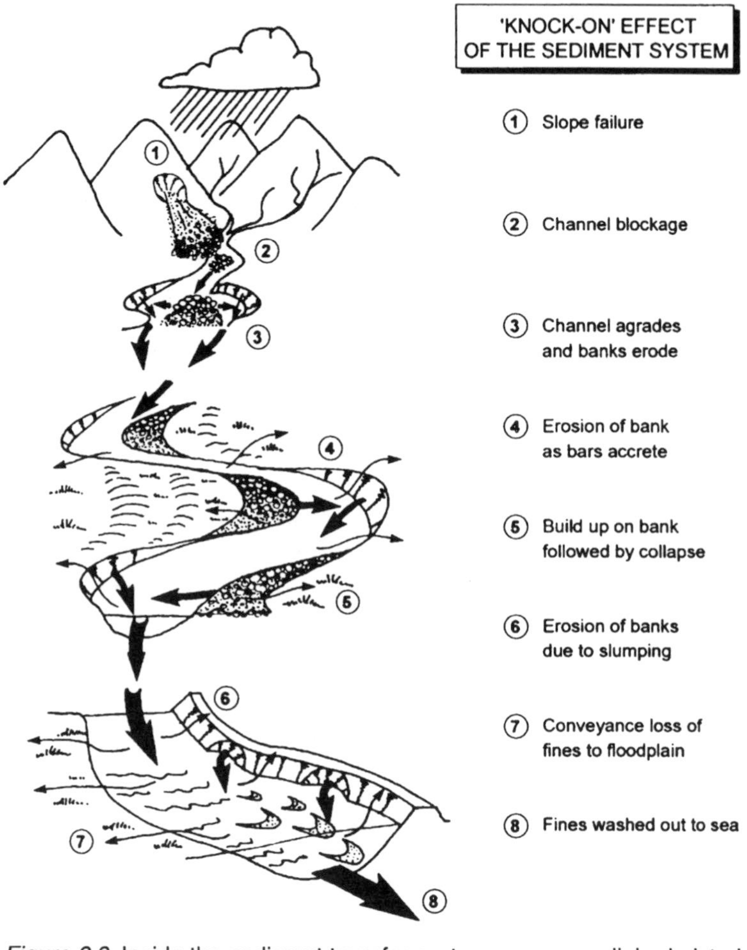

Figure 2.3 Inside the sediment transfer system: a process-linked pictorial model (after Newson and Sear, 1994)

for such a view in the high levels of predictability of flow, phenomena such as flood peaks and low flows from morphometric, cover and climatic variables measured from basin maps (see for example NERC, 1975; Institute of Hydrology, 1980).

2.1.3 Timescales of river basin development in nature

The introduction to this chapter has outlined the differences of spatial and temporal scales in the perceptions and research agendas of the professionals involved in river basin management. These differences are critical to the concept of equilibrium states in the physical development of basins via the sediment transport system. Schumm and Lichty (1965) have tabulated the pattern of status changes between variables controlling (independent) or controlled by (dependent) river basin evolution.

In a more practical form, Knighton (1984) has illustrated the time and space scales over which various features of the river basin landscape may be said to evolve (and therefore attain some sort of equilibrium or steady-state condition). Not only are conditions and concepts of equilibrium essential to successful, sustainable management techniques in natural systems but they, too, change with different timescales. Once again Schumm (1977) comes to the rescue by expressing this diagrammatically; Figure 2.4 combines the Knighton and Schumm approaches.

Traditionally geomorphologists, locked into the cyclic thinking of W. M. Davis, have considered the concept of *grade* as being appropriate to the time span over which humans study river systems, implying that 'nice adjustment' of Playfair. Engineers, too, have employed a steady-state concept in their design of new river channels (e.g. for irrigation) or their 'training' of eroding or depositing, i.e. unstable, channels. For nearly twenty years now, however, geomorphologists have moved away from notions of stability in river systems to those of *metastability,* or periods of quasi-stability interrupted by episodes of rapid change which appear to managers as challenging demonstrations of instability. The new approach is best summarised in the term *thresholds.*

Threshold phenomena are widespread in science (Newson, 1992a), particularly in materials where failure phenomena (rapid change between two stable states) abound. The importance of threshold concepts to river geomorphology is that they permit a 'middle road' between two previously dominant philosophies of landscape development – those dominated by catastrophes and, in contrast, by (slow) progressive processes. There are also practical reasons for the recent popularity of threshold phenomena; in their recent move towards process investigations, geomorphologists have inevitably worked in field situations where the engineering approach through regime designs for channels has failed (see Section 2.2.2).

What perturbations to steady state, elucidated by field studies, have become worthy of incorporation within the threshold concept? The major cases are as follows:

(a) River basins affected by artificial developments such as river regulation (Petts, 1984) or urbanisation (Roberts, 1989).
(b) Semi-arid river basins where sediment supply is little constrained by soil and vegetation cover and where the alternation of flood and drought extremes is commonplace. Basins affected by fire are also included (see Graf, 1985).
(c) River basins affected by 'rare great' floods have been said to exhibit a range of threshold phenomena.

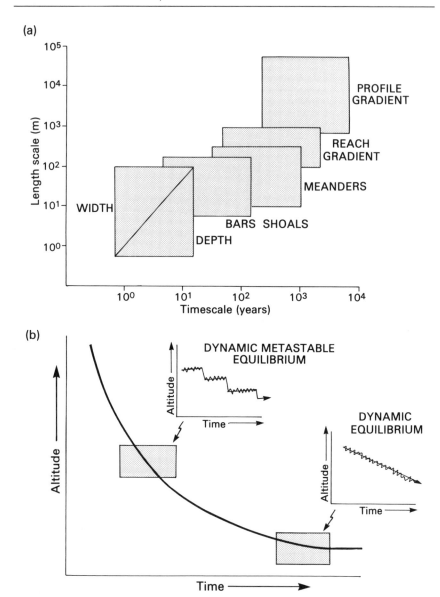

Figure 2.4 Timescales of river system sediment transport and morphological response:
(a) The characteristic length and timescales of major river forms (after Knighton, 1984)
(b) Two major forms of equilibria adopted by river systems during the geological cycle (Schumm, 1977)

(d) Model-scale experimental river basins in which input conditions are held steady (by the operation of artificial rainfall simulators) but in which sediment output occurs as a series of discrete episodes – much of Schumm's early work was on such a model system.

The important basic points for geomorphologists about threshold behaviour in river basin sediment systems are that it changes our philosophy of timescales of change (see Figure 2.4b) and that an important type of threshold is intrinsic to river systems and therefore operates without the need for a large external change such as climate change or an extreme flood. The geomorphic threshold therefore develops by the slow, progressive processes of weathering on slopes or storage of sediments on valley floors to the point where the morphology of the feature in question makes it inherently unstable even under the 'normal' range of conditions. The threshold is crossed when the weathered material on the slope fails, for example, as a rotational slump, or the valley floor steepens to the point where gullying develops. Clearly, the practical implications of geomorphic thresholds include a need to consider the flood record of a basin in relation to recent morphological change and the importance of avoiding developments such as cultivation and drainage on sediment storage zones which are in a threshold condition. Basin development occurs against a background of a spatially and temporally varied set of stability conditions in the sediment system; these need geomorphological audit (see below) as part of environmental impact assessment.

Slope angle has been used as a successful geomorphological predictor of intrinsic threshold behaviour in semi-arid terrains, notably in the west of the USA. Graf (1985) illustrates that the instability apparent to managers of the Colorado River basin derives from the superimposition of intrinsic thresholds, occasionally crossed by gullying in the headwaters, compounded by extrinsic thresholds imposed by climatic variability and traditional engineering interventions such as dam construction and floodplain development.

Clearly, Ferguson's 'jerky conveyor belt' is therefore a valuable perception of the river basin sediment transport system. While thresholds should be strictly defined and should not be applied to patterns and processes which are merely threshold-like, the educational value to river basin management systems is clear. Short-term management devices which merely 'set up' future threshold behaviour are the equivalent of sweeping dust under the carpet. Furthermore, surveys to investigate signs of symptoms of instability indicative of the state of a river basin, or river reach, are relatively easy to operationalise before inappropriate regime-based designs are employed by engineers (Lewin et al., 1988). When engineers ask, 'what is the alternative design procedure?', one is forced to admit that here the unpalatable choice may be between doing very little (or using low-cost, 'soft' engineering), implying a conservation philosophy and a retreat from the river, and placing emphasis on very indirect forms of river engineering, for example, erosion control on sensitive slopes and valley floors.

We should not escape from this extremely broad approach to river sediment systems without a view of the entire system. The workings of the

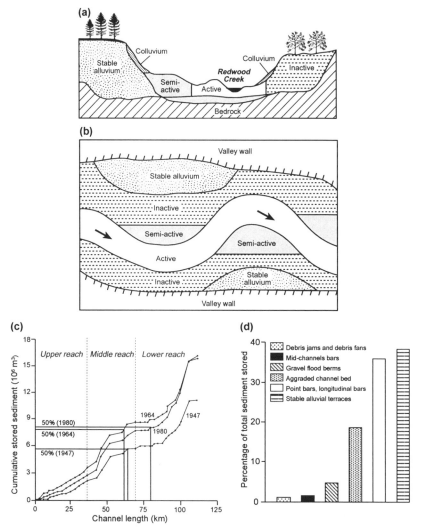

Figure 2.5 Sediment storage – a regulatory function of catchment ecosystems: Redwood Creek, California (after Madej, 1984)
(a) Schematic guide to the four main stores in cross-section
(b) Schematic guide to the four main stores in planform
(c) Cumulative stores of sediments stored at three dates (centre of mass indicated by 50% line)
(d) Volume (%) of sediment in various compartments

'conveyor belt' are perhaps best demonstrated by diagrammatic and tabular illustrations of where sediments are stored in the system and their average residence time in those stores before 'moving on' by rejoining the channel system. Once in the channel, depending on its length, sediment size is extremely important in determining residence time. In short rivers the suspended load may well reach the estuary on one flood but coarser sediments may become incorporated in the floodplain at several sites downstream. Sediments from process legacies (such as glaciation) may also leave storage to the channel and therefore complicate the assessment of contemporary sediment budgets (see Section 2.4).

Madej (1984) has compiled the spatial and temporal picture of fluvial sediments stored in the Redwood Creek basin of northern California. Figure 2.5 here shows her classification of the storages as active, semi-active, inactive and stable together with their distribution in planform, valley cross-section and with drainage area. Such an inventory is of exceptional value in assessing the likely impact of 'opening' or 'closing' storages, for example, by cultivation/drainage of hitherto stable zones or flood proofing the channel from semi-active areas. Unfortunately, the research effort to assess a semi-natural catchment in this way prior to development is prohibitive; however, it remains for the concept of sediment storages and their role in the spontaneous regulation functions of the basin ecosystem to be taken on board by developers.

2.2 CHANNEL MORPHOLOGY

Channel morphology is an output from the transfer system – it can be diagnostic of system states and geomorphologists have recently paid much more attention to channel classifications and typologies in an effort to formalise their knowledge of this diagnostic value. At every site on a river system which is transporting water and sediment the channel's morphology is adjusted, or is in the process of adjusting, to these downstream fluxes. Characteristic *planform* channel morphologies typify transport systems for different calibres of sediment (see Figure 2.6); relationships between planforms and flow regimes are known to exist but are poorly quantified because of the influence of sediments and of other factors to which we may pay only brief attention.

It is, for example, clear that channel erosion and deposition to bring about adjustment of form to flow cannot occur if the bounding materials are too resistant. Therefore, bedrock or glacially derived deposits may confine the channel to a non-equilibrium pattern. The content of silt and clay in channel boundary materials is also effective in determining the width/depth allocation in an individual cross-section (see Schumm, 1977, ch. 5). Of perhaps greater importance to modern forms of management orientated towards conservation is the importance of bank vegetation (in-channel vegetation also influences flow capacity).

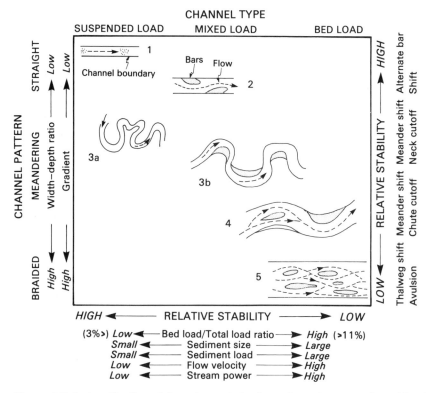

Figure 2.6 A classification of river channel planforms based upon sediment load, cross-section and stability (Schumm, 1985)

Whereas the stability and roughness of inert materials in the channel is taken as invariable, except by massive and costly works, that attributable to vegetation has traditionally been considered manageable, and a large part of river management work is to that end.

However, there is a dual attitude to the value of vegetation in rivers. On the one hand, its role in preventing scour and protecting bed and banks is recognised, both through the binding action of roots and through the streamlining of flexible leaves and stems. The UK Hydraulics Research Station (Charlton *et al.*, 1978) have established, for example, that unvegetated and short grass channels are, on average, 30 per cent wider than their tree-lined counterparts. On the other hand, there are, in lowland rivers, known flood risks resulting from additional roughness of profuse 'weed growth', the reduction of channel capacity by the bulk of plants, the possibility of increased turbulence around trees in floods, the risk of sudden bank failure if a tree falls, and the possibility of log-jams or weed-jams at bridge points damming the flow.

The balance would appear to favour removing everything except short perennial grasses – and, sadly, miles of watercourses in lowland areas have been reduced to that. But, fortunately, many river managers are glad to retain and even develop varied river vegetation, for its beauty, its landscape value and for the wildlife it harbours.

Traditional methods of river management frequently used the attributes of native plants to stabilise banks and deflect flows where needed. Routine, regular maintenance by hand labour ensured that plant cover did not degenerate. Thus, sallows and alders were planted on rivers, reedbeds along canal banks, and turf walls packed down as toe protection along dikes. However, traditional practices have, in many places, been forgotten or are seen as a 'luxury' under present financial and staffing arrangements. In the rush to capitalise (in more ways than one) on the products of the last fifty years of technology in machinery and materials, river managers have tended to forget the advantages of utilising vegetation to stabilise river sections.

2.2.1 Stable/unstable channels and channel change

It is the river engineer's job to treat problems of river instability at a specific site, yet it is important to have a synoptic view of the river from source to mouth, as a continuous transport system subdivisible, as shown above, only by sediment size and availability which, in turn, control morphology. Nevertheless, it has been traditionally thought in Britain that 'upstream' controls on river form had little influence on what happened in a given reach: average flows of both water and sediment have been considered unvarying in the long term. Consequently, the engineer has used empirical data, often from such non-dynamic environments as canals, to guide him on the selection of the basic channel dimensions of width and depth and the stream's velocity. The anticipated success of such channel designs can be judged from the terms 'regime' and 'grade' which are used to describe a 'trained' river in perfect equilibrium with its dominant flow, usually taken to be the annual flood. However, rivers respond over different timescales to different influences. Over geological timescales rivers can clearly cut down to sea level. Davis deserves credit for working out this long timescale, but it coloured much of his argument and eventual classification. At the timescale over which riparian owners these days expect the river to be stabilised, it is the lateral migration of the channel and not downcutting which is of most concern.

Furthermore, it is lateral migration which is best recorded as a pattern, if not a process, by geomorphologists. The sinuosity of river planforms has always excited the interest of a number of sciences but key factors in our present knowledge based on channel patterns have been the ability to make measurements of bank erosion relatively simply and our access to relatively long periods of data from old plans, maps and aerial photography (see compilations by Gregory, 1977 and Hooke and Kain, 1982; also Figure 2.7).

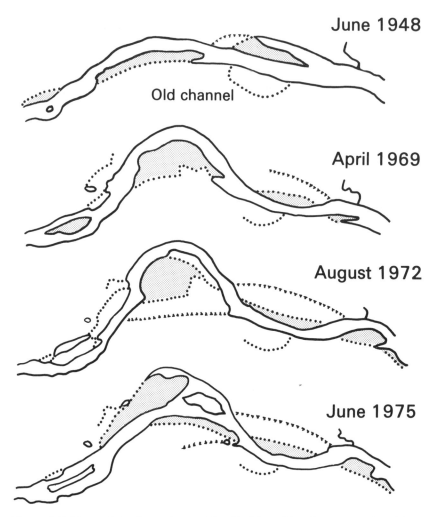

June 1948

Old channel

April 1969

August 1972

June 1975

Figure 2.7 Channel planform change identified from historic maps and aerial photographs: River Severn, Maesmawr, mid-Wales (Thorne and Lewin, 1982)

As a result of this concentration in fact we have tended to underestimate the contemporary importance, at least locally, of *vertical* morphological development. This imbalance is, however, being corrected (Schumm *et al.*, 1984; Lewin *et al.*, 1988).

The tendency of rivers to follow curved rather than straight courses is another systematic problem of river engineering. The Davis classification, by linking meanders to the 'mature' or 'old age' river, has tended to give the impression that a river wanders aimlessly without energy for erosion. This is a fallacy, since meanders develop in floods, when the river has maximum available energy. Energy not used to overcome bed and bank friction is far from equally distributed across a river. Instead, turbulence breaks up into a number of cells, producing currents which act laterally with as much force as the main thread of downstream flow. These secondary flows are generated in any stream and with a regularity which relates to the stream's width, that is, to the space available for the cells to develop.

In streams where banks are compact but where a coarse sediment load is carried, secondary flows create midstream shoals, or riffles, and pools at fairly regular intervals; where the banks are erodable, riffles and pools still occur, but the secondary cells produce a sinuous river planform by lateral erosion and deposition. Cut-banks occur opposite shoals and together they create the familiar pattern of the meandering river. Given further development, rivers may also exhibit a form in which both banks are cut and the shoals are mainly in mid-stream; this is the braided river typical of streams in recently glaciated areas.

This natural tendency to sinuosity, and the tendency for meanders and other less regular river bends to migrate by erosion and deposition, means that the equilibrium regime approach which is mostly indexed from data on straight reaches is seldom successful in coping with sinuosity. Much expense and environmental degradation is involved in forcing a river to flow where it is put when the designed planform is inappropriate to the reach's position within the system. Clearly, however, river engineers would willingly use a new empirical method of predicting river planform pattern, if one existed. The most hopeful method is that based upon the secondary flow pattern itself, because it describes the pattern with which energy is expended within the channel. Current meters now exist which can measure cross currents such as these; even simple photography will identify the locus of maximum stress on the banks and reveal just where any necessary structural or vegetative bank protection can be positioned.

In river systems there is still argument over the form of equilibrium adopted by rivers (see Section 2.1). Formerly Schumm's concept of threshold change was widely applied to river planforms, especially to the segregation of meandering and braided forms on the basis of channel slope and discharge, which was originally proposed by Leopold and Wolman (1957) and widely used as an indication as to 'what the planform should be' in a given reach. However, the boundary line has been regarded as a threshold condition implying costly instability in planforms. This view is widely challenged at present (Carson, 1984; Ferguson, 1987), with the result that managers will need new guidance as to the natural stability or instability of channels.

There are form guides to river instability (Lewin et al., 1988) but there are also good grounds for assuming that man-made changes to flow and sediment systems evoke morphological change which may occur over relatively short timescales.

Table 2.2(a) Identifying sediment sources: indicators of stability

Categories	Upland zone	Transfer zone	Lowland zone
Potential sediment sources	Slope debris Peat slides Alluvial fans Boulder berms Channel bars Bank erosion Forest ditches Forest roads Moorland 'grips' Mining	River cliffs Terraces Bank erosion Channel bars Field drains Field runoff Urban runoff Tributaries/upstream Mining	Upstream Bank erosion Bed movement Field drains Field runoff Wind blown material Tributaries Urban runoff Tidal sediments Mining
Evidence of erosion	Perched boulder berms Terraces Old channels Old slope failures Undermined structures Exposed tree roots Narrow/deep channel Bank failures both banks Armoured/compacted bed Gravel exposed in banks	Terraces Old channels Narrow/deep channels Undermined structures Exposed tree roots Bank failures Armoured/compacted bed Deep gravel exposed	Old channels Undermined structures Exposed tree roots Narrow/deep channel Bank failures Deep gravel exposure
Evidence of aggradation	Buried structures Buried soils Large uncompacted bars Eroding banks at shallows Contracting bridge space Deep fine sediment in bank Many unvegetated bars	Buried structures Buried soils Large uncompacted bars Eroding banks at shallows Contracting bridge space Deep fine sediment in bank Many unvegetated bars	Buried structures Buried soils Large silt/clay banks Eroding banks at shallows Contracting bridge space Deep fine sediment in bank Many unvegetated bars
Evidence of stability	Vegetated bars and banks Compacted weed-covered bed Bank erosion rare Old structures in position	Vegetated bars and banks Compacted weed-covered bed Bank erosion rare Old structures in position	Vegetated bars and banks Weed-covered bed Bank erosion rare Old structures in position

Source: Newson and Sear (1994)

Table 2.2(b) Qualitative models of channel metamorphosis, illustrating the direction of morphological response to particular combinations of changing discharge and sediment yield (after Schumm, 1969)

(a) Increase in discharge alone
$\quad Q^+ \quad w^+d^+F^+L^+s^-$

Decrease in discharge alone
$\quad Q^- \quad w^-d^-F^-L^-s^+$

(b) Increase in bed material discharge
$\quad G_b \quad w^+d^-F^+L^+s^+P^-$

Decrease in bed material discharge
$\quad G_b \quad w^-d^+F^+L^+s^-P^+$

(c) Discharge and bed material load increase together; e.g. during urban construction, or early stages of afforestation
$$Q^+G_b \quad w^+d^=F^+L^+s^=P^-$$

(d) Discharge and bed material load decrease together; e.g. downstream from a reservoir
$$Q^-G_b \quad w^-d^=F^-L^-s^=P^+$$

(e) Discharge increases as bed material load decreases; e.g. increasing humidity in an initially sub-humid zone
$$Q^+G_b \quad w^=d^+F^-L^=s^-P^+$$

(f) Discharge decreases as bed material load increases; e.g. increased water use combined with land-use pressure
$$Q^-G_b \quad w^=d^-F^+L^=s^+P^-$$

It is perhaps time for geomorphologists to be less exceptionalist in their treatment of instability; in all systems characterised by dynamic, rather than static, equilibrium stability is an illusion – particularly if engineers feel that they can impose it. In considering the river channel and catchment system in relation to the potential impacts of development we should perhaps move to Barrow's (1995) use of the terms *sensitivity* (the degree to which an ecosystem undergoes change) and *resilience* (its ability to change without breakdown) as sub-components of 'stability'. These terms could be applied to the sediment transfer system via, for example, different classes of channel (note that 'relative stability' is an axis of Figure 2.6). In view of the many successful attempts to restore river channel morphology now recorded (see Chapter 6) it is perhaps resilience which is the more important character-istic, though channels of high sensitivity should be protected *ab initio* from negative development impacts. We also need to consider that metastable equilibrium conditions are quite common in the fluvial system (see Figure 2.4b) and field indications of the state of the catchment (such as those provided by Sear *et al.*, 1995; see Table 2.2a) should be used in conjunction with Schumm's brainstorming model in Table 2.2b.

Flow changes can be expected following changes in land use, land management (e.g. drainage) and the manipulation of river flows by the water industry (e.g. abstraction for water supply). The developments occurring in upland areas today – afforestation, timber harvesting, and farm improvements – mean that changes in sediment supply can also be expected. Such changes are already being monitored in a few catchments and give some guide to the scale of change to be expected in others.

However, the complexity of interrelated factors suggests that accurate predictive models for a particular system cannot yet be confidently extrapolated from data derived elsewhere. Erosion rates and stream loads need to be monitored as well as water flows so that in river management and project planning the system's degree of metastability can be taken into account.

2.2.2 Hydraulic geometry, regime and channel design

While the trajectory of much current geomorphological research favours a generally metastable interpretation of channel form developments, it is essential to review briefly the equilibrium (steady-state) approaches which continue to guide management of the world's river channels.

Hydraulic geometry and *regime* approaches represent, respectively, the geomorphological and engineering approaches to the adaptation of channel width, depth and velocity of flow to water discharge. The former is empirical, involving data collection foremost; the latter is theoretical and therefore more flexible and capable of extension to include channel slope, roughness, and sediment load. Both assume that in the longer term (timescales are not specified) there is an adjustment of channel dimensions to flow which is predictable. Thus regime theory is ideal for the construction of new conveyance channels (needed for flood relief or irrigation) and was largely 'proved' in practice with low-gradient canals lined by cohesive materials.

Because hydraulic geometry research represents an important phase in the quantification of fluvial geomorphology there are many compilations. After its widespread popularisation in the formative text of Leopold *et al.* (1964) numerous studies were carried out, either of changes at-a-station with varying flows or downstream as flow builds up.

Park (1977) compiled the first international review of the results. This indicated little consistency in the exponents (rate of change) reported for width, depth and velocity changes with discharge at-a-station or downstream; the downstream problem appeared to be inconsistent choice of an indicative discharge for the basin channel network. Park concluded that alternatives to simple linear relationships should be explored or the conclusion drawn that 'maladjustment could not be proven for flow transmission'.

Regime channel design begins with a clear problem – that of producing an active channel which may well scour or fill temporarily but which, during its design lifetime, will transfer its charge of water and sediment without profound morphological change. The design lifetime of an engineered channel varies but may be up to a century and it should be emphasised that the nature of river engineering schemes presents opportunities to reinforce or maintain regime channels as they adjust; geomorphologists argue, therefore, that the steady-state assumption of such designs is seldom fairly stated or tested. The regime approach selects a stable width and depth and, therefore, a velocity (together making up water discharge); the hydraulic resistance, slope and sediment transport properties are also chosen from deterministic equations (Table 2.3).

Slope is rarely adjustable in channel design since bridges, sewer outfalls and other structures cannot (except in new irrigation schemes) be adjusted *a posteriori* to channel slope. Here regime theory is at its most vulnerable and there is abundant evidence that straight channel planforms built in good faith to provide flood protection are liable to adjust to a meandering or even to a braided pattern without considerable structural protection; the paradox is that, by selecting the 'wrong' assessment of a channel's dynamics engineers can increase damage in the longer term. Figure 2.8 illustrates how the planform of the Rhine has reacted to 'training' and shows how the changed gradient which accompanies a cut-off (natural or engineered) promotes

Table 2.3 Regime approaches to stable channels, geomorphological approaches to changing channels

Regime equations for engineering design

Equations for straight sand-bed channels Blench (1952)	Equations for straight gravel-bed channels Kellerhals (1967)
$w = (F_b/F_s)^{0.5} \, Q^{0.5}$ $d = (F_s/F_b)^{1/3} \, Q^{1/3}$ $s = \dfrac{F_b^{\,5/6} \, F_s^{\,1/12} \, v^{1/4}}{3.64g Q^{1/6}(1 + C_b/2330)}$ $F_b = 0.58 D_{50}^{0.5} \, (1 + 0.012 C_b)$ $F_s = 0.09$ (sandy loam) to $\quad\quad 0.028$ (clay) $v =$ velocity of flow $R =$ hydraulic radius (cross-sectional area/wetted perimeter) $C_b =$ bankfull cross-sectional area $Q =$ discharge $w =$ channel width $g =$ gravitational acceleration $s =$ slope of water surface $\tau_{0c} =$ critical shear stress for sediment transport $\rho_w =$ density of water $d =$ depth of flow	$w = 3.26 \, Q^{0.5}$ $\dfrac{v}{\sqrt{(gRs)}} = 6.5\left(\dfrac{d}{D_{90}}\right)^{1/4}$ $\tau_{0c} = \rho_w gRs = 15.79 D_{90}^{0.8}$ $d = 0.1830 Q^{0.4} \, D_{90}^{-0.12}$ $v = 1.68 Q^{0.1} \, D_{90}^{0.12}$ $s = 0.086 Q^{0.1} \, D_{90}^{0.92}$
(F_b and F_s are Blench's bed and side factors)	$\left.\begin{array}{l}D_{50}\\ D_{90}\end{array}\right\}$ are sediment sizes at which 50%/90% are coarser

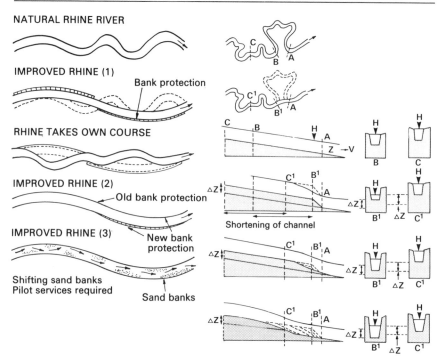

NATURAL RHINE RIVER

IMPROVED RHINE (1)

Bank protection

RHINE TAKES OWN COURSE

IMPROVED RHINE (2)

Old bank protection

IMPROVED RHINE (3)

New bank protection

Shifting sand banks
Pilot services required

Sand banks

Shortening of channel

Figure 2.8 Engineering problems of meandering channels:
   (a) The response of the Rhine to 'training'
   (b) Reaction of a meandering channel to cut-off (Ryckborst, 1980)

erosion and deposition. 'Natural' channel morphology may be considered as part of the spontaneous regulation function of the transfer part of the basin ecosystem; it can only be sacrificed at a cost.

Recognising this interdependence of form and process in fluvial systems, streams in North America have been experimentally designed to form and maintain riffles, pools and meanders where desired (principally by fishing interests). No structures were used; control of bank slopes and variation in channel cross-sections resulted in the emergence of point bars and pools after the first above-normal flow, in the required locations, even in straight reaches. It is too much to expect that there could be similar research before all new channelisation projects in the UK. However, computer modelling has enabled more complex designs to be attempted and tested, with encouraging results in the field for 'natural' habitats and landscapes.

2.3 FLOODPLAINS

It is perhaps invidious to select only one piece of characteristic river morphology for separate treatment in this chapter. The floodplain, and a set of both narrower and wider valley features, are, however, of critical importance to a range of human river-use systems and to the conservation of nature (Newson, 1992b). They represent a further component of the spontaneous regulation function in the sediment transfer system. Therefore, it is essential to know how floodplains form, what are their natural regulatory functions and how our use of them involves costs as well as benefits.

A cross-section through a river valley in the transfer zone indicates several important features (Figure 2.9). The *valley floor is* the broadest definition, encompassing all the landforms dominated by processes of deposition, including legacies such as those of glacial deposits. At the narrowest level is the *river corridor,* recently defined officially for conservation purposes in the UK and governed by interactions with channel flow and sediment regimes via the agency of flora and fauna dependent on the channel itself (we can also include human recreation). In other countries the concept of *riparian zones* is becoming part of river management culture, again emphasising the interdependence of the channel and its surrounding features.

The floodplain represents that area across which the river escapes during floods and therefore may be subdivided by the frequency of the flood concerned. Since the discharge which fills the channel *(bankfull discharge* of hydraulic geometry; *dominant discharge* of regime) occurs between once a year and every other year, floodplains appear captivatingly suitable for human settlement, agriculture and communications. There is considerable surprise and anxiety when, subsequently, damaging, costly and fatal inundations occur.

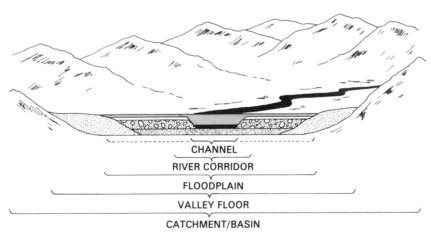

Figure 2.9 Zones of the valley floor within the basin

2.3.1   Floodplain formation and functions: floods, aquifers

Much of the damage caused by inundation of floodplains occurs by relatively concentrated flows down features resembling infilled remainders of river channels; these so-called palaeochannels also provide the best indication of floodplain formation processes. These processes are closely linked to lateral mobility of active river channels. As sinuous patterns evolve and migrate, valley sides become eroded; the active channel therefore creates a wider valley in which both flood waters and sediments can be stored. The storage of sediments is, as we observed in Section 2.1, a salient property of the transfer system and in confined river reaches there is clearly a throughput of both flow and load until spreading can occur. In regions which have experienced climatic fluctuations, particularly glaciation, valleys tend to have adjusted to the flow of ice or to substantial meltwater streams. The present channel is therefore often a 'misfit' in a wide valley. This makes the confinement which our occupation of floodplains usually implies (by flood- or erosion-protection engineering) even more intensive.

Sediments enter floodplain storage in two principal ways: by deposition in the channel (e.g. as bars and shoals) followed by abandonment, through migration, of that channel, or by out-of-bank flows. Clearly, the former mechanism favours coarse clasts of sediment and the latter fine clasts; the result is that floodplains are frequently composite, with coarse material below, topped by fines. A further environment for the deposition of fines is the 'backwater', partially abandoned, channels of alluvial (silt/clay) reaches.

The coating of fines, often carrying organic and chemical nutrients, makes a fertile parent material for soil development – one of the original reasons for the 'hydraulic civilisations', notably that of the Nile, whose floodplain is seasonally inundated and 'fertilised' with silt. There are two dimensions in which floodplain sediments act as aquifer deposits. First, the down-river flow of water is seldom restricted to the channel itself; large volumes can move 'invisibly' close to the open channel but having leaked into the coarser deposits of the floodplain, often following the palaeochannels described above. There are two notable demonstrations of this phenomenon: the successful abstraction of relatively large amounts of quite pure water for supply from floodplain deposits (including those of former river courses in the semi-arid zone) and the 'loss' through temporary leakage of volumes of water released from regulating reservoirs (Chapter 6).

In the other dimension, at right-angles to the down-valley flow, all the runoff contributed from valley-side slopes enters the channel via the floodplain deposits. Therefore, considerable modification of flow patterns occurs across the floodplain and we now know that this is accompanied by beneficial chemical changes such as nutrient stripping (Pinay and Decamps, 1988).

2.3.2   Floodplain modifications by Man

The typical river valley of the developed world is now a wide corridor of intensive land use and water use. Since settlements and infrastructures cannot continually move in response to flood or drought, floodplains have become regulated with more or less respect for natural dynamics according to culture, cost and severity of natural impacts. In a polarised case the following developments may have occurred:

(a) Channel straightened and erosion-proofed.
(b) Extensive flood protection structures.
(c) Extensive irrigation and/or drainage.
(d) Removal of natural vegetation and wetlands.
(e) Encroachment of buildings and structures towards the channel.
(f) Use of floodplain and channels/palaeochannels for waste disposal.

These all result from the exploitative nature of the development process and it is only in recent years that a more cautious approach has been necessitated by the competing objectives of conservation and recreation and by the high costs of the more heroic defence of our occupation of floodplains against natural processes.

There is now a developing field of 'fluvial hydrosystems' centring on the natural relationship between the flow and sediment transfer functions of the active channel in terms of interrelationships with floodplain processes. French geomorphologists and biologists have championed this development (see Amoros *et al.*, 1987 and Figure 2.10) and are supporting it with research into the timescales of stability necessary for valley floors to carry out their natural functions of wildlife corridors, water storage and filtration. In Figure 2.10 the biological populations A, B and C are sequentially incorporated in the stream ecosystem, as are those at depths X, Y and Z, providing natural stream migration rates are permitted. Biota have high survival rates in fluvial storage zones. The human imprint on developed floodplains has, however, consistently sought to separate the areas of settlement or agriculture from the water and sediment transfer system, isolating and destroying the important source of genetic information which would normally be stored, supplied and stored again. In the USA there is also a trend towards the zoning of floodplains so as to permit a graded land use with less vulnerable uses adjacent to the channel and more vulnerable uses at a 'safe' distance (see Chapter 4). In the UK new research points to the desirability of considerable floodplain 'retirement' to allow nutrient-stripping from the effluent of productive agriculture on valley sides and interfluves. There is also a considerable interest in the natural use of floodplains to store flood waters, a use to which farmers have apparently become adjusted as in the case of Lincoln (see Chapter 7).

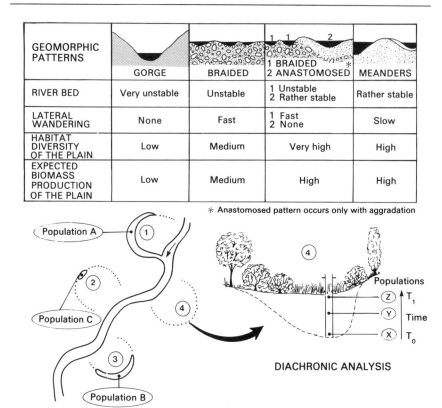

GEOMORPHIC PATTERNS	GORGE	BRAIDED	1 BRAIDED 2 ANASTOMOSED	MEANDERS
RIVER BED	Very unstable	Unstable	1 Unstable 2 Rather stable	Rather stable
LATERAL WANDERING	None	Fast	1 Fast 2 None	Slow
HABITAT DIVERSITY OF THE PLAIN	Low	Medium	Very high	High
EXPECTED BIOMASS PRODUCTION OF THE PLAIN	Low	Medium	High	High

＊ Anastomosed pattern occurs only with aggradation

SYNCHRONIC ANALYSIS

DIACHRONIC ANALYSIS

Figure 2.10 The importance of 'wild' river channel migration for the creation of a variety of habitats in space and through time (Amoros *et al.*, 1987)

2.4   BASIN SEDIMENT SYSTEMS

It is perhaps surprising that even in those nations which face severe river basin management problems as a result of soil erosion, river instability, or other factors promoting high rates of sediment transport, measurement networks are generally sparse or absent. Consequently, it is virtually impossible to calculate either average or extreme transport rates other than by compiling lists of those measurement results which are available (e.g. Holeman, 1968). This assemblage of data leads to attempts to predict the sediment *yield* of rivers (i.e. the rate of output per unit area) on the basis of climate or geology (e.g. Jansen and Painter, 1974). Notwithstanding such predictive tools it is usually essential to make on-the-spot measurements in those cases where a specific problem of erosion or sedimentation is present.

The available techniques are far from satisfactory, particularly because none of the routine ones produces a continuous record of either suspended or bedload yields, especially during flow extremes. The need for simultaneous measurements of river flow also conspires to make sediment yield and transport rate measurements difficult. Progress is, however, under way towards, for example, the use of turbidity sensors to measure suspended sediment loads (Walling, 1977) and the installation of pressure-sensing or weighing traps for bedload (Reid and Frostick, 1986).

2.4.1 Sediment budgets

The last thirty years of research in drainage basin sediment systems have been a period of process studies backed by less detailed calibrations of the magnitude of inputs and outputs of the system, paralleling the measurement of water balances in hydrology. A recent international symposium (Bordas and Walling, 1988) indicates a considerable geographical spread of research effort but a continuing anxiety over techniques and over the ability to construct satisfactory budgets given the 'jerkiness' of the sediment conveyor from source to sink and the fact that many sinks are, albeit temporary, in mid-basin. The much shorter turnover times in hydrology make budgeting on an annual basis much more satisfactory.

The most severe practical problem confronting the measurements of sediment budgets for management (e.g. to determine the rate at which a new reservoir will fill with sediments) is that temporal and spatial discrepancies in throughput of sediments lead to mismatch of, for example, soil erosion rates and sediment yields further down the basin (see also Chapter 6, pp. 216–17). This discrepancy is, however, considered to have certain regularities and the basin *sediment delivery ratio* defines the proportion of headwater sediment supply which reaches the outlet of river basins of increasing area. The delivery ratio requires much more research, particularly because it implies a smooth operation for processes we know to be disjointed (see Section 2.1). Nevertheless, for small basins it is now common to see the sediment budget compartmentalised into all the relevant sources and sinks (see Foster *et al.*, 1988 and Figure 2.11). Such sediment sourcing is an ideal corollary to mapping of storage zones as a preface to basin development (Section 2.1).

The problem of delivery processes becomes much more severe when in certain glaciated or eroded regions a considerable legacy of stored sediments becomes re-mobilised after hundreds or even thousands of years. On the eastern seaboard of the USA the era of European settlement led to massive soil erosion; much of the resulting sediment was, however, stored as *colluvium* (a basal slope deposit – shown in Figure 3.13b).

2.4.2   World sediment yields

Given the requirement posed by the concept of sediment delivery to assess sediment yields much more thoroughly than has been the case to date, it is essential to conclude this chapter with a brief coverage of the major factors which control sediment yields and of how, within a typical large basin, we may refer yields back to their controlling processes.

Jansen and Painter (1974) provide a simple assessment of 'the global' denudation rate: 26.7×10^9 tonnes per year; the range of other published estimates which they review is from 12.7×10^9 tonnes to 58.1×10^9 tonnes. Such estimates are of most use to geologists making comparisons with estimated rates of 'new rock' production. For river basin managers it is necessary to assess the geographical variability of yield for standard areas. Holeman (1968) does this by continent:

	$tonnes. \, km^{-2}. \, yr^{-1}$
North America	97
South America	63
Africa	27
Australia	33
Europe	35
Asia	600

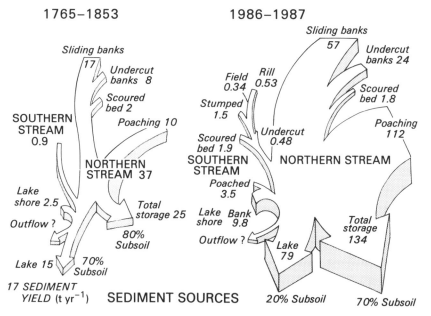

Figure 2.11 Source components of the sediment inflow to two Midland lakes (Foster *et al.*, 1988)

Jansen and Painter (1974) by climate:

	$tonnes. km^{-2}. yr^{-1}$
Tropical rainy	71.5
Dry	169.0
Humid, mediterranean	714.4
Humid, cool	46.5

and Fleming (1969) by vegetation type. Schumm (Figure 3.13a) also provides a broad-brush link to precipitation and temperature.

A major review of all the component processes of continental denudation, produced by Saunders and Young (1983), concluded that acceleration of natural erosion rates by human activities ranges from two to three times with moderately intense land use to ten times with intense land use. However, in any such estimation it is necessary to recall the local, partial and unsophisticated nature of most geomorphological measurements; bedload transport is frequently merely 'added on' as a token 10 per cent and solutes, ignored in this chapter (see Chapter 3), are likely to equal or exceed sediment yields in many large river basins.

The most integrative concept with which to leave this chapter on the basic physical stability of the river basin is perhaps that of sediment residence times. Figure 2.12 shows a section through a typical humid, cool river basin and indicates the location of the key sediment storages and an estimate of the average duration of that storage. This illustration stresses:

(a) The division of the transfer system into supply and transport.
(b) The importance of storage of sediments.
(c) The vulnerability of stores to mobilisation through river instability or land-use change.
(d) The importance of taking longer timescales into account in basin management.

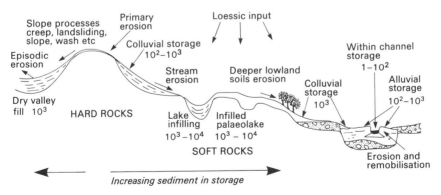

Figure 2.12 Duration and location of long-term sediment storage in the fluvial landscape (Brown, 1987)

2.5   SUMMARY: KEY ELEMENTS OF THE NATURAL SYSTEM, A SENSITIVITY ASSESSMENT

We have seen at many points of this chapter the central role of the basin transfer system in the ecosystem model and, therefore, in any assessment of sustainable basin development. Marchand and Toornstra (1986) tabulate in detail the implications of loss of spontaneous regulation functions. The confidence of the geomorphologist in seeking a role in river basin management springs from the fact that, when the sediment transport system of any basin becomes destabilised, that basin 'falls apart' in more than one sense (see White, 1982). The advantages of the basin to human settlement and resource-utilisation patterns disappear as the system degrades. In many senses river sediments, once the source of wealth for the hydraulic civilisations, are the worst pollutants. Their sources are so widespread that control is far less rapidly achieved than for most chemical pollutants, not that they can be divorced from chemical pollutants since many of the latter are bound to sediment particles (Slaymaker, 1982).

While the river floodplain has many advantages to human resource utilisation and newly discovered values for conservation and for moderating and modulating the transport system, a destabilised river system will suffer from irregular and largely unpredictable extremes of flow which in some environments can remove the entire floodplain. At a much more modest level the expenditure necessary to 'clean up' such a system by removing the sedimentation from reservoirs, navigation-ways and even nature reserves puts an unbalancing burden on the broad base of holistic river basin management. Our new knowledge of the prevalence of threshold reactions by the river basin system, especially the potential of intrinsic thresholds, further adds to the dismal recovery prospects of a basin destabilised by poor land management, poor flow management or over-development.

As will be revealed in Chapters 5 and 6, the most sensitive world environments for such destabilisation to occur are precisely those in the semi-arid and mountainous marginal environments where population and development pressures are most likely to produce it.

While the river managers of moderate humid-temperate and long-developed nations are learning only slowly the importance of the system of fluvial geomorphology, it is imperative that those in charge of the process of development programmes in these sensitive environments do not engineer the ultimate failure of their ambitious programmes.

A considerable problem exists in convincing responsible agencies and institutions that fluvial geomorphology is relevant to river basin management (see Brookes, 1995a); a ready alternative exists in applying the results of engineering science, which takes a different route towards similar ends (Figure 2.13) but as a higher professional standing simply because geomorphology is, in most countries, an academic, not an applied subject.

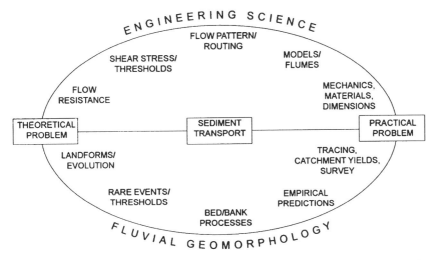

Figure 2.13 Fluvial geomorphology and engineering science: separate research agendas but common origins and aims for research questions (after Sear and Newson, 1994)

Table 2.4 Impact matrix for basin development – encompassing longer-term geomorphological viewpoint

Development problem	Geomorphological status of basin	Outcome/ risk
Dam construction	(a) Soil erosion active	Sedimentation in upland basins/dams**
	(b) Holocene/historical erosion	Sedimentation from stores***
	(c) Soil erosion controlled at source	Degradation of sediment stores***
River regulation	(a) Dam above alluvial zone	Channel change below dam*
	(b) Dam with flood control/hydro-power	Habitat effects via floods/bed structure**
	(c) Dam near coast	Coastal erosion*
Land-use development	(a) Land on slope/active tectonics	Failure unless small-scale***
	(b) Land on active/semi-active stores	Changes to hydrology release sediment***
	(c) Floodplain land	Losss of habitat/ spontaneous regulation***

Source: after Newson (1994a)
Note: Asterisks in third column indicate seriousness of risk.

Nevertheless, fluvial specialists are now making a considerable contribution to river management and it is often the simpler classificatory and observational materials which are of most help to river managers. For example, the river channel classification by Rosgen (1994) adds considerably to that provided here in Figure 2.6 and can be used as the basis both for judging adverse impacts and for schemes of restoration based on the 'natural' dimensions and features to be expected for a particular type of channel in a particular location in the basin.

Preservation of river systems is impossible because of their dynamism; however, conservation of a few wilderness rivers is equally misguided. The fundamental physical integrity of river basins must be respected by any management scheme which claims to be long term (see Table 2.4); against this claim some of the most immediately alarming effects of chemical pollution pale into insignificance.

Chapter 3

Land and water
Interactions

The sediment transfer system is normally laid bare for us to see; regulators of the system, both spontaneous (natural) and imposed are observable even if we choose to ignore their influence, but it is more difficult to apply the ecosystem concept to patterns and processes of runoff. Hydrology is a young science and, while its early days were preoccupied with surface processes, hydrologists now admit that many of the key points of the runoff system are hidden – in soil, in rock or in plants. Hydrological process studies are a feature of the last thirty years and we remain in ignorance about many of the key controls; we may assert the obvious – that land use and land management influence runoff and water quality – but proving the point remains difficult, especially when another human influence (the direct one of damming and piping water) is less subtle. Few hydrologists are prepared to take an ecological or 'land/water' approach to their publications (with some exceptions, for example, Falkenmark and Chapman, 1989).

As we have previously observed, 'water and land' best represents the driving force behind the hydraulic civilisations, their use of the land being controlled by the availability of water and the efficiency of the distribution network. Just as the controlling variables of fluvial geomorphology may change their status as dependent or independent according to timescales (see Chapter 2), so land and water have become reversed in order of influence – with our use of land having first priority, at least in the humid zones in which development has been rapid.

An example will illustrate the rapidity with which the balance may change. One of the preoccupations of water resource managers in the UK between 1930 and 1970 was that human recreational pressure might damage reservoir and river water quality and lead to the spread of disease, e.g. typhoid. During the 1970s recreational facilities were slowly developed on both reservoirs and rivers, with the purification costs borne by the water suppliers. By 1990 the situation had reflexed totally and recreational water users were being warned that reservoir and river water threatened them as the result of algal blooms!

The need now to address 'land and water' as a prelude to sustainable basin management puts pressure on scientists to identify and quantify those relationships of cause and effect which link the vast majority of the area of a river basin – its land surface – to the highly significant minority area – its rivers. At the same time, there is a danger that we neglect groundwater; while not 'organised' around surface river basins, the importance of groundwater supplies for human communities and some habitats makes that component of runoff an increasing concern of those pressing for sustainable basin development. Management units for pollution control now include 'groundwater protection zones' (see Section 3.2). The need for information and data on the influence of land use and land management on surface and subsurface runoff is urgent. The problems of environmental science in this situation are considered in Chapter 8; for the moment we must proceed as though hydrology were an exact science and its axioms worthy of operationalising. The hydrological cycle, quantified for river basin units, is an ideal starting point.

3.1   VEGETATION, SOILS AND HYDROLOGY

The UK is a humid, temperate land surface with a dense network of surface stream channels. However, even in mountainous parts of the UK, less than 2 per cent of the surface area of a river basin is occupied by stream channels. Clearly, therefore, with the exception of some lake and reservoir catchments, the huge majority of rainfall is translated to river flows via a canopy of vegetation and a mantle of soils, weathered bedrock or drift and via underground routes.

While it is now known through our experience with acid rain that precipitation itself may carry the pollution resulting from poor environmental management, we may assume for most of this chapter that vegetation and soils have had their naturally benign influences on the quantity and quality of groundwater and river flow severely corrupted by their development for productive uses. While it may 'stand to reason' to the non-scientist that this is so, it has taken hydrological experiments more than a century to elucidate the detail and to extend initial rural preoccupations to an urban and suburban context.

Chapter 1 relates the slow historical progress towards a cyclic concept of balancing components of the global water mass. For three hundred years, however, it has been axiomatic to hydrology that the hydrological cycle (Figure 3.1) links the important storages and fluxes of global moisture. The land surface is one of the important switching points in the cycle and a large number of land surface variables act to control the routing of precipitation through the land phase of the cycle (see Figure 3.2). Although major 'fixed' controls are operated by relief and climate, the hydrological cascade of storages and flows in the surface zone is important, critically important in

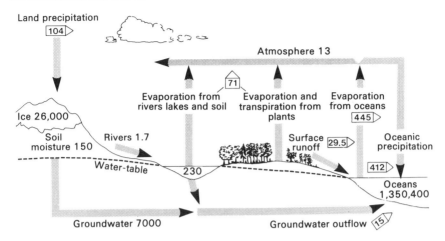

Figure 3.1 The global hydrological cycle; all volumes in thousands of cubic kilometres

many relief and climate zones. In fact land-use hydrology is very much a regional science, with the exact effects (particularly their magnitude) of land use dependent largely on regional conditions; for instance, precipitation and evaporation volumes and seasonality in the case of forest effects on water quantities reaching rivers (Newson and Calder, 1989 – see also Table 3.1).

3.1.1   The hydrological cycle in nature and the role of vegetation

The most telling demonstration of the relationship between fluxes and storages in the land phase of the hydrological cycle is that, while biological, domestic and industrial demand for water is almost constant, precipitation occurs for a relatively small proportion of the time; even in the humid conditions of the UK there are few places in which rain or snow falls more than 15 per cent of all time and many important regions where the figure is less than 5 per cent. 'Storage is the answer', both *artificial*, in obvious masses called reservoirs and, more importantly, *natural*, in the pores between soil and rock particles, faults and joints in aquifers, in ponds and lakes and in the channel network. These storages have been calculated as shown in Table 3.2; they can be represented at the river basin scale as a hardware model or as a conceptual model such as that shown in Figure 3.2. The latter device can be used to illustrate the potential impacts on the volume and quality of storages and the flows between them of land-use developments and land management techniques.

Land-use hydrology has at its core the solution of the water balance equation for a particular unit of land, normally a river basin but often at

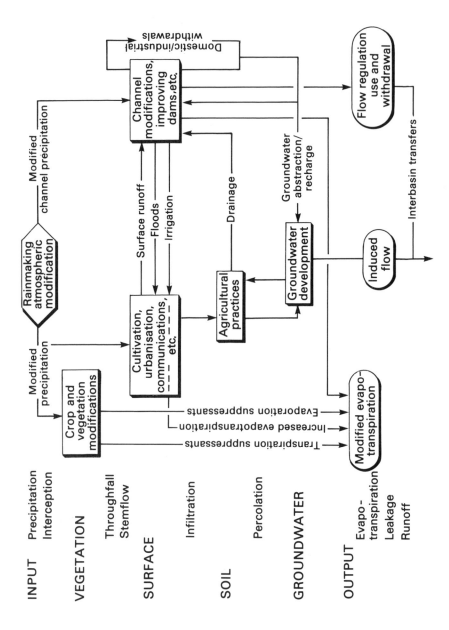

Figure 3.2 The river basin hydrological cycle shown as storages and flows, with artificial influences identified (original by J. Lewin)

Table 3.1 Regional process controls in forest hydrology

Regional element	Effects on processes
CLIMATE	
Rainfall regime	Low intensities/long durations favour interception if other conditions are right
Temperature/sunshine	Higher radiation inputs stimulate physiological water use via transpiration. Radiation climates essentially differ from advection climates
Wind speed	Advection energy and ventilation favour interception but also high rates of physiological stress and hence high transpiration where trees deeply rooted in wet soils
TREES	
Species	Often irrelevant but fast growers use more water if available
Stomatal control	Ofen superior to genetically produced agricultural crops
Rooting depth	Critical control on transpiration rate in dry conditions, especially if soil has good moisture storage (e.g. phreatophyte)
Spatial scale of forest	Controls edge effects v. core effects
SOIL/GEOLOGY	
Permeability	Controls water use and need/effectiveness of cultivation and drainage
Chemistry	Controls impact of acid deposition and need to apply nutrients/other chemicals
MANAGEMENT	Virtually complete control except over crop height and climate

much smaller 'plot' scales where measurements can be more accurate and control more secure:

$$P \qquad -Et \qquad +/-S \qquad =Q$$
Precipitation Evapotranspiration   Changes in storage Discharge in stream

It can be solved for any time period but generally an annual or flood hydrograph base is chosen so that initial and final conditions are similar, reducing the need to make comprehensive storage measurements. Even without this difficulty the measurement of evapotranspiration has proved very difficult to make directly and climatic calculations of the potential rate are much more common than those of actual rate. The ability to measure actual rates of evapotranspiration on small numbers of plants in plots or lysimeters has made these small-scale experiments popular but more recently methods of direct measurement of the vapour flux from plant surfaces have been devised.

Table 3.2 Storages and fluxes in the global hydrological cycle

	Values (km³ × 10³)	Percentage of total
STORAGE		
Ocean	1,350,000.0	97.403
Atmosphere	13.0	0.00094
Land	35,977.8	2.596
Rivers	1.7	0.00012
Freshwater lakes	100.0	0.0072
Inland seas, saline	105.0	0.0076
Soil water	70 0	0.0051
Groundwater	8,200.0	0.592
Ice caps/glaciers	27,500.0	1.984
Biota	1.1	0.00008
ANNUAL FLUX		
Evaporation	496.0	
Ocean	425.0	
Land	71.0	
Precipitation	496.0	
Ocean	385.0	
Land	111.0	
Runoff to oceans	41.5	
Rivers	27.0	
Groundwater	12.0	
Glacial meltwater	2.5	

The other dilemma in evaporation studies is that of how to separate two controlling processes in the loss of moisture from vegetated surfaces:

(a) *Interception* is the detention of precipitation on plant surfaces and its re-evaporation from that location.
(b) *Transpiration* is the 'use' of water by the plant physiological system linking roots and leaf stomata (variable openings).

It was Dalton who, in 1801, demonstrated that evaporation rates from containers of warmed water were proportional to the vapour pressure difference between water and air; this resistance or aerodynamic approach to the phenomenon became less easy to solve for field sites where climatological data made the energy balance approach through measurements of radiative power for evaporation more popular. In 1948 the late Howard Penman combined the two approaches in an equation which was to facilitate vastly estimation of water balances, particularly in the humid temperate zone.

To allow for reduced actual evaporation via the transpiration process when soil moisture is depleted, Penman introduced a concept of the *root constant*, fixed differently according to crop rooting depth. We may take simplistic guidance from this concept that deeper-rooted plants (such as trees) 'use' more water than shallower-rooted plants (such as grasses), a difference which may well hold for warm, dry climates with a plentiful supply of water to roots (e.g. riparian trees or phreatophyte species).

Calder (1990) traces the isolation and solution of a more difficult problem with the original Penman equation – one of canopy height, not rooting depth. Penman's aerodynamic calculations are appropriate for short, but not tall crops, especially in windy climates when the canopy is wet. Under dry conditions the Penman approach has some mutually compensating errors between aerodynamic and radiative terms but the errors are so large ($\times2$) for wet, tall crops such as forests that a separate approach to evaporation under these conditions (i.e. of interception) is now the norm for evaporation equations and models.

The issue of relative rates of evaporation achieved under different conditions in different climates by interception and transpiration by species of different heights has dominated land-use hydrology this century.

3.1.2 Important canopy processes

In terms of our knowledge of their hydrological effects, the above-ground portions of natural plant, crop and forest covers have grown in importance since the processes they control have been investigated more closely. As Figure 3.2 demonstrates, the canopy is itself a temporary store in the hydrological cycle. While not constituting a large depth of storage in terms of incoming precipitation (commonly <5 mm), the lateral extent of plant cover is considerable and therefore volumes are by no means negligible. While storage is only temporary the individually small volume of each storage site and its comparatively large surface area mean that both evaporative and chemical (solute) transfers are efficient in the canopy under most conditions.

It is important to list the properties of vegetation canopies which influence their hydrological behaviour (Table 3.3). In the early days of qualitative observations on, for example, the hydrological influence of natural forests (see Chapter 8 and Kittredge, 1948) canopies were seen as exercising a simple, single, umbrella-like influence on rainfall and a generally protective, beneficial outcome was assumed for river behaviour. Nevertheless, even at that stage, failure to separate the precise canopy processes and their variability with canopy properties led to controversy and misunderstanding. Issues of canopy effects were often confused with those of soils and soil management (see Section 3.1.3). No rapid improvement was produced by quantified observations; here catchment experiments involving a manipulation of canopy cover, such as afforestation or deforestation/harvesting, were found to produce different effects on streamflow (or different magnitudes of effect), depending on plant species, climate and other variables.

Table 3.3 Vegetation canopy properties having an influence on the hydrological performance of vegetation cover

Interception (a)	Transpiration (b)	Evaporation (a + b)
'ARCHITECTURE'	'PHYSIOLOGY'	'SITE'
Crop height	Stomatal cover	Exposure – regional
Canopy depth	Seasonal growth	– local
Leaf area (per plant)	Growth stage of plant	Ventilation
Leaf shape	Health of canopy	Albedo
Crop spacing		Radiation climate
		Seasonality of site

It is perhaps not surprising that ignorance prevailed so long in the field of canopy hydrology; one of the dominant processes operating at canopy level, evaporation, is the most difficult to measure or predict in the hydrological cycle. Ultimately, therefore, clarification has come from a wide variety of scales of research carried out by several disciplines in numerous climates (even so, neglecting the importance of vegetation management). Vegetation changes may be accidental (fire, disease) or deliberate, but in each case the 'before and after' covers may be different (e.g. trees may follow bracken, or heather, or scrub rather than short grass).

It has been clear that during the short life of the International Hydrological Decade and Programme (1965–74 and subsequently) forest hydrology has dominated studies of land-use effects. In the proceedings of the first international symposium on the subject in this period, held in Wellington (IAHS-UNESCO, 1970), it is clear that, while a very large majority of papers are on techniques and problems of extrapolation (indicating a youthful stage?), a significant number reported results from forest studies. Ten years later in Helsinki the organisers were confident enough to call the meeting 'The influence of man on the hydrological cycle' (IAHS, 1980); again forests dominated with fourteen pages (regulation effects were next with nine) and by 1981 the Vancouver Symposium was devoted entirely to 'Forest hydrology and watershed management' (Swanson *et al.*, 1987). Table 3.4 shows the categorisation and country of origin of papers in this volume. It is also indicative of the maturing state of forest hydrology that watershed management appeared in the title of the Vancouver Symposium.

Despite the difficulties of extrapolating data for applications in a regional science (see Chapter 8), it is now generally accepted that in rain-dominated climates tall vegetation makes a net demand for water from a river basin. Work by Bosch and Hewlett (1982) has drawn together the results of catchment studies round the world (Figure 3.3) and may be used as an approximate guide to the trade-offs between a canopy cover and the water balance of the catchment. Their review results from ninety-four catchment experiments involving timber harvesting from climates ranging from

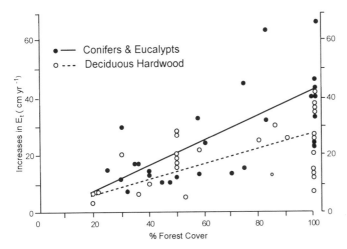

Figure 3.3 Forest cover and increased evapotranspiration: an international review (after Bosch and Hewlett, 1982)

<300 mm to >3,000 mm of annual precipitation. There is a consistent increase in catchment yield when tree cover is removed, though the magnitude varies (positively) with annual rainfall and, obviously, the proportion of the forest cover removed. Forest types are difficult to compare within the same climatic zone but, broadly, conifers and eucalypts cause 40 mm change in annual yield for the loss of 10 per cent cover, deciduous hardwoods 25 mm and brush 10 mm.

Figures 3.10 a and b show similar conclusions for two studies in the UK. More practical guidance, as applied to forest management in the USA, comes in Chapter 4, while the pressing problem of the water use by eucalypts in the developing world is covered by Calder *et al.* (1992) – see Chapter 5. Perhaps the major outstanding forest hydrology problem facing river basin development is in the humid tropics where timber extraction is popularly associated with enormous problems of land degradation. The lack of controlled catchment experiments and poor access in the tropics means that we have yet to get a broad regional picture of the subtleties of impacts on rivers (Bonell and Balek, 1993). The picture is, however, complex because moisture is not limiting in the soil but the impact of raindrops and cultivation techniques at the soil surface probably controls both runoff and erosion processes post-felling (Lal, 1993 and Figure 3.4). An additional argument about flood flows from tropical humid forest watersheds is the controlling role of the rainfall intensities, which are high enough to minimise the land cover (canopy) impact on runoff in the biggest storms.

Bonell and Balek review the important East African catchment experiments (Blackie *et al.*, 1979) in which changes in land use from forest to productive systems led to a variety of runoff yield responses. Changing

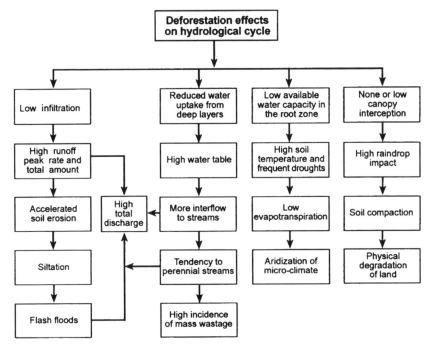

Figure 3.4 Deforestation: hydrological impact assessment, with special relevance to the tropics (after Lal, 1993)

from rain forest to tea plantation reduced runoff yield, changing from evergreen forest to smallholder cultivation much increased it, while changing from bamboo to pine had no effect.

3.1.3   Patterns of soil hydrological processes

One of the major pieces of empirical progress in twentieth-century hydrology has been in determining the spatial pattern of runoff processes. Geographers have played a prominent part in the retreat from a concept of catchment surface runoff to a variety of spatially orientated runoff zones in which permutations of surface and subsurface runoff occur at various times during a runoff event, for example the flood hydrograph. Unfortunately, much of this research has been carried out in the humid temperate climate zones of the world; thus problems of development (often focused in semi-arid or humid tropical zones) cannot easily be incorporated in the applications of our recent knowledge. The climate variables which make this constraint are similar to those in Table 3.1 for regional controls in forest hydrology; they include rainfall seasonality, intensity and presence or absence of snow. Runoff processes are also controlled by such factors as surface crusting in semi-arid soils, cracking, animal burrowing/treading and amount of ground cover by vegetation.

Two prominent variants of the newer conceptual framework are proposed: the *partial contributing area* (PCA) and the *dynamic contributing area* (DCA). Their strengths and validity are both determined by the pattern of rainfall (or snowmelt) interaction with the surface of the soil. Figures 3.5 a and b illustrate an earlier different view of this interaction, assembled from observations in the semi-arid rangelands of the USA by R. E. Horton (1933). The Horton model at its simplest states that runoff occurs from the surface of the soil when rainfall intensity exceeds the infiltration capacity of that soil. It is a rate-exceedance rather than a capacity-exceedance, an important distinction to make when considering the regional and local application of the model. 'Overland flow' or 'surface runoff' was assumed to have universal applicability in hydrology and had the further advantage, as conceived by Horton, of explaining many of the geomorphological features of the drainage basin by focusing attention on soil erosion (see Figure 3.5b). Under overland flow conditions (i.e. at the critical rainfall intensity) we assume that river flow is derived from every part of the river basin, a concept largely supported by the predictive power of basin area data.

The Horton runoff model went largely unquestioned during thirty important years in the development of applied hydrology in which techniques of prediction such as the unit hydrograph were developed. However, field measurements of infiltration capacity in soils proved difficult to make compared with the relative ease of rainfall intensity gauging; the model was therefore difficult to apply in detail at a river basin scale. Those attempts to link infiltration capacity to soil type suggested that, for the humid temperate zone, rainfall intensities seldom exceeded infiltration capacities for the majority of soils. Direct observations of runoff phenomena on slopes in the field also revealed an important role, in terms of both volumes and rates of runoff, for subsurface flows within the soil profile.

Table 3.4 The broadening scope of forest hydrology as represented by the topic and origin of papers in *Forest Hydrology and Watershed Management* (IAHS, 1987)

Topic	Papers	Nations
Acid precipitation	10	4 (USA 4; UK 3)
Erosion and sedimentation	14	10 (none dominant)
Snow in forested basins	4	2 (USA 3)
Evapotranspiration	7	7 (none dominant)
Soil moisture	7	4 (USA 3; UK 2)
Whole watershed studies	11	7 (USA 3; New Zealand 3)
Simulation models	6	5 (USA 2)

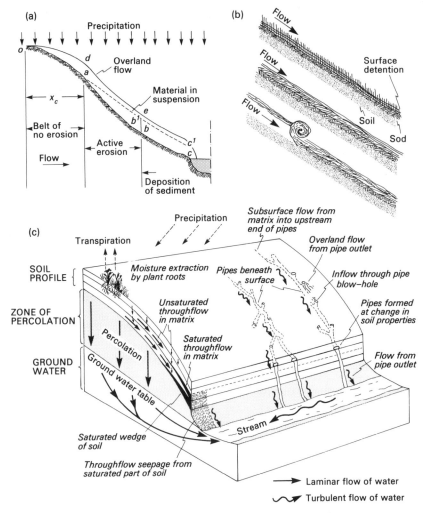

Figure 3.5 The overland flow model of R. E. Horton and its replacement by a
view incorporating a much larger role for soil throughflows:
(a) The basic Horton model showing the profile of overland flows on
slopes (Horton, 1945)
(b) Breakdown of grass cover during intense rains (Hortonian flow)
(Horton, 1945)
(c) Slope section and plan to emphasise saturated zones built up by
subsurface flow processes (after Atkinson, 1978)

As the section of hillslope shown in Figure 3.5c demonstrates, we now
understand a variety of routes for runoff on slopes, especially in vegetated,
humid landscapes with a good development of soil horizons. Control passes
from the soil surface in Horton's model to the state of soil moisture storage

in, and rate of filling of, successive layers of soil. Vegetation and crop covers, together with management practice, therefore assume an even greater significance than in the Hortonian model; surface flows remain effective but mainly where soils are saturated, a phenomenon typical of the base of slopes (as shown) and one which can clearly extend upslope in any soil horizon during a rain storm. This extension is the basis of the dynamic contributing area concept. Figure 3.6 illustrates how the saturated areas of catchment at the start of a small flood event expand during the event.

A further feature of Figure 3.5c worthy of note in connection with land use and management is the existence in many soil profiles of natural 'pipe' networks which clearly promote efficient soil drainage; subsurface runoff or *throughflow* is not, therefore, an invariably slow phenomenon.

The progress made in understanding subsurface runoff processes in the last thirty years illustrates the benefits (and expense) of research in hydrology; guidance to those concerned with basin management includes an appreciation of the (spontaneous) regulatory role of natural vegetation covers, soil organisms, soil chemistry (in the case of stream acidification and eutrophication) and soil erosion.

If canopy processes in hydrology largely underlie the importance of crop choice, dominated by research on forest hydrology, soil processes justify the

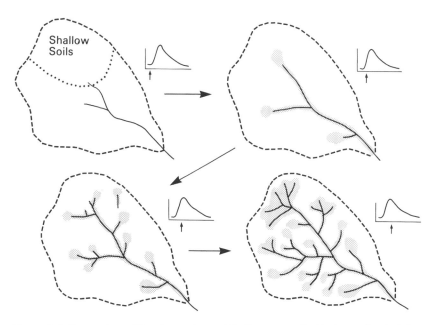

Figure 3.6 Expansion of the dynamic contributing area of a drainage basin into ephemeral channels and areas of saturated soils during the passage of a flood hydrograph (Hewlett and Nutter, 1970)

attention of river basin managers to surface treatments, of which urban covers have achieved a similar dominance in research terms.

Figure 3.7 illustrates the major river basin influences of forest, bare, urban and cultivated covers. Table 3.5 confirms a considerable variability of impermeability of urban covers but the most obvious urban effect on the natural system of Figure 3.5c is that of re-routing precipitation at the surface.

Table 3.5 demonstrates a measure of flexibility in urban design and it has become standard practice in North America to mitigate the urban influence on both runoff and sediment yields by innovating with materials to reduce urban impermeability. Another aspect of urban hydrology which promotes hydrological change is the channelling of the extra runoff from an impermeable surface into an efficient system of drains and sewers; this drainage system is also being modified with the introduction of storage tanks, hydraulic 'brakes' in sewers and surface balancing ponds.

Clearly, city planners need an approximation of the runoff impacts of their developments and simple predictions at least allow a precautionary approach to likely downstream increases in runoff (see Figure 3.10e).

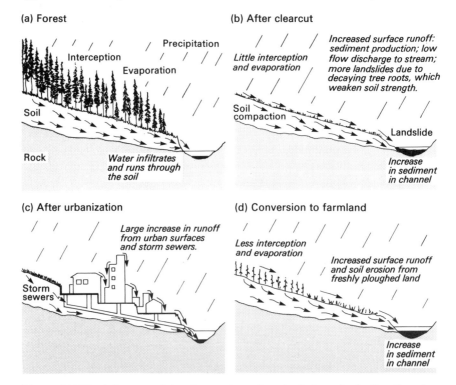

Figure 3.7 The influence of development on slope hydrology, indicating the role of agriculture and urbanisation

Table 3.5 Approximations of the impermeability of various surfaces

Type of surface	Impermeability (%)
Watertight roof surfaces	70–95
Asphalt paving in good order	85–90
Stone, brick and wooden block pavements:	
with tightly cemented joints	75–85
with open or uncertain joints	50–70
Inferior block pavements with open joints	40–50
Macadam roads and paths	25–60
Gravel roads and paths	15–30
Unpaved surfaces, railway yards, vacant lot	10–30
Parks, gardens, lawns, meadows – depending on the	
surface slope and character of the subsoil	5–25

Research in urban hydrology has been dogged by the sheer complexity of the system (much of which disrupts the flow paths of the natural surface catchment), the need to develop special instrumentation and the difficulty of finding experimental sites for 'urban v. rural' comparisons or 'before and after' studies at a catchment scale (see Figure 3.10f). The most recent phase of urban hydrology has, therefore, tended to concentrate on smaller scale investigations of elements of the urban runoff system, such as roofs and gutters, pavements and streets (Hollis, 1988).

Agricultural disruption to soil processes in the runoff cascade is also imperfectly understood; again the difficulties of finding experimental sites and of innovating techniques abound – to which one may add the problem of generalising between national and even regional differences in cropping and cultivation techniques. As an example of the problems of generalising land-use effects on hydrology we may cite the long debate in the British Isles about the effects of farm and forest drainage on the flood hydrograph. Are streams in heavily drained areas more prone to rapid and high flood hydrographs? Empirical studies yield both 'yes' and 'no' answers. At first a separation appeared between sites on peat soils (enhanced response) and mineral soils (reduced response); see Newson and Robinson (1983). Subsequently, for clay soils, detailed investigations revealed that two conditions exist within the same drained area: under wet conditions undrained land responds more rapidly to rainfall than drained land because soil cracks become closed (Robinson and Beven, 1983; Figure 3.10d). Furthermore Robinson *et al.* (1985) revealed the importance of mechanically cracking the soil when draining land and Reid and Parkinson (1984) the importance of landform in controlling soil moisture and hence drainage flow response. There is clearly not one single answer!

Knowledge of runoff zones, such as those proposed by protagonists of PCAs, DCAs, or even the traditional surface runoff models, is essential to those who wish to control land use and land management in support of river basin management (Dunne and Leopold, 1978). For example, surface drainage channels installed in areas of a catchment prone to either Hortonian surface runoff or seasonal saturation are unlikely to be effective in increasing soil moisture storage. Instead these drains merely channel surface runoff more quickly to natural channels, increasing flood peaks by acting as extensions to that natural network.

Guidance to planners and others without hydrological knowledge can be obtained from patterns of natural vegetation which often mark very clearly the zones of runoff generation in a catchment. Plants tolerant of saturated soils clearly delimit the partial contributing area. Mapping vegetation communities also reveals variations which indicate soil piping zones, seeps and other diagnostic features.

3.2   GROUNDWATER EXPLOITATION AND PROTECTION

Because our perception of river basin development is dominated by images of surface water management (and mismanagement) – dams, canals, pipelines, treatment works, polluted waters – the impact of surface activities on subsurface hydrology is often neglected. Groundwater is important to river basin development for several reasons:

(a) Groundwater forms the major water supply source in many parts of the world, notably in rural areas, especially in the semi-arid and humid tropics.
(b) Where not the dominant source, the inherent values of groundwater – purity, constancy and extent (beneath the land area) – make it an important resource to moderate and modulate the surface water regime and surface ecosystems.
(c) Many of the infrastructure projects of development require foundations or storage capacity underground – thus groundwater is an influence on engineering, construction and waste disposal.

All of the hydrological influences reviewed in this chapter apply to subsurface waters but there are particular sensitivities and technical problems with the groundwater system which are worth drawing out separately. Generally, groundwater systems react slowly to surface impacts but more profoundly and there are greater costs in rectification when damage is done. While groundwater hydrology has a long history and good theoretical support, and the world's more important aquifers have been explored with geophysical techniques to the point where they can be hydrologically modelled, a major problem of lack of knowledge of subsurface processes often exists at the local scale.

Few parts of the Earth's land surface are truly impermeable and subsurface waters exist in both shallow and deep aquifers. Surface rivers flowing across their own alluvium often interchange water with the alluvial aquifer, with important benefits for abstraction by both human and plant communities. The inter-relationship between aquifers and rivers can be demonstrated even more clearly where groundwater abstraction rates have been allowed to grow to a point where spring-fed rivers all but dry up (as has been the case in England during recent summers). Deeper aquifers in sedimentary rocks are the best known contemporary subsurface reservoirs for water; sustainable exploitation entails balancing abstraction with the rate of recharge from the surface and protection against contamination. The more difficult and dangerous situation is where such an aquifer received good rates of recharge in past, humid climates, making the 'reservoir' essentially one of 'fossil' water. Fossil groundwater basins are prominent in the water resources of North Africa and the Arabian peninsula (Simpson, 1991). In Saudi Arabia non-renewable groundwater provides over 75 per cent of the nation's water needs; the resource is dwindling and saline intrusion of seawater is occurring and the aquifer is drained by abstraction (Al-Ibrahim, 1990). In such cases the American terms groundwater 'mining' and groundwater 'overdraft' are both expressive and relevant; there is little hope of sustainable development based on this water if contemporary recharge is reduced, for example, over much of the semi-arid and arid world. Figure 3.8 shows those areas of the USA where groundwater overdrafts have been incurred (that is, where withdrawal is greater than recharge). Foster and Chilton (1993) remind us of the importance of groundwater in the humid tropics where waterborne disease and sedimentation are problems faced by surface water exploitation and where groundwater is generally available where it is needed (rather than needing transfer).

A further category of aquifer is worthy of mention because of the popular impression that underground water is naturally filtered and therefore always of high purity: crystalline and fissured hard-rock aquifers, for example, granites and hard limestones, transmit water through routes much larger than the pores more typical of sediments. The most spectacular example is that of karstic areas where public water supplies emerge from caves in limestone; 'filtration' is virtually non-existent in such systems except in the overlying soil layers. Very little research has been conducted on the relationship between surface land use and groundwater resources – this has been concentrated in the realm of surface water and carried out through catchment experiments. However, a very direct example comes from a popular Australian technique for controlling the seepage of saline groundwater into agricultural lands by reducing groundwater recharge (and thereby lowering the water table) through the growth of forest plantations (George, 1990). The original cause of the salinisation problem was the destruction of the natural forest cover in the name of agricultural development – a strongly practical demonstration of the effects of the interception process.

Much more concern is now being expressed about the impact of surface land use and land management on groundwater contamination and pollution

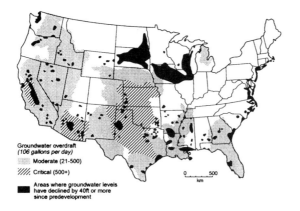

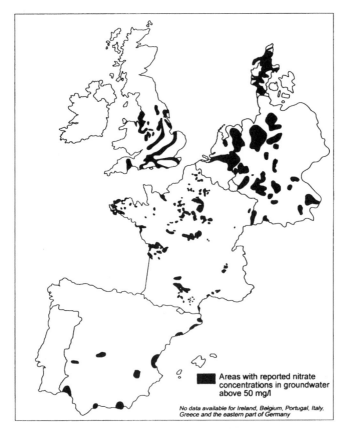

Figure 3.8 Unsustainable management of groundwater:
(a) 'Overdraft' (withdrawal > recharge) and excessive depletion rates in the USA
(b) Serious nitrate pollution of groundwater in the EU (after RIZA, 1991)

(United States Environmental Protection Agency (EPA), 1990; RIZA, 1991). The EPA divides the threats according to whether they are on the land surface, subsurface (but above the water table) and below the water table. They are:

- On the land surface: contaminated surface water, land disposal of wastes, including sludge from water/sewage treatment plants, stockpiles and tailings, salt on roads, animal feeds, fertilisers and pesticides, airborne chemicals and accidental spills (e.g. road and rail freight).
- Below the surface: septic tanks, landfills, cemeteries, leakage from underground tanks and pipelines, artificial recharge waters, mines and saline intrusion (sea water entering coastal aquifers).

The RIZA report, for the European Communities, divides between rural and urban threats. The most significant rural threat in Europe is from agricultural practices, particularly the use of fertilisers and pesticides. In about 70 per cent of the EU's agricultural soils, nitrate concentrations are above the Communities' target value of 25 mg/l (Figure 3.8b shows those which exceed the EU Drinking Water Standard); in 65 per cent of the area the standards for pesticides are exceeded or will be shortly. The National Rivers Authority in England and Wales (now the Environment Agency) has set out proposals for Groundwater Protection Zones (which already exist for protection against nitrate contamination – see also Chapter 8). Maps of vulnerability to groundwater pollution are being published (NRA, 1992) and there are models for the calculation of the land areas at the surface needed to protect water sources (Adams and Foster, 1992).

3.3   RUNOFF MODIFICATIONS IN DEVELOPED RIVER BASINS

Despite the evidence offered in Section 3.1 of the potential for hydrological effects of land use, it is essential to be highly specific when suggesting links between land use or land management and properties of the river hydrograph: merely 'that land use has altered the river flow' will not suffice without a wealth of detail about both the land use and the flow properties which are alleged to have been affected. Clearly, coincidental changes in both must not be ruled out and statistical 'proof' will be of limited application without a knowledge of causal processes. A further problem of interpretation is the direct impacts on river flow behaviour and groundwater fluctuations already made by human exploitation in the developed world. Only against a background of 'naturalised flows' (with artificial impacts on volume and timing by dams and structures removed) can the effects of land use and land management be judged.

In land-use hydrology we need to subdivide crop (also stages of the *crop cycle*) from the infrastructure to grow and harvest it and the cover effects of the urban surface from those of infrastructure, for example, sewerage.

Effectively we can divide between land *use* and land *management*; through management practice we can seek to mitigate undesirable impacts and enhance desirable effects. In terms of the river flow properties influenced by land use, these may be extreme and obvious in some climates (e.g. a dry river bed) but in many cases the influences are very much more subtle, possibly seasonal and possibly variable between patches of the same land use because of physiographic, climatic or management differences.

One must not neglect the operation of natural controls on the runoff process; for example, basins may differ profoundly in their flood or drought

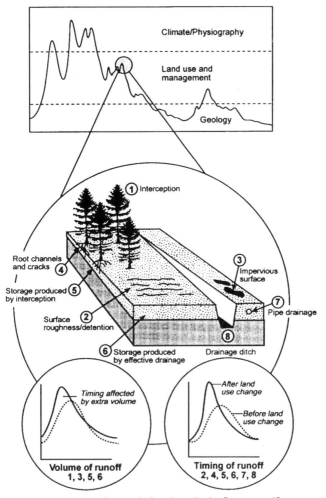

Figure 3.9 The annual hydrograph and (outset) influences of land use and management on the volume and timing of individual flow events

behaviour because of physiographic or geological differences. Figure 3.9 offers a simple guide, based on the annual hydrograph of a hypothetical humid zone surface river, to those aspects of flow liable to be influenced by land use; clearly, the land-use and management influence may make its biggest impact in the mid-range of flows where subtle changes in volume or timing (Figure 3.9 – outset) may become exaggerated over longer timescales.

3.3.1   Modifications to runoff volume

The dried-up river bed resulting from over-abstraction is a clear and extreme version of an artificial volumetric change in river flows! Nevertheless, changes of river flow could also be the result of timing changes – a less equable temporal spread of an unaltered volume. The two properties are obviously linked but the emphasis in this and the following sections will be on those links with land use where the major influence is on one of these properties (see also Table 3.6).

Land-use and land management effects on runoff volume are clearly most likely where a change is brought about in:

(a) Evaporative loss from an identical precipitation volume.
(b) Surface characteristics of a basin which influence the detention and storage of runoff.

Both cropping (including the crops) and urban/industrial land uses may be expected to cause such changes – at key switching points in the land phase of the hydrological cycle – at the canopy level and the soil surface level. At the canopy level, there is now an emerging empirical consensus among hydrologists that a forest cover reduces runoff volume under most climatic conditions.

The intensive research programme on forest hydrology in Britain has yielded further details of the interception process and of the essentially exclusive operation of the two components of forest moisture loss: interception and transpiration. Calder and Newson (1979) conclude that in the British uplands interception ratios for plantation conifers converge at 30 per cent of annual precipitation (Figure 3.10a). During periods in which the forest canopy is moist, interception entirely dominates transpiration but when the canopy is dry the reverse is true. Consequently, in drier climates the tree's own physiological control of transpiration may come to dominate its comparative water usage in relation to adjacent land covers; in practice understorey vegetation in forests also uses moisture and Roberts (1983) has concluded that overall forest transpiration is essentially a conservative process with similar annual rates across Europe (approximately 333 mm/yr).

Outside catchment experiments very little work has been performed on catchments monitored for water supply or pollution control purposes. Figure 3.10b reveals the utility of such a sample, from Wales, in which

Table 3.6 Land-use/management effects on the flow of rivers

	Volume of runoff	Timing of runoff
FORESTS		
Planting	Increase if previous cover cleared	More rapid if cultivation/drainage required
Mature cover	Decrease in climates favouring interception or transpiration; increase where snow trapped in forests	More rapid if cultivation/drainage required
Harvesting	Increase until cover re-established	Location and extent of harvesting critical
URBAN		
Surface	Increase in surface volume and totals where replacing crop/forest cover	More rapid as more precipitation retained on surface
Drain/sewer systems	No influence except on groundwater volumes	Most flood flows made more rapid; extreme floods may be ponded in urban area
AGRICULTURE		
Drainage	May reduce long-term storages and reduce low-flow volume	Effect depends on primary soil permeability
Cropping	Evaporative use by irrigated crops reduces net runoff volume	Little direct effect in humid, temperate climates; elsewhere critical to surface runoff and erosion
QUARRYING/MINING		
Surface water	Little affected	Restored open-cast often yields rapid runoff
Groundwater	Stored/pumped; boosts low-flow volumes	Little affected

there is a clear negative relationship between the volume of annual runoff (as a proportion of rainfall) and the forest cover of the catchment. Nevertheless, results of this type, together with those of Bosch and Hewlett (1982), are much more rare than those from process-based or catchment studies and it should be emphasised that predictive techniques for runoff volume used at national and international scales do not, as yet, incorporate forest or other land-use variables – physiographic and climatic variables are much more effective in statistical predictions of runoff volume. This apparent mismatch

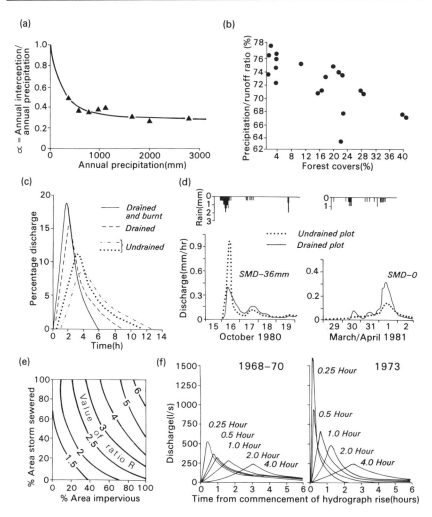

Figure 3.10 Hydrological effects of rural land use and land management:
- (a) Interception ratios of mature forest canopies in the UK (Calder and Newson, 1979)
- (b) Reduction of annual runoff coefficient with increasing forest cover, Wales (Mas'ud, 1987)
- (c) Moorland drainage and the flow hydrograph (Conway and Millar, 1960)
- (d) Farm (under-) drainage and the flow hydrograph, indicating the role of antecedent conditions as soil moisture deficit (SMD) in routeing rainfall through/over the soil (Robinson and Beven, 1983)
- (e) Mean annual flood increments (cf. natural floods) with increasing urban cover and sewerage (Leopold, 1968)
- (f) Flood hydrograph changes after urbanisation (Walling, 1979)

between small-scale findings on process and large-scale, statistical predictions is followed up in Chapter 8.

3.3.2 Modifications to runoff timing

Figure 3.4c shows the multitude of processes and routes by which precipitation reaches the humid zone surface stream. We have already considered this as a cascade of storages and clearly timing of runoff is proportional, in a simplistic system, to the number of storages through which each molecule of runoff passes. Response times vary accordingly as Figures 3.10 c, d and f demonstrate. Any land-use or management strategy which significantly re-routes runoff will alter its timing: the simplest example is surface drainage of the soil. It should once again be emphasised, however, that timing changes will also bring about volume changes in another part of the system, for example, urban storm sewers will deprive the underlying aquifers of recharge.

Table 3.7 illustrates the range of velocities characteristic of different runoff routes in the river basin; however, it would be simplistic to consider changes in runoff timing to be merely a function of re-routeing between these categories: the volumes re-routed and the destination of the re-route are also important. Two examples illustrate this point: soil pipes (and drainage pipes) appear to be rapid routes for runoff but often small volumes only are involved and, in the case of soil pipes, the destination of the runoff involves storage delays and slower routeing; in extreme floods urban drainage systems overflow and the resulting inundation tends to reduce the velocity of the floodwave as a whole.

Despite these caveats, artificial processes which gather runoff efficiently into hydraulically smooth channels and which reduce the distance to the stream or increase the pathway gradient will promote faster flood responses. This is particularly true if some other change has also occurred to promote a

Table 3.7 Estimated flow velocity of various hydrologic processes

Flow medium	Velocity range (m/h)
Open channel	300–10,000
Overland flow	50–500
Pipeflow	50–500
Matrix throughflow	0.005–0.3
Groundwater flow:	
Sandstone	0.001–10
Shale	0.00000001–1.0
Jointed limestone	10–500

Source: Weyman (1975)

higher volume of runoff, for example, in Figure 3.10c where a small peat catchment has been both burned (sealing the surface) and drained. In the case of subsurface drainage the condition of the soil between surface and drainage pipes has a subtle control on the timing of responses. As already described above, hydrologists in the UK have discovered that a dry, cracked clay soil promotes a more rapid response by drained farmland but when the same soil is wet (and only slowly permeable), undrained farmland responds more quickly because it routes more rainfall across the surface (Figure 3.10d).

The combination of increased volumes of runoff and more efficient drainage from traditionally planned urban areas is shown by Figure 3.10e, in which the peak of the annual flood is increased in proportion to both the impervious area and the extent to which it is drained. The evolution of the more rapid urban response through time is shown in Figure 3.10f.

3.3.3   Regulated rivers, an introduction

It is a paradox of our subject area in this chapter that, while the intention is to substantiate a planned approach to the joint management of land and water, it is water engineering which has produced a more identifiable, and in many cases more marked, effect on river flows than has land use or management. The regulation of river flows, either by design to ameliorate the natural regime, or as a secondary impact of water use, dates back to the first dams and irrigation schemes.

The extent of river regulation by dams has been gauged by several authors; it extends into areas outside the effects of land use and management for development because it is to pristine, remote, wilderness or mountainous sites that dam builders look for good foundations, abundant runoff and good water quality.

Beaumont (1978) plots the growth of international dam construction since 1840 (Figure 3.11); building continues undiminished (Lvovitch and White, 1990) although rather fewer, but larger schemes are now the norm. Between 400 and 700 dams are currently being built every year around the world.

To the obvious effects of damming rivers on their downstream flow patterns we may add the following water uses which effectively alter the volume and timing of river flows downstream in ways which resemble damming:

(a) Major industrial users of river water; few are consumptive but even a use for cooling will modify river flows.
(b) Irrigation use of river water is both consumptive and switches the river water to a slower return flow route.
(c) Water use by human settlements both imposes demands on flow from dams by direct supply, from natural or regulated rivers, and returns this water as sewage which places quality demands on the downstream river.

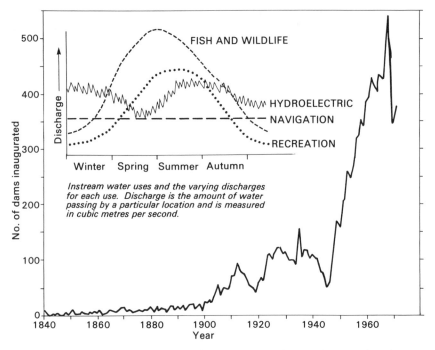

Figure 3.11 Direct control over river flows: river regulation. Growth of world dam
construction (Beaumont, 1978) and (inset) the seasonal flow
regime suitable for natural and exploitable purposes

Some indication of the extent of the last of these categories can be gained
from the fact that up to 34 per cent of the flow volume of the River Trent
(England) during average conditions is provided from sewers, not by
'natural' sources (Farrimond, 1980). Because the sources of water for human
consumption are often remote from the settlement and yet sewers discharge
to the nearest river, such systems often effect a transfer between basins.

The major technical problems raised by river regulation from reservoirs
are given special attention in Chapter 6. For the moment we remain purely
with the magnitude of artificial influences on the regime of river flow
(volume, timing) in order to contrast those of land use and direct regulation.

Figure 3.11 also presents the conflicting requirements of different river
users for particular patterns of flow. To these may be added the needs of
riparian consumers and the need to protect settlements from both floods and
droughts. It is clearly wrong to consider that the 'best' river regime for
settled, civilised human life is a constant flow! Ecologically the pattern of
fluvial systems and habitat derives just as much from flow variability
through time as from the physiographic and nutritional features of river
systems.

Table 3.8 lists the types of changes in volume and timing of flow occasioned by a variety of regulating activities both direct and indirect (cf. Table 3.6, land-use effects).

One of the grave difficulties in making quantitative assessments of regulating effects is that many of the causes are of long standing and pre-date our official flow measuring networks. Thus, for example, many UK upland reservoirs date from the nineteenth century but, despite protests over their 'compensation flows' (see Chapter 1), their precise effect on downstream volumes and timing could not be assessed effectively until recently. In the case of less localised effects such as sewerage outfalls, they are so numerous in developed regions that even with flow measurement stations installed, reduction of total flow to 'naturalised' flow is difficult; flow measurements are made on the larger sewers but, for many, flow rates are an approximation.

Clearly, too, impacts become reduced and blurred in terms of a unique hydrological signal as one moves further downstream from the point of maximum activity, be it dam site, plantation or city. Scaling these decay factors (which operate in time as well as space) is as difficult with direct regulating influences as it is with those of land use/management. As Table 3.8 shows, the location of unregulated tributaries in the system makes

Table 3.8 River regulation: magnitude of effects on flow regime

	Volume	Timing	Downstream distance
DIRECT EFFECTS			
Direct supply reservoir	Reduces	Delays flood peak by storage	
River regulating reservoirs	Reduces and increases according to conditions	Tends to delay natural floods	
Hydro-power (dam)	Diurnal or seasonal pulses	Towards artificial regime	Depends on position and size of unregulated tributaries
Run-of-river	Little impact	Little impact	
Flood control	Reduces/ removes peaks	Delays	
INDIRECT EFFECTS			
Irrigation return flows	Reduces	Little effect	
Sewerage return flows	Redirects/ transfers	More rapid if includes storm drain	

almost every case unique but also offers a new site factor for consideration in locating major river control structures (Newson, 1994a).

3.4   VEGETATION, SOILS AND WATER QUALITY

Knowledge of the hydrological cycle (Figure 3.1) prepares us for the links between land use/management and changes in runoff timing and volume. During the last decade, however, possibly reflecting an increased human perception of environmental degradation and of mankind's controlling role, hydrologists have developed detailed programmes of research on river basin water chemistry and its variation with land use. They have been joined by ecologists anxious to investigate links between water quality and habitat; much of this latter research has been done at smaller scales and, once again, we find forests, both natural and exploited, exciting the interest of many research programmes.

In Chapter 2 we emphasised the role of the basin sediment system in controlling the basic physical systems of river basins, particularly in the longer term. Sediment transport is also a feature of water quality changes brought about by land-use activity; in this case it is fine sediments (sands, silts and clays) which we consider. Nevertheless, the emphasis in this section will be away from sediments and towards water chemistry.

Before continuing it is appropriate to make a simple subdivision between two types of basin effect in this category. Whether the problem is *contamination* (the introduction of new materials and compounds to the system) or *pollution* (the introduction of damaging loads or concentrations of material or compounds), we separate two sources:

(a) Point sources are identifiable, such as the obvious outfalls of irrigation return flows or sewerage systems.
(b) Diffuse sources are much more difficult to identify, such as nutrients and pesticides added evenly to agriculture or forestry.

The source may also be continuous or intermittent (Figure 3.12) and there are difficult intermediate categories such as the farm slurry-store which, if it is emptied carefully and spread on the land, becomes a diffuse source of nutrient chemicals but which, if it spills to a stream, becomes a point source. Pollution incidents are also, therefore, land-use related. It is obvious to those managing river basins that scientific detection of, and legal controls on, diffuse pollution are much more problematic than the equivalents for point sources.

The difficulties associated with diffuse sources mean that proving links between rural land use/management and water quality are much greater than for urban links. Furthermore, scientific studies will find difficulties of extrapolation beyond the boundaries of research sites, in this case far more insuperable than those of scaling a universal physical process such as

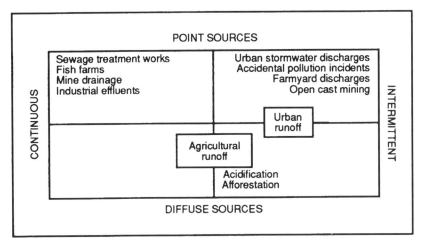

Figure 3.12 Sources of pollution: a simple classification (after British Ecological
Society, 1990)

interception. Studies of river sediment systems or water chemistry systems
have additional problems of downstream changes in processes and of the
importance of local conditions of culture, climate, soils and rock types.

3.4.1   Land use and the fine sediment system

In Chapter 2 we were concerned with the formation and maintenance of
river channels, mainly the realm of coarse sediments; here coverage shifts to
the finer, suspended or 'wash' load which often originates in soil erosion and
constitutes a physical pollutant (or chemical pollutant carrier) downstream.

Archaeologists have revealed in many regions of the world an association
between the human artefacts of a datable civilisation and characteristics of
the sediments which contain them, indicating a major erosion phase
associated with typical land use of the era. In semi-arid lands the land-use
effect is via settled irrigated agriculture. In humid regions of North America,
Europe and Asia there are clear deforestation- and overgrazing-related
erosion horizons. Often a rival climatic explanation is available but, with
close analysis of how basins route the products of soil erosion, it is clear that
in certain major areas of basins, notably in hollows and at the base of slopes,
land-use practices themselves can account for impressive thicknesses of
deposit. We may therefore hypothesise a link between 'land-use' sediments
in colluvium and 'climate' sediments in alluvium (see further debate in
Chapter 6).

A much more detailed analysis than this is available for the eastern
seaboard of the USA. Meade (1982) describes how soil erosion was
increased tenfold by European settlement; large quantities of the resulting

sediment are still stored on hillslopes and valley floors, exaggerating contemporary sediment yields. The impact of modern man in the region has been to trap the same sediment in reservoirs; the remainder is stored in estuaries and coastal marshes. Knox (1989) has pointed to the fact that river valleys act as huge stores of alluvium from such periods, demonstrating the evidence of prehistoric clearances in Wisconsin (see Figure 3.13c).

On a world scale the interplay of climatic and land-use (in this case natural vegetation) covers is available through the work of Schumm. Figure 3.13a shows how, because transport processes in river systems are critical for moving fine sediments in suspension from river basins, the broad characteristics of hydrology determine sediment yields at a given time. However, the feedback of moisture availability to vegetation cover means that humid climates have a far lower yield than the potential indicated by their high runoff.

Not surprisingly, therefore, it is disruption to vegetation covers in climates of intense rainfalls which leads to rapid deterioration of water quality and river habitats through soil erosion. Urbanisation produces the worst impact during construction and the urbanisation of semi-arid lands is especially damaging; however, tropical deforestation exposes the most sensitive soils on earth to some of the most intense rainfalls on earth and has a very serious impact on water quality downstream.

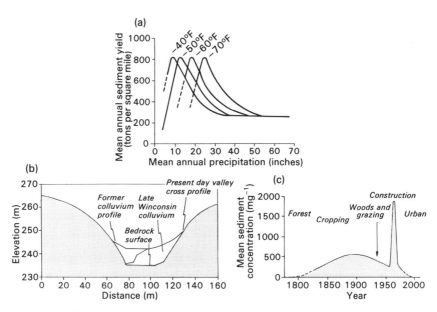

Figure 3.13 Controls on sediment transport:
(a) Climate, on a world scale (Schumm, 1977)
(b) Storage of historically eroded material (Knox, 1989)
(c) General development processes on pristine landscape
    (Wolman, 1967)

For those particles of soil released from soil erosion the route to the ocean may occur as a complex and lengthy set of paths via storages (see also Chapter 6 for technical discussion). The path lengths between storages, once entrained by channel flows, are much greater. Lambert and Walling (1988) conclude that channel storage of fine sediment in the Exe basin, in the UK, is minimal and the channel efficiently conveys suspended material direct to the coast.

The question to pose next is whether fine sediments entrained by rivers represent a significant contamination or pollution of the environment (implications for reservoir sedimentation, flood control, etc. are discussed in Chapter 6). The concentration of fine sediments in suspension is referred to, and can be measured as, turbidity – an occlusion of the natural clarity of the pure substance to various degrees. Turbid water requires expensive filtration for all human users of water.

At Holmestyles Reservoir in Yorkshire, forestry ploughing and draining operations created severe pollution by suspended sediments, costing £23,000 to cure because filtration equipment needed replacement. At Cray Reservoir in Wales a public water supply was out of commission for some days as the result of forestry operations.

Turning to biotic conditions in turbid streams, some research indicates very little damage from, for example, quarry waste materials. However, fine waste material has geomorphological implications, as Richards (1982) demonstrates for Cornish streams contaminated by china-clay waste; width/depth ratios of polluted channels are under half of those for 'natural' streams, showing an adjustment to the finer load carried. A series of laboratory experiments conducted by Alabaster (1972) indicates that the presence of fine materials in suspension seriously exacerbates the toxic effect of certain chemicals in water. Cross-linkages between solid and soluble phases of river flows are emerging from many land-use related river pollution studies. For example, phosphates become preferentially adsorbed to fine sediments (Walling, 1990). The loss of organic and mineral nutrients from eroded soils can become an accumulating pollution hazard in lakes and reservoirs.

In studies of sedimentation from 'hydraulic mining' (in which water is used to expose and sort ores and dispose of waste), the fine fraction of the sediment mix released appears to be an important carrier of the metal pollution downstream (Macklin and Dowsett, 1989) to lakes, slack-water sites on the river and to floodplains where poisoning of vegetation and cattle may continue for decades.

3.4.2 Solute processes, mineral and nutrient

A protestor arguing against the fluoridation of public water supplies was once heard to object to 'the addition of a chemical to pure water'. Nature conspires to add many chemicals to river water; in some cases the concentration or loading of wholly 'natural' chemicals pollutes water to the extent that it is undrinkable and the stream lifeless. In the case of crystalline limestone terrains the chemical solution of the landscape actually creates the typical river network (an underground one flowing through caves).

Figure 3.14 shows both systematically and pictorially the origins of natural stream water quality (in rural areas). There are both mineral and nutrient cycles, often largely controlled by soil/vegetation systems and agricultural practices, which explain chemical content at any point in the hydrological cycle. Two elements of this system are often neglected by field monitoring campaigns: the chemical characteristics of precipitation and the chemical modifications produced by 'in-stream' processes, often controlled by stream biota. Indeed it is upon the in-stream processes of purification that we depend across the world when using rivers as conveyor belts for the waste products of housing, farming and industry.

'Natural' water quality is governed by two great sources of mineral salts, the ocean and ocean-deposited sediments containing abundant sodium and chloride, and by rock types consisting of lithified marine organisms, rich in calcium and carbonate (Walling and Webb, 1986). These ions are merely discriminators; the full list of major ions in river water is shown in Table 3.9.

Whether introduced by 'natural', processes or by land-use related modifications, the soil is a primary site for geochemical reactions and the concentration of ions is influenced by both the rate of solution (or production of soluble chemical) and the rate of removal by soil moisture movement. An understanding of hydrological patterns, flow paths and flow

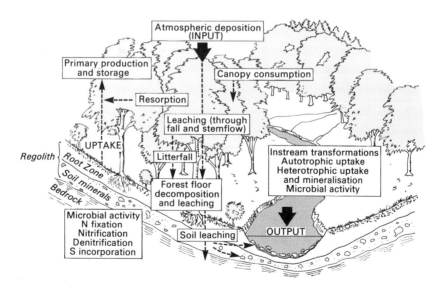

Figure 3.14 Processes leading to the production of stream solute loads in a natural catchment

Table 3.9 Average composition of world river water

Author	Concentration (mgl^{-1})								
	Ca^{2+}	Mg^{2+}	Na^+	K^+	Cl^-	SO_4^{2-}	HCO_3	SiO_2	*Total*
Livingstone (1963)	15.0	4.1	6.3	2.3	7.8	11.2	58.4	13.1	120[a]
Meybeck (1979)	13.4	3.35	5.15	1.3	5.75	8.25	52.0	10.4	99.6[b]
Meybeck (1983)	13.5	3.6	7.4	1.35	9.6	8.7	52.0[c]	10.4	106.6[b]

[a] Total content of all solutes.
[b] Total listed constituents.
[c] HCO_3 not listed in Meybeck (1983); Meybeck (1979) value used.

velocities is therefore of immense importance in managing water quality problems resulting from land use. For example, nitrate pollution arising from intense production on well-cultivated, freely draining soils can be mitigated as the throughflow drains to the channel via saturated floodplain soils where de-nitrification occurs (Pinay and Decamps, 1988; Hill, 1990).

In the case of urban and industrial land use, the soil plays a much less prominent role unless waste products are applied to land – an increasingly popular option for sewage sludge, the solid product of urban sewage purification. Mostly, however, urban industrial water quality problems are not routed by the national hydrological pathways but originate in an industrial process at a site with a clear point discharge direct to a stream (see Table 3.10).

However, it is perhaps by the location of urban and industrial developments in relation to the river network and sensitive aquifers (and the timing of discharges) that point sources of pollution will be controlled by dilution. There can never be an end to water pollution but it is part of arguments based around ecosystems and their 'carrying capacity' for human resources and wastes that society should plan the location of potentially damaging activities. Figure 3.15 shows that urban and industrial planning policies can be backed, in other parts of a developed basin, by management of atmospheric deposition (calculating the critical loads for ecosystems), rural land-use policies, protection of aquifers and implementation of rural 'buffer zones' of natural vegetation next to channels. Protection zones can also be applied to industrial developments where strict controls on emissions and protection measures against pollution incidents are applied. The critical policy need is for land-use planners and water (catchment) planners to coordinate their approach to the carrying capacity of each local freshwater environment, as well as to the needs of the whole basin (Newson, 1994b).

Table 3.10 Some sources and causes of water quality deterioration

Effluent	Factors affecting water quality deterioration
Domestic sewage	BOD, suspended solids, ammonia, nitrate, phosphate
Vegetable processing	BOD, suspended solids, colour
Chemical industry	BOD, ammonia, phelons, non-biodegradable organics, heat
Iron and steel manufacture	Cyanide, phenols, thiocyanate, pH, ammonia, sulphides
Coal mining	Suspended solids, iron, pH, dissolved solids
Metal finishing	Cyanide, copper, cadmium, nickel, pH
Brewing	Suspended solids, BOD, pH
Dairy products	BOD, pH
Oil refineries	Heat, ammonia, phenols, oil, sulphide
Quarrying	Suspended solids, oil
Power generation	Heat

Table 3.11 Point sources and the expected contamination of groundwater

Source	Inorganic contaminants	Organic contaminants
Urban areas	Heavy metals and salts	Oil products, biodegradable organics
Industrial sites	Heavy metals	Chlorinated hydrocarbons, hydrocarbons, oil products
Landfills	Salts and heavy metals	Biodegradable organics and xenobiotics
Mining disposal sites	Heavy metals, salts and arsenics	Xenobiotics
Dredged sediment and sludge disposal	Heavy metals	Xenobiotics
Hazardous waste sites	Heavy metals (concentrated)	Concentrated xenobiotics
Leaking storage tanks	–	Oil products (petrol)
Line sources (motorways, railways, sewerage systems, etc.)	Heavy metals, salts	Oil products, pesticides

Returning to diffuse-source water quality changes related to land use or land management, these may occur as the result of the following processes:

(a) Hydrological changes occurring as a consequence of canopy processes, resulting in, for example, increased evaporation and therefore concentration of solutes trapped from the air, excreted by the plant or released by weathering from the soil below.

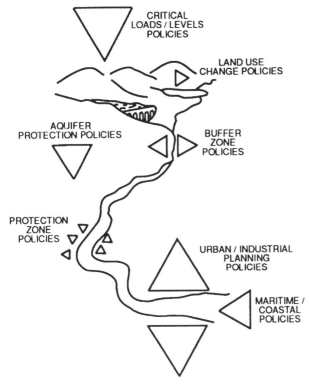

CRITICAL
LOADS / LEVELS
POLICIES

LAND USE
CHANGE POLICIES

AQUIFER
PROTECTION POLICIES

BUFFER
ZONE
POLICIES

PROTECTION
ZONE
POLICIES

URBAN / INDUSTRIAL
PLANNING
POLICIES

MARITIME /
COASTAL
POLICIES

Figure 3.15 A simple typology of planning policies in relation to river and groundwater pollution

(b) Hydrological changes resulting from the route(s) taken by water through the soil column, e.g. towards or away from reactive zones such as weathering horizons or via more or less rapid routes downslope to the nearest collecting channel itself; that channel may itself be an artefact of the land management system.

(c) Chemical changes resulting from the burning or decay of a crop, the removal of a crop or stronger or weaker livestock pressure on land.

(d) Chemical changes resulting from the use of artificial chemicals such as fertilisers and pesticides.

(e) Chemical changes occurring in transit through the stream channel system because that system mixes reactive chemicals or because reactive sites such as metabolising organisms are encountered in the stream ecosystem.

It is, however, the diluting power and biological vitality of streamflow from the upper parts of river basins which allows such point discharges to be made. Thus it is not merely the practice of land-disposal of sewage sludge which links urban to rural water quality; the loss of 'purification power' which occurs as the result of rural pollution is highly significant to urban land use since the world's popular city sites are often riparian and downstream.

Rural water quality changes are caused by a number of natural phenomena, including the seasonal march of temperature and moisture, operating through their effects on micro-organisms and, as already demonstrated, on flow routes and velocities of runoff. A graphic demonstration of change in a non-polluting chemical characteristic of water quality is shown in Figure 3.16, derived from a felling experiment performed on a natural forest at Hubbard Brook in the USA. Felling produces a change in runoff volume and soil temperatures (due to loss of canopy). As a result, both the production

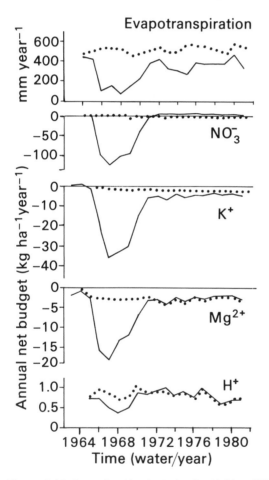

Figure 3.16 Annual net budgets for the Hubbard Brook forest catchment after clear felling (Likens *et al.*, 1978)

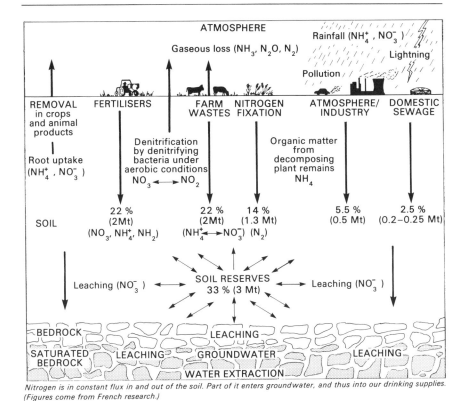

ATMOSPHERE

Gaseous loss (NH_3, N_2O, N_2)

Rainfall (NH_4^+, NO_3^-)

Lightning

Pollution

| REMOVAL in crops and animal products | FERTILISERS | FARM WASTES | NITROGEN FIXATION | ATMOSPHERE/ INDUSTRY | DOMESTIC SEWAGE |

Root uptake (NH_4^+, NO_3^-)

Denitrification by denitrifying bacteria under aerobic conditions $NO_3 \leftarrow\rightarrow NO_2$

Organic matter from decomposing plant remains NH_4

SOIL

| 22% (2Mt) (NO_3, NH_4^+, NH_2) | 22% (2Mt) ($NH_4^+ \leftarrow\rightarrow NO_3^-$) | 14% (1.3 Mt) ($N_2$) | 5.5% (0.5 Mt) | 2.5% (0.2–0.25 Mt) |

Leaching (NO_3^-) ←→ SOIL RESERVES 33% (3 Mt) ←→ Leaching (NO_3^-)

BEDROCK   LEACHING

SATURATED BEDROCK   LEACHING   GROUNDWATER   LEACHING

WATER EXTRACTION

Nitrogen is in constant flux in and out of the soil. Part of it enters groundwater, and thus into our drinking supplies.
(Figures come from French research.)

WINTER CEREALS

Fertiliser added (40kg/ha) (140 kg/ha)

Harvest   Drilling

Risk of leaching organic nitrogen

Small winter risk   *Small risk*   *Small risk*

POTATOES

Fertiliser added (c 200 kg/ha)

Irrigation   Irrigation   Harvest

Leaves killed off by spraying

Planting in ridges

Long period Risk of leaching organically derived nitrogen   *Small risk*   *Small risk*   *Small risk*

Nov   Dec   Jan   Feb   Mar   Apr   May   June   July   Aug   Sep   Oct

Leaching is influenced by the crops grown on the land, climatic conditions and according to the season.

Figure 3.17 The nitrogen cycle for developed land surfaces and the seasonal risk of nitrate leaching for two crops, winter cereals and potatoes

and transport of ions such as nitrate, potassium, magnesium and hydrogen are increased (though the phenomenon is transient, declining as forest regrowth occurs).

Figure 3.17 illustrates a much more pressing rural water quality problem, that of the increasing activity of nutrient cycling (in this case nitrogen) which is an inevitable corollary of optimised agricultural production systems. Industry and domestic waste are also implicated as Figure 3.17 shows but the major and obvious source illustrated for the developed humid temperate zone is 'bag nitrogen' or artificial fertiliser. Nevertheless, the relative role and time-dependence of fertiliser inputs are not fully understood; Figure 3.17 indicates considerable scope for altering crops, crop patterns or crop cycles to minimise the seasonal mobilisation and loss of nitrate to water courses.

The loss of pollutants from the surface of urban areas is a more difficult problem. As Figure 3.18 shows, pollutants collect on urban surfaces such as roads as the direct result of 'normal' urban activity. The British proclivity to dog ownership, for example, leads to an annual deposition of 17 g per square metre of dog faeces which, together with waste oils, litter, de-icing salt and other pollutants, finds its way to streams by virtue of the need to provide efficient road drainage.

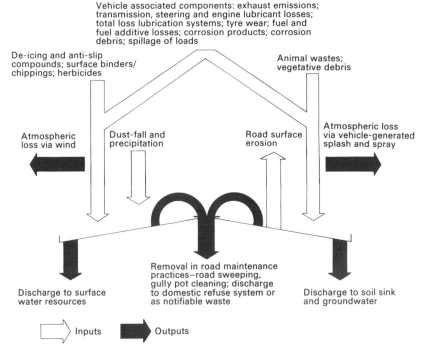

Figure 3.18 Chemical influences on runoff from urban surfaces (Pope, 1980)

The developed and fast-developing economies of the world are currently experiencing very extensive problems of water quality, not only from the perspective of purification needs for human consumption – a measurable cost if health standards are to be maintained – but also in relation to the recreational use of freshwaters, either directly by bathing and boating or indirectly via fisheries. Witter and Carrasco (1996) describe the water quality implications of rapid development in the Rio Yaque del Norte

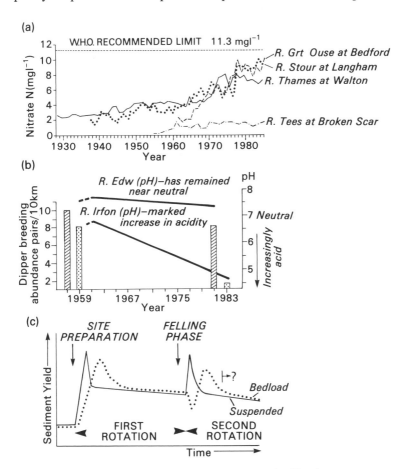

Figure 3.19 Changing river water quality as a result of land use:
    (a) Increasing nitrate concentrations in England and Wales (Roberts and Marsh, 1987)
    (b) Decreasing pH (increasing acidity) revealed by bird numbers breeding by two streams in Wales; vertical bars refer to inferred pH for the two streams (Tyler, 1987)
    (c) Variations in sediment discharge during forest rotation (coniferous plantations, upland UK) (Leeks, personal communication)

watershed (Dominican Republic) as 'a development bomb waiting to explode'. A loss of fisheries also bespeaks a further indefinable cost to species diversity, that is, declining water quality is a conservation threat. In addition to the impact on surface waters it is now concluded that, despite the filtration capacity of many aquifers, groundwaters are now suffering extensive pollution (Section 3.2); furthermore, most rivers discharge into the world's oceans whose capacity to absorb polluting loads is huge but finite.

Two time trends illustrated in Figure 3.19 are becoming relatively common in Europe and North America. In the richer agricultural and highly populated areas nitrate concentrations are increasing at an alarming rate; in upland areas of low agricultural productivity (especially after afforestation in the UK) pH measurements are declining. In the case shown by Figure 3.19b, the decline of pH is indicated by a decline in the numbers of a bird of conservation significance whose food consists of invertebrates which are intolerant of industrially derived acidity which is scavenged from the atmosphere by recent conifer plantations. In both these cases – of eutrophication and acidification – a complex chain of policies involving the whole basin, including its aquifers below and air pollution above it, is necessary to effect control. As in the case of damage to the sediment transfer system (Chapter 2), pro-active attention to the ecosystem's spontaneous regulators, such as wetlands for nutrient removal and base-rich subsoils to neutralize acidity, is an essential part of environmental impact assessment.

3.5   CONCLUSIONS

The severe problems encountered in constructing a generalised body of scientific results, operable in river basin management, from a series of mainly small-scale, short-term empirical studies of land-use effects are featured in Chapter 8. In an inexact, environmental science such as hydrology we must at the very least be sure about the errors involved in field studies. Of overriding concern and crucial to the design of catchment experiments, particularly those concerned with the water balance, is the requirement that the experimental error attached to the effect being studied is not larger than the effect itself. The level of investment needed for scientific programmes in this topic means that we remain in ignorance of key relationships in certain types of land-use practice such as those in the developing world. It has taken decades before a proper hydrological evaluation has occurred of the growth of eucalypt tree plantations in many regions of water shortage (see Calder et al., 1992) and of the impact of development in the humid tropics (see Bonell et al., 1993). The humid tropics, like the world's drylands, are in desperate need of hydrological study in advance of rapid urbanisation, deforestation, metal mining and dam building. They comprise nearly a quarter of the world's land area, will soon

Table 3.12 Removal mechanisms in wetlands for the contaminants in wastewater

Mechanisms	Contaminant affected							
	Settleable solids	Colloidal solids	BOD	Nitrogen	Phosphorus	Heavy metals	Refractory organics	Bacteria and virus
Physical								
Sedimentation	P	S	I	I	I	I	I	I
Filtration	S	S						
Adsorption		S						
Chemical								
Precipitation					P	P		
Adsorption					P	P	S	
Decomposition							P	P
Biological								
Bacterial metabolism		P	P	P			P	
Plant metabolism				S			S	S
Plant absorption				S	S	S	S	
Natural die-off								P

Notes: P = primary effect, S = secondary effect, I = incidental effect (effect occurring incidental to removal of another contaminant).

house a third of its population, have most of the world's uncut forests, most of the unharnessed hydroelectric power and most of the planet's genetic riches. An added hydrological component is the critical role of basins such as the Amazon in controlling the water balance of the planet as a whole. In some nations there are particular reasons for acting quickly and providing 'watershed protection' by varying natural plant covers, deliberate choice of crops in the interests of water management or protective laws against urbanisation. In as many others, however, doubts as to whether science can correctly identify and measure causes are used as a perfect recipe for policy delay.

The river basin ecosystem and its aquifers operate a series of natural controls on the runoff process (both volume and timing) and on the chemical exchanges occurring as water passes through storages such as the vegetation canopy and soil. Table 3.12 illustrates, as an example, the chemical and biological purification operated by wetlands on the water which passes through them. Development often proceeds by modifying these regulators and the result is a different, often hazardous, pattern of runoff and water quality. Costly regulators, such as water purification plants, are then required to 'put back' the regulators, but not before biota and habitat diversity have been damaged. Land-use hydrology seeks to expand our knowledge of the processes which govern the operation of these spontaneous regulators. Land use and land management are also options for managing the impacts of point sources of the polluting byproducts of development; perhaps the visionary view of a balanced approach is that where the sewage from a settlement or chemical plant passes into a conserved or restored wetland for purification, with the wetland simultaneously providing flood protection and a haven for wintering birds. Before such a vision can be realised it is essential that researching the hydrological impacts of land use and land management comes to be seen by national governments as a strategic environmental assessment procedure rather than merely good basic science. In the international review of basin development which now follows one of the clearly differentiating features between the developed and developing world is the level of capability to carry out such assessments.

Chapter 4

Managing land and water in the developed world

An international survey

4.1 DEVELOPMENT AND THE RIVER BASIN

Geographers and politicians argue over definitions of development, over maps of where it has occurred, is occurring and will occur and over rates of development. Perhaps the most commendable classification to emerge in recent years separates off the least developed countries (LDCs). However, virtually all definitions imply some form of 'cultural colonialism' by implying that development is a process and pattern laid down immutably by the 'First World' (Bissio, 1988). The trade system directed by world capitalism confirms an international homogeneity; in the field of water projects the hegemonies of finance and technology transfer are also forces.

To anyone based in the UK, a small nation on an island, developed for habitation over thousands of years, the two most startling aspects of river basin development are the size of the river basins tackled by both ancient and modern management schemes (consider those shown on Figure 4.1) and the rapidity with which technology transfer is now leading to convergence of schemes. Two important new questions then emerge: Are large schemes good schemes? Does convergent technology negate local variation in environment? These cannot be answered until the end of Chapter 5.

An even more basic question arises as to the precise role of water development in the development process as a whole. As Cox (1988) puts it, 'much of the history of civilization involves a continuing effort to enhance the positive aspects of the water resource and to control the negative aspects' (p. 91).

However, Cox goes on to reveal that empirical studies find it difficult to link water development to the location and intensity of private economic activity – water has not been a high priority in industrial or urban location in the developed world (see also Rees, 1969). In the developing world, however, the predominating agricultural or primary industrial economy is much more likely to be boosted by deliberate investment in water. The history of water development in the USA demonstrates that a continuation of government support for 'cheap water' after the initial agricultural period leads inevitably to

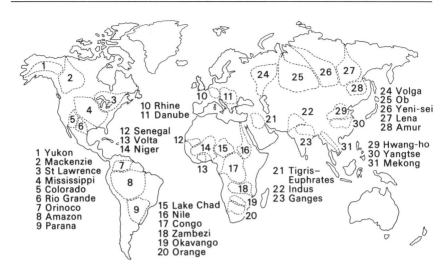

Figure 4.1 The world's largest river basins

the low ranking of water as a location factor. Sustainability arguments and economic instruments for environmental management suggest that pricing drastically influences attitudes to resource use (and hazard avoidance).

A simplistic division of development in river basin terms (presented in Table 4.1) establishes the priority of life-permitting water schemes over life-enhancing ones. Ideally we would move towards fully comprehensive schemes

Table 4.1 River basin development: prioritising the issues of water-based schemes

	'Developed'		'Developing'
Life-permitting	Priorities often domestic and industrial supplies	WATER RESOURCES	Priorities often irrigation and hydro-electric power
	Priorities normally urban centres	FLOOD PROTECTION	Priority is food security
	Major influence associated with property rights	FISHERIES	Subsistence only: little enhancement
	Reacts to 'chemophobia'	POLLUTION CONTROL	Eradication of disease
Life-enhancing	Increasingly Increasingly	RECREATION CONSERVATION	Little known Little known

across the globe but one aspect of local variability is that politicians with budgets are prone to listing priorities. One contribution made by Table 4.1 is that, for the 'developed world', the subject of this chapter, the list represents an approximate chronological sequence of river basin management themes.

The basis for a selection of nations in this chapter inevitably involves literature in the English language but also includes the potential for the 'national experiences' of river basin development in the USA, Canada, New Zealand and Australia to exemplify both physical and cultural influences on public policy with regard to rivers.

4.2   RIVER BASIN MANAGEMENT IN THE USA

The USA is essentially two nations in terms of water resource development; although the Mississippi basin (Figure 4.2) appears to integrate a vast internal drainage system, the division between 'humid' and 'dry' America is too profound in terms of history, attitude, law and contemporary problems to be ignored. The problems of water-based development in the West are well and colourfully documented from the days of the earliest exploration of 'The Interior'. Government finance was used to hold and populate the West against natural hazards and indigenous cultures, making a study of river basins in that region highly instructive. By contrast, in the humid East the water agenda is dominated by issues of industrial planning and environmental protection.

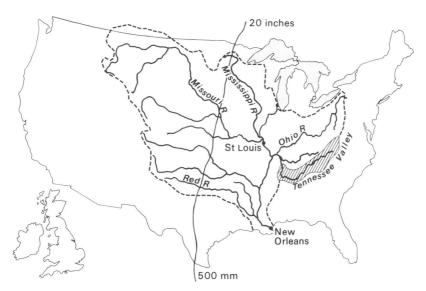

Figure 4.2 The USA, with the British Isles at the same scale, showing the extent and central position of the Mississippi basin

4.2.1 Exploration, development, destabilisation

In 1803 the USA completed the diplomatic coup of doubling its land area by the purchase from Napoleon of Louisiana; 'at four cents an acre' (Cooke, 1973), the opening of an interior for the young nation was a critical development for the President, Thomas Jefferson. We know that by 1804 Jefferson had despatched Lewis and Clark on an expedition, largely routed along rivers, to explore the new hinterland, to survey and to make records of all natural resources (de Voto, 1953).

While Lewis and Clark's three-year expedition contributed an essential knowledge base to the 'go West' mentality which continues to pervade the USA (*vide* the continued growth of the 'Sun Belt'; see below), a more important journey in terms of water development was that made by the Civil War veteran John Wesley Powell. It is a tribute to the Powell expedition's use of photography that Stephens and Shoemaker (1987) have recently been able to assemble a compelling visual record of a century of environmental changes in the basins of the Green and Colorado Rivers. Powell it was who first spoke of 'reclamation' for the harsh habitats of the West; he favoured a careful and conservational application of damming and irrigating but the US Government reacted slowly. However, in the year of Powell's death (1902) the Reclamation Act was passed by Congress, paving the way for Federal intervention in the financing of water schemes in the West. The Mormons' settlement of Salt Lake City provided, perhaps, too inspiring a vision of what could be achieved by irrigation. James Michener's epic historical novel *Centennial* (1975) contains abundant factual evidence of the impacts of 'reclamation' on the lives of settlers in the Colorado township which gives the book its name. The development of water resources is at first a private venture, steered by Russian immigrant Brumbaugh; he immediately concludes that the eastern seaboard's law of *riparian rights* is totally inappropriate for his grandiose scheme for diverting water from the River Platte ('too thick to drink; too thin to plough') on to his land so as to convert from an extensive livestock to an intensive arable economy.

> The average rainfall in parts of England where Riparian Rights were codified was more than thirty-five inches a year, and a farmer's big problem was getting excess water off his land ... 'We must have a new law', Brumbaugh grumbled ... And [Beck] the seedy lawyer proceeded to do so. He devised a brilliant new concept of a river ...: 'The public owns the rivers and all the water in them. The use of that water resides in the man who first took it onto his land and put it to practical purposes .... First-in-time, first-in right.'
>
> (Michener, 1975, pp. 683–4)

Michener colourfully collapses the evolution of the law of *Prior Appropriation* in the American West but correctly identifies the attitude and environment behind it; as noted below it is the current battle between attitude and environment which disfigures water planning in relation to land resources in the West. The triumph of prior appropriation was not inevitable, because much of the land in the West was in Federal ownership; however, the earliest rulings were made in the legacy of mineral rights and a 'gold rush' mentality. The translation to water 'capture' for irrigation was easy under the prevailing political ethos of expansion. Bates *et al.* (1993) quote the Colorado Supreme Court in 1872:

> In a dry and thirsty land it is necessary to divert the waters of streams from their natural channels, in order to obtain the fruits of the soil, and this necessity is so universal and imperious that it claims recognition of the law.

A Secretary of the Interior described irrigation in 1893 as 'the magic wand' and there was a common impression of the climate of the West that 'rain follows the plough'. In fact the outcome was put in doubt. The USA's colonisation of the more extreme climatic zones of its land led in the 1930s to disaster: successive droughts led to the Dust Bowl which forced millions of farming people to migrate, often to other dry areas without a soil erosion problem.

4.2.2   The Colorado basin

Reisner (1990) has labelled the Colorado an 'American Nile' because the river, despite its moderate size (Figure 4.3), is 'the most legislated, most debated, and most litigated in the entire world' (p. 125). It has the largest population and industrial base of other world rivers of a similar level of exploitation and management. It is, in most years, all used (Figure 4.3e), its delta being dry. Its unhappy record of allocation to competing states resembles that of the Nile and the growth of the American West this century, superimposed upon considerable environmental change (both natural and resulting from water management itself) makes a fascinating, if distressing, geographical study (see Graf, 1985).

The waters of the Colorado were first used for a major irrigation project in the Imperial Valley of California in 1901. Dams were built on tributaries from 1910 but by 1922 there was sufficient competition for the relatively modest annual flow of the Colorado itself that the Colorado River Compact was arranged to split this resource between the upper and lower (altitude) states along its steep, often gorge-like course. The Compact assumed an average flow for the river which was 33 per cent higher than the true value (despite flow gauges set up on the river system since 1895 (Graf, 1992)), a mistake which works to the disadvantage of the Upper Basin which is obliged to deliver a set quantity to Lee Ferry, the 'hinge' point with the Lower Basin. The Compact also failed to consider the water entitlements of

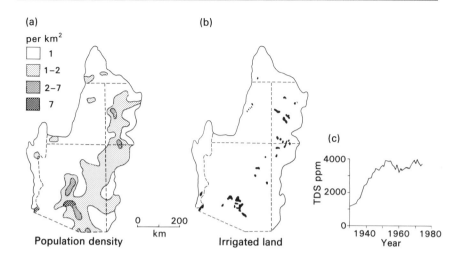

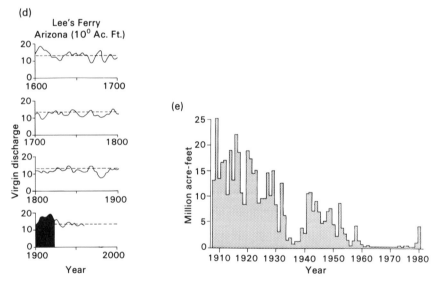

Figure 4.3 Features of the Colorado basin, an 'American Nile':
  (a) Population density
  (b) Irrigation schemes
  (c) The rise of total dissolved solids (TDS) in the river as irrigation
      schemes have developed
  (d) Time series of natural ('virgin') discharge, illustrating the wetter
      period preceding the negotiation of the Colorado Compact (all
      from Graf, 1985)
  (e) Annual flows of the Colorado below all major abstractions,
      1910–80 (Englebert and Scheuring, 1984)

Indian tribes and did not fully address the water needs of Mexico. There are recent and continuing implications for both errors (e.g. Burton, 1991, on the restoration of Indian rights). In 1928 approval was given for the Boulder Canyon Project in which the Hoover Dam was built by 1935. Since that moment, and largely as a result of Federal activity (by the Bureau of Reclamation or US Army Corps of Engineers), the river has been comprehensively dammed (Table 4.2) and diverted, at first to feed the irrigation of high-value arable crops in the south-west but also in the interests of 'high altitude' irrigation (mainly for livestock farming) in the upper basin; clearly the result of Federal projects is cheap water. Reisner (1990) remarks that 'in the West water flows uphill towards money' – and in this case the money is that of the taxpayer or of those paying for the expensive electricity produced by the dams. Graf (1985) likens the Colorado to 'the world's largest plumbing system' because of the dams.

In the lower parts of the Colorado a continuing battle occurs between California and Arizona over rights to the Colorado. Arizona, inhabited by a successful Indian culture based on irrigation (the Hohokam) between 300 BC and AD 1400, experienced a quadrupling of population between 1920 and 1960. Growth of the 'sun-belt' continues and puts further pressure

Table 4.2 Major high dams in the Colorado system

River	State	Dam	Reservoir	Date	Height (m) completed	Capacity (acre-feet)[a]
Colorado	AZ CA	Parker	L. Havasu	1938	98	807,615
Colorado	AZ NV	Davis	L. Mohave	1950	61	2,267,487
Colorado	AZ NV	Hoover	L. Mead	1936	221	37,103,436
Colorado	AZ	Glen Canyon	L. Powell	1964	216	33,668,298
Colorado	CO	Granby	L. Granby	1950	91	673,218
Los Pinos	CO	Vallecito	Pine L.	1941	49	161,523
Blue	CO	Green Mtn	Green Mtn L.	1943	94	155,359
Aqua Fria	AZ	Waddell	L. Pleasant	1927	78	203,445
Gila	AZ	Coolidge	San Carlos L.	1928	76	1,506,726
Verde	AZ	Horseshoe	Horseshoe L.	1949	43	173,853
Verde	AZ	Bartlett	Bartlett L.	1939	87	224,406
Salt	AZ	Stewart Mtn	Saguarto L.	1930	63	87,543
Salt	AZ	Mormon Flat	Canyon L.	1938	68	72,747
Salt	AZ	Horse Mesa	Apache L.	1927	93	305,784
Salt	AZ	Roosevelt	Roosevelt L.	1911	85	1,723,734
Green	WY	Flaming Gorge	Flaming Gorge	1964	153	4,724,856
Strawberry	UT	Strawberry	Strawberry L.	1913	22	286,056
Price	UT	Scofield	Scofield L.	1946	38	690,480
San Juan	NM	Navajo	Navajo L.	1963	123	2,130,624
					Total storage	86,721,822

Source: Graf (1985)
[a] 1 acre-foot = 1,233 m^3.

on the entire Colorado resource system (Sommers and Lounsbury, 1988). Graf (1985) undermines the basis of allocating the average flow of the Colorado between competing states on the grounds of a number of natural instabilities in the basin, many of which are exacerbated by the existing management (largely based on dam construction). There is inherent climatic variability in the south-west USA, the result of oscillating positions of the jet stream (Figure 4.3d). For example, the period for calculating discharge averages prior to the Compact was wetter than long-term figures now available; the 1940s–60s, a period of the urbanisation of floodplains, were largely flood-free. As Graf puts it, 'Planning for variability rather than for averages is therefore the most likely route to success' (p. 154).

Among the variability detailed by Graf, the Colorado suffers from:

(a) The periodic formation of deeply incised (21 metres deep) channels known as *arroyos* which severely damage agricultural land and release tonnes of sediment downstream to hinder basin management.

(b) Channel changes which threaten irrigated agriculture, settlements and communications, with narrow, single-thread channels widening and braiding after 'rare' floods (e.g. those in the late 1970s and 1980s in Arizona which produced channels 1.5 kilometres wide).

(c) Colonisation of floodplain and channel bars by phreatophyte vegetation such as willow, cottonwood and tamarisk which evaporates much-needed river water and produces more extensive flooding than would occur with a 'clean' floodplain. There is, however, controversy about the benefits of removing phreatophytes (Graf, 1992).

Each of these inherent variabilities may be climatically driven or may result from threshold behaviour of the geomorphological system (see Schumm, 1977) or else is conditioned by factors such as overgrazing, the introduction of non-native species such as tamarisk or thoughtless 'channelisation' of problem rivers.

The dams and irrigation systems are also responsible for unwelcome changes such as sedimentation (not only in reservoirs but also in upstream channels), habitat changes (especially below dams with a heavily controlled flow regime) and salinisation (Figure 4.3c). The latter problem threatened to wreck the political structure of basin management when polluted, saline water, the result of crop and reservoir evaporation, was supplied over the border to give Mexico its legitimate share of the Colorado's water.

Despite these problems, Graf (1985) asserts that 'a factor common to the ancient and modern approaches is a perspective on the river basin as an holistic entity, a complex grouping of individual parts that function together' (p. 3).

There is, in fact, little evidence in the Colorado that such an holistic outlook is substantiated by institutions and technology. Indeed, it was the continuing inability to coordinate the exploitation of the Colorado and to reconcile it with conservation that forced the US Government to set up the National Water Commission in 1968. Another retrospective action was the commissioning by the US Bureau of Reclamation (in 1986) of an environmental assessment of Glen Canyon Dam (built in 1964); concluding the Symposium at which results were presented, Leopold (1991) stresses the need for interactive management of the Colorado in view of its inherent changeability, for more data collection and monitoring and for different management scenarios to be explored to give flexibility in response to natural and artificial hazards. In March 1996 an artificial flood was released from Glen Canyon Dam to redistribute channel sediments through the Grand Canyon, re-establishing bars and shoals 'lost' since the regulation regime from Glen Canyon began.

4.2.3   The Tennessee Valley Authority (TVA)

The Tennessee River is a tributary of the Ohio, itself a tributary of the Mississippi which occupies half of the continental USA (Figure 4.2). The Tennessee is itself huge, comprising an area of 80 per cent of that of England and Wales, with a flow which is twenty-four times that of the Thames or 70 per cent of that of the Nile.

There are many tensions in American public policy but two of the greatest, between Federal and State action and between private and public investment in projects, have joined to threaten the success of a model river basin project, much imitated the world over, begun in south-eastern USA in 1933. A hydro-electric power dam, on the River Tennessee at Muscle Shoals, was to be the first act of the Tennessee Valley Authority (TVA), a powerful river basin institution, charged with both land and water management, which President Theodore Roosevelt called 'a corporation clothed with the power of government, but possessed with the flexibility and initiative of a private enterprise' (Palmer, 1986, p. 34). As Palmer writes, 'New dams were at the cutting edge of the TVA and the national attitude about rivers was reflected and shaped by the agency' (p. 35 and see Figure iii, in the Prologue, this volume). Later, President Truman tried to create similar authorities for the Missouri and Columbia rivers but Congress refused. The dams on the Tennessee (nine major sites) and on tributaries (forty-two sites) created a longer lake shoreline than on the Great Lakes; however, despite the heroic achievements of this conservation scheme (in the sense of the word 'conservation' in vogue at the time), it has been much criticised.

Chandler (1984) points out that the costs were far more than the $1.5 billion spent on capital schemes: 10 per cent of the Valley's best farmland was drowned and most of the benefits of flood control were enjoyed by a single city – Chattanooga. Saha and Barrow (1981) dismiss the TVA as 'a massive electricity generating utility' and have drawn attention to the use of that power for polluting activities such as fertiliser manufacture and to the subsequent generation by the TVA of coal-fired and nuclear power. The availability of local power has, however, led to a regeneration of regional prosperity; the navigation improvements built into each dam scheme have also aided the connectivity of a 'backwoods' area (originally isolated by the 37-mile-long rapids at Muscle Shoals).

However, the TVA had become an exploratory concept in soil and water management because of its approach to erosion control; 'watershed management' was a term first used there and an agency spawned by the approach – the US Soil Conservation Service – was to have a major influence throughout the world, often by reversing TVA scales of operation and concentrating on small dams and on-farm techniques. The reach of the TVA, via government agencies, to farmers has been impressive by way of:

(a) Land reclamation from gullying.
(b) Soil conservation.
(c) Use of fertilisers (many produced at Muscle Shoals).
(d) Demonstration farms – crop diversification.
(e) Cooperatives for marketing.

In addition, the TVA needed to manage reservoir water levels to control malarial mosquitoes; this it does by keeping water levels raised during the period April to June when plant growth and stagnant shallows would encourage breeding. The results have been good and further success has come from reducing water levels in winter so as to store floodwaters in the system (see Figure 4.4); despite drowning land beneath its dams the TVA

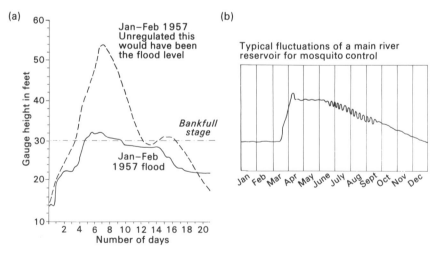

Figure 4.4 Tennessee Valley Authority reservoir regimes for:
(a) Containing flood runoff
(b) Controlling mosquito life cycle (malaria control) (TVA)

has, however, freed much good land from flooding and yet more through erosion control in the uplands.

In a recent review of the environmental effects of dams, Brown and Shelton (1983), both employees of the TVA, strongly support the record of their dams in making the Tennessee River 'one of the most useful in the world' (p. 139). The TVA 'means as much in meeting the challenge of the future as it did in helping to overcome the problems of the 1930s' (p. 150). However, they also admit to problems in responding to 'environmental and societal effects in the planning and decision-making process' (p. 150), a point taken up in Chapter 7 here. Feldman (1991) points out that the TVA continued to promote dams (such as Tellico – see below) in the 1960s 'even though few economic or environmental justifications could be offered on its behalf'.

4.2.4   Dams and river basin management

There are around 90,000 dams 'of any consequence' in the USA (Feldman, 1991); Texas and Kansas lead the nation in the number of dams, each with over 5,000 (Shuman, 1995). Beaumont (1983) lists 5,609 'large' (over 15 metres high) dams in the USA, the majority of which are single purpose, mainly for flood control. Palmer (1986) takes the environmentalist position against the USA's talent for dam-building:

> Reservoirs have flooded the oldest known settlement in North America, the second-deepest canyon, the second most popular whitewater, the habitat of endangered species, one of the first national parks, tens of thousands of homes, rich farmland, desert canyons and virgin forests. In building dams we have blocked the best runs of salmon and broken the oldest Indian treaty – one that George Washington signed.
>
> (Palmer, 1986, p. 1)

Palmer describes the USA as having a 'culture of rivers, free-flowing and luxuriant with habitat'; an attitude which, while latent all this century, peaked in the successful campaigns to prevent dams in Grand Canyon and the Green River Canyons in the 1960s. He sets up a chronology of river abuse up to this time, including the activities of the US Army Corps of Engineers in the East through their flood-control dams as well as the irrigation schemes of the Bureau of Reclamation in the West. By 1980 all agencies and private developers had built 50,000 dams of 25 feet or more. Conservationists had begun to oppose dams in the celebrated Hetch Hetchy of Yosemite National Park; at this stage they faced other conservationists who interpreted the term as meaning careful, rational exploitation of resources (American civil policy enshrined this latter definition throughout the early years of this century). The protests failed; in 1908 the Hetch Hetchy dam was approved.

In the 1950s the biggest environmental conservation issue in the USA concerned a proposal under the Upper Colorado River Storage Project to dam the canyons on the Green River in Dinosaur National Monument. In this case the objectors won, only to lose the battle for Glen Canyon, upstream of Grand Canyon on the Colorado in Arizona, as the Bureau of Reclamation took up their charge to 'build it elsewhere'.

Palmer writes that in the 1960 and 1970s:

> A revolution in attitudes about rivers moved through the country and touched every stream. The late 1960s and early 1970s brought powerful ingredients for change: a growing sense of scarcity, the environmental movement, activism by conservationists and landowners, applications of science and economics coupled with publicity, recreational use, and tight money – all contributing to a national movement to save threatened rivers.
>
> (Palmer, 1986, p. 93)

The new spirit had an unexpected outcome in the case of the Tellico Dam, proposed for the Little Tennessee by the TVA in 1966 but delayed by protests from fishermen; in 1973 a zoologist discovered a rare fish, the snail darter, in the river near the dam site and it was placed on the 'endangered species' list. Progress on the dam was halted but, despite an excellent case against the dam for many other reasons (President Carter had long opposed dam construction), the conservationists' case based on the snail darter was lost and the Tellico Dam was completed.

In 1965 the Wild Rivers Act was passed to bring a measure of protection to the rivers of the West; by 1968 the National Wild and Scenic Rivers Act extended protection for eight rivers, with a reserve list of twenty-seven. The Act prohibits dams or other Federal projects which would damage listed rivers; the list has been extended through the surveys of the Department of the Interior.

A long-running act of contrition about the impact of dams in the USA has been the effort to restore the migratory fishery on the Columbia River in the North-West. The Columbia has sixty-six major hydro-power dams and barrages, the power-base of the region's economy but one which has, since 1931, ruined the rights of both Native Americans and sport fishermen to catch the river's indigenous Chinook salmon. In 1980 the North-West Power Council was formed to mitigate the barrier and regulation effects of the dams; progress has, however, been slow enough for environmentalists to invoke the Endangered Species Act for the salmon. The Columbia River crosses the boundary with Canada, and Hume (1992) has listed the objections of Canadian environmentalists to the river's dams.

Another problem with the American proclivity to dam-building has been the relatively scant attention paid to safety, despite the excellence of the national standing in hydrological research. By 1984, in spite of variable uptake of legislation, 2,900 non-Federal dams had been declared unsafe, most because of poor spillway design. The American Society of Civil Engineers responded by establishing a Task Force which first classified dam safety into three levels of importance depending on the damage caused by dam failure; design is relatively straightforward in the very large and very small damage categories but a cost-benefit approach, with much consideration of social and environmental consequences of failure, needs to be taken in-between these extremes (ASCE, 1988a). In the same year the US Committee on Large Dams prepared an update of its survey of the type, frequency and trends in US dam failures (ASCE, 1988b). The 'narrative description' of incidents makes exciting, if distressing, reading. Over 500 incidents are on record, including 125 major failures and 125 'accidents' (Figure 4.5). The failure of the Teton Dam in Idaho in 1976 killed fourteen people and cost $400 million.

The USA's experience with dams and dam disasters draws out the following features to be borne in mind when dealing with river basin management schemes elsewhere, particularly in the developing world:

(a) Dams have a finite lifetime and a risk assessment is a necessary part of any scheme, for both social and environmental reasons.

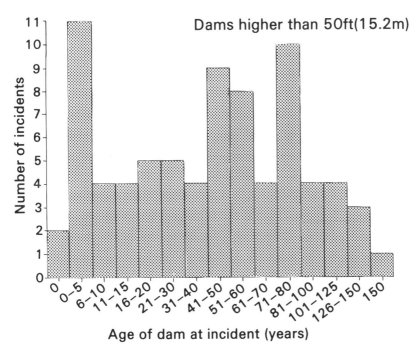

Figure 4.5 Dam incidents, USA, including damage and disaster (American Society of Civil Engineers, 1988b)

(b) Dam-building is a scale element of river basin development and can be aimed at small, e.g. farm, scale developments as well as the larger, riskier, but more prestigious structures.

(c) Dam construction at any scale bespeaks maintenance, safety checks and a well-resourced system of disaster warnings if downstream communities are in danger.

A major change of outlook on dam-building and one which also affected land-use policy occurred in 1968 with the passing of the National Flood Insurance Act. This Act sought to change the balance of the Federal approach to flood protection from one reliant upon dams, levees, diversion channels, and so on (structural approaches) to one seeking to identify the use of flood-prone land and to charge an insurance premium consistent with the risk of flooding. In 1973 the Flood Disaster Protection Act forced developers to arrange such insurance cover before becoming eligible for any federally related finance for development. By 1980 over $95 billion of insurance had been taken out but with an emphasis on coastal, not river, flooding (Platt *et al.*, 1983). There were also land-use planning implications (see below). Shuman (1995) reveals that the current focus for dam-

Plate 4.1 Devastation resulting in the valley of the Roaring River, Colorado, USA from the failure of the Lawn Lake Dam, July 1982 (photo M. D. Newson)

related environmental impact studies is the removal of dams which have served their purpose, are unsafe or have had undesirable effects on migratory fish (e.g. dams on the Elwha River in Washington state). Shuman lists six dam removals between 1962 and 1992 and four current proposals. Dam removals can be controversial and sediment-related problems from the release of stored silts can also have environmentally harmful impacts.

4.2.5   Land-use issues in American river basin management

It is typical of water management in a developed nation that there have been resources to put into data collection for monitoring and research. Information is not a panacea, however, and knowledge for management requires that data are collected with a theme in mind. The ecosystem model for an holistically managed river basin makes certain demands about the format and location for data collection; however, the tradition in the developed world is normally to let the intensive exploitation functions of the basin define the data programmes: data are 'good' if they serve development, not protection.

We have not, so far, mentioned an important US agency which maintains the network of basic hydrological observations in the country as a whole – the US Geological Survey (USGS). The Survey exists to:

(a) Collect and analyse data for water resources planning and control.
(b) Assess the physical, biological and chemical characteristics of surface and ground water.
(c) Coordinate all Federal activity in water-resource data gathering and provide advice to other agencies.
(d) Operate State institutes and grant programmes for water resources research.

The Survey's data-gathering activity dates from John Wesley Powell's day; from 1888 Powell sought to train Survey staff to make the necessary measurements to enable satisfactory irrigation and flood control schemes to be devised. The USGS is also technically well-equipped to carry out research on the effects of land use.

The research catchment programme, run at a much smaller scale, is very active in the USA; the USGS participates largely through involvement with State water resource institutes. Additionally the US Department of Agriculture, especially the Forest Service, carries out its own research into farming and forestry effects. Because such agencies also own and manage a large amount of land there is a good deal of immediate practical benefit from land-use hydrology.

As an indication of the major land-and-water interactions in the US research programme, in the American Society of Civil Engineers'

Watershed Management 1980 volume (see Snyder, and McClimans references below), the following farm and forest issues are covered:

Farming $\begin{cases} \text{Soil erosion and tillage} \\ \text{Range management and stocking rates} \end{cases}$

Forestry $\begin{cases} \text{Water yield} \\ \text{Erosion} \\ \text{Roading} \\ \text{Fire} \end{cases}$

This spread and balance of issues changes little in six other symposia of the Society's Watershed Management Committee (see the most recent: Ward, 1995), though there is an increasing emphasis on the translation of hydrological data into 'multi-stakeholder management plans', particularly in the forested headwaters of the Pacific Northwest region.

Satterlund and Adams (1992) have assembled a guide entitled *Wildland Watershed Management*, essentially a headwater hydrology manual, from which Figure 4.6 is taken to show the proven water resource values of brush management in semi-arid Arizona and the importance of patterns of land management in felling forests to increase runoff. The explanation of the effect of these patterns lies in the fact that interception losses from forest canopies require ventilation around the edge of the canopy – the more edge the greater the losses during and after rainfall.

Watershed management principles have also, albeit slowly, entered the control of water quality – particularly of non-point pollution – in the USA. The US Environmental Protection Agency (1986) has divided the nation into seventy-six 'ecoregions', representing statistically significant combinations of natural controls on water quality. State quality control programmes then monitor the physical, chemical and biological background quality of indicator watersheds in order to set local standards. In some states (e.g. Illinois; Polls and Lanyon, 1980) homogeneous land uses receive similar monitoring. The Illinois study has tabulated the characteristic pollutant loads contributed from each major land use.

The US approach to forest management and water resources has the benefit of many years of continuous detailed research dating back to the early years of the century, through one or more forest cycles (American forestry has an added environmental issue to address on felling since much of the forested land is pristine). By now forest management is very carefully controlled to meet water resources and pollution objectives. For example, the US Environmental Protection Agency requires the Forest Service to carry out evaluations of non-point sources of pollution including soil erosion/man movement, water temperature, dissolved oxygen, nutrients and pesticides. Snyder (1980) summarises these impacts while McClimans (1980) develops the practical aspects of forest management via 'best

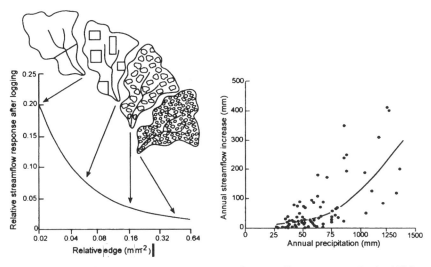

Figure 4.6 Detailed canopy management and streamflow response in the USA
(after Satterlund and Adams, 1992):
  (a) Pattern of tree cover removal (25 per cent removed) and
      streamflow response
  (b) Removal of brush cover on semi-arid watersheds in Arizona and
      streamflow response

management practices' for forestry activities (for UK parallels, see Chapter
8); these were instituted by amendment to the Federal Water Pollution
Control Act 1972. The BMP procedures link pollution or other damage
hazards to slope angle and distance to streams (Figure 4.7).

Heede and King (1990) address the problem that to harvest timber as part
of watershed management may lead to more problems (of pollution and
sedimentation) than it solves (in water yield). In their study area of Arizona
the use of 'state of the art' machinery and of strict rules on its use created no
extra surface runoff or soil erosion problems. In the East, Douglass (1983)
has demonstrated how the use of simple predictive models can coordinate a
felling programme to augment water yield steadily, though he claims that
interventionist policies to bring about such a programme are not yet needed
in the eastern states.

The ill-fated National Water Commission (1968–83) carefully considered the benefits of the land-use management option in a national water
resource strategy (US National Water Commission, 1973). The major
control options available were said to be:

(a) Forest and brush management, critical because of the 1,000 mm average
    rainfall of the US forests (cf. 610 mm on other land uses). There is, for
    example, a strong case for a 'joint product' (timber and water) from the
    sub-alpine forests of the Upper Colorado.

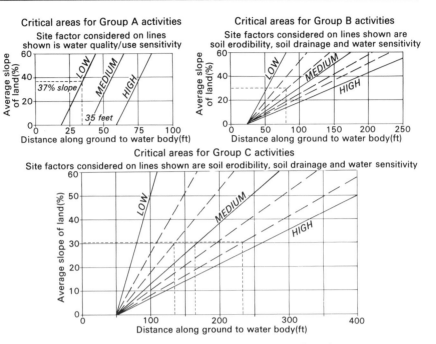

Critical areas for Group A activities
Site factor considered on lines
shown is water quality/use sensitivity

Critical areas for Group B activities
Site factors considered on lines shown are
soil erodibility, soil drainage and water sensitivity

Critical areas for Group C activities
Site factors considered on lines shown are soil erodibility, soil drainage and water sensitivity

Figure 4.7 Assessment graphs for mitigating the damage from forestry operations close to streams. Group A activities include aerial spraying, Group B cultivating and road work, Group C ground skidding (McClimans, 1980)

(b) Phreatophyte management, already referred to in connection with the Colorado but comprising 6.5 million hectares in the south-west. (For comparative evaporation rates see Table 4.3.)

(c) Management of snow packs by vegetation (largely forest) management; in Colorado 20 per cent of the state's water resources come from snow and in California the proportion stands at 51 per cent.

(d) Soil surface treatments to improve runoff characteristics.

Since the National Flood Insurance Act of 1968 and the Flood Disaster Protection Act of 1973 a major change of approach to land use has occurred in relation to floodplains. One-sixth of all urban land in the USA lies within the 100-year flood line. Since the 1937 Flood Control Act the Federal government has spent $25 billion on structural flood control works but annual flood losses now amount to $2 billion. More than 20,000 communities in the USA are practising land-use regulations under the National Flood Insurance Program (Muckleston, 1990); $162 billions' worth of property on floodplains was insured by the late 1980s. The Federal Government is allowed to purchase land with persistent flooding problems, rather than protect it at great expense and to the detriment of downstream settlements. A

Table 4.3 Measured evapotranspiration rates reported
for phreatophytes and other riparian
vegetation in the south-western United States

Species	Rate (mm water per year)
Bare Ground	640−810
Phreatophytes	
Cottonwood	1,540−2,480
Mesquite	1,014
Tamarisk	340−2,800
Willow	770−1,390
Shrubs	
Arrowweed	2,440
Fourwing Saltbush	1,090
Grasses	
Bermuda	1,070−2,760
Blue Panic	1,240−1,330
Alta Fescue	1,820
St Augustine	1,660
Alkalai Sacaton	430
Saltgrass	540
Crops	
Alfalfa	1,750−2,130
Barley	300−1,120
Cotton	650−1,050
Wheat	660−1,120

Source: Graf (1985)

national Program for Floodplain Management has been slow to be realised in practice but decision-making software abounds for achieving optimum uses (including flood storage, wetland conservation and groundwater recharge). In this case the Federal influence has been too gentle in the face of state, community and private involvement. Montz and Gruntfest (1986) claim that the Flood Insurance Act has only worked well where success is easiest; flood damage continues to rise and floodplain encroachment has not been halted in growing urban areas. Here is a real case for a single authority at the basin scale; possibly the Mississippi floods of 1993 will encourage such an approach for the nation's most important river.

Prince (1995) makes a useful summary of the most recent and most influential flood in US flood history − that on the Mississippi during July and August 1993; peak flows were the highest on record at forty-two gauges and in excess of the 100-year recurrence interval at forty-six. The loss of life totalled forty-eight people (Table 4.4), 42,000 homes were destroyed and 21,000 square miles went under water. 'Verdicts' on the flood pointed to the implications of the almost total loss of wetland flood storage in the upper basin (Illinois, Iowa) and to the impact of stream channelisation. Almost

Table 4.4 The 1993 Mississippi flood: flood damage statistics

State	Flood-related deaths	Property damage ($ million)	Agricultural loss ($ million)
Minnesota	4	51	865
North Dakota	2	100	420
South Dakota	3	26	725
Iowa	5	1,250	450
Nebraska	2	50	292
Kansas	1	160	434
Wisconsin	2	101	800
Illinois	4	930	565
Missouri	25	2,000	2,000

Source: *The Flood of '93*, Associated Press, New York

half of the Corps of Engineers' 229 'federal' levees were overtopped, bringing the Corps to the view that the lack of a comprehensive floodplain plan for the Upper Mississippi basin results in a wasted investment; state and private levee districts have taken the same view. As one American engineer put it, 'The river owns the land. We should not forget that fact. When we do it targets our error.'

4.2.6   Land-use planning in river basin units

The Commission's other recommendations include institutional changes to permit more rational planning but in practice most of the compliance by land owners and managers is coming from the threat of pollution legislation rather than from proactive rural planning. While California has watershed planning systems to coordinate the work of agencies in support of the State Wild and Scenic Rivers Act (Watson, 1980), other states have not followed suit and the other examples of basin planning are either urban or part of the brief of interstate and interagency planning for large basins (see below).

There have been several phases of interest in river basin planning authorities subsequent to the Colorado Compact in 1922 and the Tennessee Valley Authority in 1933. In the 1940s the Federal Interagency River Basin Commission provided advice based upon 'layering' the individual activities of a number of agencies in some of the larger basins (Figure 4.8). This scale and type of activity was abolished in 1966; the Water Resource Planning Act 1965 (Black, 1982) had established River Basin Commissions as much more proactive in planning; these paid the penalty of being recommended by the National Water Commission as 'new and unique regional institutions' and clearly regional institutions figure very little in American policy. Much more enduring have proved to be the river

Figure 4.8 Agency structures surrounding the management of major US river basins

basin compacts between states where the Federal government is also involved, for example, the Delaware River Basin Compact (DRBC) (see Majumdar *et al.*, 1988). The Commission comprises the Governors of Delaware, New Jersey, Pennsylvania (and New York, which has created 90 per cent of the water storage capacity as dams in the upper basin), together with the Secretary of the Interior. Goodell (1988) describes its functions as:

> to encourage and provide for the planning, conservation, utilisation, development, management and control of the water resources of the basin and to apply the principle of equal and uniform treatment to all water users who are similarly situated and to all users of related facilities, without regard to established political boundaries.
>
> (Goodell, 1988, p. 286)

It successfully saw the basin through the drought of 1961–7 when it took over the operation of several agency schemes (having been instituted after devastating floods in 1955!). It has also arranged referenda on further dam projects and cancelled two following adverse votes. Patrick (1992) considers that the DRBC has had major successes in cleaning up water quality in the Delaware estuary – largely through using a system of tradeable permits to discharge pollutants up to the point where 90 per cent of the water's natural capacity to dilute and disperse is utilised. However, DRBC has failed to reconcile differences in the planning styles and structures, for example, between New Jersey and Pennsylvania. Polhemus (1988) concludes:

> the Commission is tightly controlled by the states and has not been permitted to develop an innovative operating programme. Although the Commission could evolve into a significant control factor in the basin, influencing growth, economic development and the balance of power in the region, it has been compromised to a limited water management role.
>
> (Polhemus, 1988, pp. 317–18)

It may well be that concerns with water quality will be paramount in driving river basin management policy in the USA towards further basin-wide authorities. Patrick (1992) describes the Federal Water Pollution Control Act (the 'Clean Water Act') as having the twin goals of making rivers 'fishable and swimmable'; making them fishable requires allowance for fish migration and implies treatment of the river as a system. The Act has been amended to allow an attack on non-point pollution sources and this could further strengthen the case for basin authorities. However, there are severe political obstructions to be tackled first.

4.2.7   Cultural, political, legal attitudes

Feldman (1991) considers that a fundamental constraint on the work of the National Water Commission was that it was expressly forbidden to consider interbasin transfers of water because this would have created an environment in which interstate, basin-wide authorities were inevitable. Feldman goes on to say that the whole problem of river management in the USA has two main components: 'the fragmented nature of natural resources policy making and the venerable tradition of antistatism in American political culture' (ironic since one of the main enduring problems of water in the West has been federal control/subsidy for unsustainable water-based development).

As Reisner (1990) puts it, the venture West by high density settlement creates the dilemma for water distribution there:

> Any place with less than twenty inches of rainfall is hostile terrain to a farmer depending solely on the sky, and a place that receives seven inches or less – as Phoenix, El Paso and Reno do – is arguably no place to inhabit at all.
>
> (p. 3)

Worster (1992) describes three stages of the growth of unsustainable water-based growth in the American West, beginning with the Mormon settlement of Utah (1847: 'incipience'), boosted by the Bureau of Reclamation (1902: 'florescence') and finally 'empire' from the 1940s onwards when 'the two forces of government and private wealth achieved a powerful alliance, bringing every major western river under their unified control and perfecting a hydraulic society without peer in history' (p. 64).

How can the USA now escape the ethos which led to the situation west of the Mississippi? It will be extremely hard to break the law of prior appropriation in the West. Even though the law allows rights to be established only to the extent of 'beneficial use' of the water abstracted (Gangstad, 1990), only the most profligate uses are frowned on by the courts (Bates *et al.*, 1993, quote the use of a river to flood a field in order to drown gophers as being non-beneficial use). The sale of water rights is perhaps the major way in which current patterns of water requirements can be represented in property rights but it gives no guarantee that ecological uses of water will be respected, other than in wild and scenic rivers. Pricing and legislating water exploitation is something which successive US Governments have found desperately hard to do; a number of Presidents have begun their political careers with 'pork barrel' water projects won in Washington for local electors. President Jimmy Carter was extremely bold for a US President in standing up to the Bureau of Reclamation over its budget. Ingram (1990) offers a detailed political critique of the project-based water development syndrome which now needs correction.

Reisner is fatalistic:

> the tragic and ludicrous aspect of the whole situation is that cheap water keeps the machine running: the water lobby cannot have enough of it, just as the engineers cannot build enough dams and how convenient that cheap water encourages waste which results in more dams.
>
> (Reisner, 1990, p. 494)

Not just surface water resources are involved. The Ogallala aquifer underlies the High Plains between South Dakota and western Texas. From the close of the Second World War, in an era of cheap energy, its waters have been pumped on to the drought-prone lands; by now, as Reisner (1990) puts it, the aquifer is overdrawn by an amount equivalent to the flow of the Colorado.

Groundwater protection and management are likely to be high on the agenda of US water management policy-making in the next decade (Figure 3.8). Already twenty-seven states have statutes allowing intervention to control 'groundwater mining' but, significantly, those states with a large irrigation need have relied on a 'good neighbour' policy among local users (Bowman, 1990). In New Jersey, by contrast, the state has a groundwater strategy, targeted at pollution control as well as water resource allocation (Whipple and Van Abs, 1990).

The most recent estimates (Solley, 1989) are that withdrawals of water declined by 10 per cent in the USA between 1980 and 1985. The USA's five-yearly survey, begun in 1950, had hitherto always shown a rise in consumption; the downturn is ascribed to depressed commodity prices in agriculture and increased use of recycled water in industry. Nevertheless, average public and domestic use remains extremely high, at over 350 l/person/day when the water is privately supplied and almost 480 l/person/day under public schemes (illustrating the effects of costs and pricing).

Given that the 1980s have witnessed seven droughts in the USA, the discrepancy between supply and demand in some regions (Table 4.5) is likely to grow. The most extensive water scheme of all for the western USA remains a gleam in the eyes of dam-builders; the North American Water and Power Alliance (NAWAPA) was devised by a Los Angeles water and power engineer in the early 1950s. Environmentalists are appalled by its scope to bring water south from the Canadian Rockies; Canadians feel threatened by it, especially in the light of their own shortages in British Columbia and Alberta. Reisner (1990) remarks that, if built, the NAWAPA network will destroy what is left of the natural West and require the taking of Canada by force!

Clearly, some form of national planning is needed for water resources in the USA, particularly when options other than further dam-building or NAWAPA are so obvious but require institutional changes, for example, the land-use options referred to in Section 4.2.4. These would help to yield more water and, importantly, help to raise water quality. Approaches to water quality control are little and late in the USA. The wider, ecological concerns about US freshwaters have been raised in a document which sets a new, holistic, research agenda (Naiman *et al.*, 1995). *The Freshwater Imperative* lists the priority research areas as ecological restoration and rehabilitation, maintenance of biodiversity, analysis of modified flow patterns, assessment of ecosystems goods and services, predictive tools for management and solution of future problems. Even with such a knowledge base, however, the political culture of the USA mitigates against the right institutional framework for ecological management of river systems. The US Government's attitude to institutional change can be summarised by the action of Secretary of the Interior James Watt's dismissal of the NWC in 1983; one of his associates described the Commission as 'just another layer of government that you didn't need' and 'another review board that could never make up its mind on anything'. Watt reinstated many agency water projects, especially in the West, linking these to employment needs. Despite the jibes over indecision, the major findings of the Commission (US NWC, 1973) are salutary in the light of our survey here of America's dilemma:

(a) Develop an adequate data base.
(b) Conduct further research into the environmental impacts of water resource development.
(c) Utilise planning techniques which are sensitive to ecological processes and environmental values.

Table 4.5 Streamflow compared with current withdrawals and consumption (billion US gallons[a] per day)

Region	Mean annual runoff	Runoff in 95% of years	Freshwater consumptive use 1970	Projected total consumptive use		Withdrawals 1970	Projected total withdrawals	
				2000	2020		2000	2020
North Atlantic	163.0	112.0	1.8	5.0	8.5	55.0	113.9	236.3
South Atlantic–Gulf	197.0	116.0	3.3	5.7	8.3	35.0	87.4	130.2
Great Lakes	63.2	42.4	1.2	3.2	5.5	39.0	96.6	191.0
Ohio	125.0	67.5	0.9	2.5	3.6	36.0	65.1	90.2
Tennessee	41.5	24.4	0.24	0.8	1.1	7.09	13.9	18.1
Upper Mississippi	64.6	28.5	0.8	0.8	2.6	16.0	30.6	41.3
Lower Mississippi	48.4	24.6	3.6	4.5	6.3	13.0	28.0	39.4
Souris–Red–Rainy	6.17	1.91	0.07	0.5	0.5	0.3	2.0	2.8
Missouri	54.1	23.9	12.0	15.0	16.4	24.0	27.9	31.6
Arkansas–White–Red	95.8	33.4	6.8	10.6	12.3	12.0	25.3	31.6
Texas–Gulf	39.1	11.4	6.2	10.9	12.3	21.0	57.3	92.6
Rio Grande	4.9	2.1	3.3	5.0	5.5	6.3	9.5	11.7
Upper Colorado	13.45	7.50	4.1	3.1	3.1	8.1	6.6	6.7
Lower Colorado	3.19	0.85	5.0	4.6	5.3	7.2	8.4	8.9
Great Basin	5.89	2.46	3.2	3.6	3.8	6.7	7.6	7.8
Columbia–North Pacific	210.0	138.0	11.0	17.3	21.6	30.0	90.1	156.7
California	65.1	25.6	22.0	32.7	38.2	48.0	120.5	244.8
Alaska	588.0		0.02	0.1	0.2	0.2	0.9	4.2
Hawaii	13.3		0.8	1.0	1.4	2.7	4.7	8.6
Puerto Rico			0.17	0.5	0.6	3.0	8.3	13.7
Total United States	1,793.7		86.5	127.4	157.1	370.59	804.6	1,368.2

Source: US Nationd Water Commission (1973)
[a] 1 million US gallons = 3,785 ML.

(d) Develop rigorously and present as clearly as practicable the environmental impacts associated with a proposed water resources project and the available alternatives.
(e) Reach a decision.
(f) Monitor environmental consequences.

Muckleston (1990) labels the most recent phase in American legislation over water as 'counterenvironmentalism, devolution and dismemberment of institutions, facilitated by fiscal austerity for domestic programs' (p. 29).

Lack of integration in the Federal water programme also angers Palmer (1986) from the conservation standpoint. He writes:

> Twenty-five government agencies now spend $10 billion a year on water but they do not work in unison. The Department of Agriculture drains wetlands while the Fish and Wildlife Service of the Department of the Interior tries to preserve them. The Bureau of Reclamation in the Department of the Interior irrigates new farmland while the Department of Agriculture pays farmers to leave the land idle. The Fish and Wildlife Service tries to halt channelization while the Federal Emergency Management Administration pays for bulldozers to plough through streams in attempts to push gravel away after floods.
>
> (Palmer, 1986, p. 40)

It would appear that America has thrown the coordination baby out with the Federal bath water!

Rogers (1993) requests nothing less than a President's Water Commission to address the foundation of a new water policy for the USA. Black (1982) points out the paradox that, while deeply in need of a powerful and centralised agency for coordinating water *and land* planning, the people of the USA have come to distrust such large and powerful agencies through the historical activity of some of the autonomous contemporary agencies which might participate. It is perhaps the combined attitude to land and water which prevents the wider application of the river basin system argument; Caldwell and Shrader-Frechette (1993) have illustrated that the absolute rights of possession in the USA, granted at times of colonisation, are no recipe for land-use planning and that powerful political lobbies want nothing to do with interference with land, from whatever institution it comes. Quite possibly the Mississippi floods of 1993 have begun to educate millions of Americans about the mutual interaction of land and water and the connectivity between upstream cause and downstream effect. Thus a land/water phenomenon large enough to be viewed from satellites may have as much impact in Washington as political pleading by scientific commissions and experts.

4.3   CANADIAN RIVER BASIN MANAGEMENT

Canada is a land of rivers and lakes, surface water comprising 8 per cent of Canadian territory. Rivers, referred to as 'lordly' in the national anthem, were the routeways through which the nation was explored and integrated (unlike the waggon trains of the USA to the south). Nine per cent of global runoff serves a population of only 25 million people. Nevertheless, Canada has problems of water being abundant in 'the wrong places' (e.g. the cold North) but scarce by season in the West, on the prairies and in the urban agglomerations of the East (Figure 4.9). Like the USA, Canada has the problem of having developed water supply (and more notably hydropower) schemes at a large capital scale but, unlike the USA, there have been no 'pork barrel' politics with powerful Federal agencies. The provinces have controlled water development, though Federal agencies have more say in land issues. Federal interests have retained control of fisheries and navigation. There is, between federal and provincial interests, a culture of negotiation and compromise which has at times been less productive than a standoff.

In 1970 the Canada Water Act sought to provide integrated planning between Federal and provincial agencies; one of the aims was to improve public consultation and awareness. The enormous scale of Canadian water projects made this a dire necessity. For example, the La Grande project in Quebec involved a work site the size of England, under almost tundra conditions, lasting over thirteen years during which four dams were built to divert and double the flow of the La Grande in order to generate

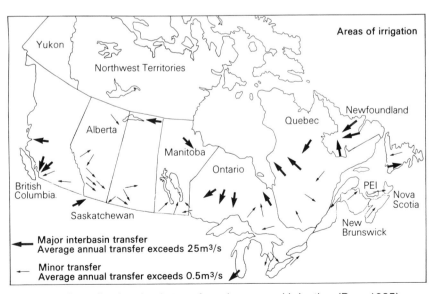

Figure 4.9 Canadian interbasin transfer schemes and irrigation (Day, 1985)

70,000 GWh of electricity. By excluding the St Lawrence and Mackenzie Rivers, roughly one-third of all Canadian flow is transferred across basin boundaries (see Figure 4.9).

4.3.1   Water transfers: a Canadian speciality

As an indication of the scale of projects for which the Water Act was implemented we may select two further examples. Table 4.6 lists the water transfer schemes under construction in 1980 and indicates the domination of hydro-electric power generation as a driving force. The dominant schemes comprised the La Grande, the Churchill Falls project ($665 million) completed in 1974, using two diversions , and the Lake Winnipeg, Churchill and Nelson project completed in 1977 ($1.36 billion). By this time Canadian water transfer flows had exceeded those of the USA and the Soviet Union combined.

These schemes had a number of undesirable social and environmental effects. For example, the Lake Winnipeg scheme threatened the environment of the Lower Churchill and South Indian Lake; a considerable change of lifestyle was implied for the communities affected. A representative body was established for discussion of implementation of the scheme, a northern development programme was established and a wide range of ecological monitoring was carried out.

Day (1985) concludes that yet more attention is needed on biophysical and social questions in Canadian transfer schemes. 'Enormous overbuilding of diversion-related hydro-electricity capacity has been costly to the Canadian public' and native groups need more equitable treatment, he claims. Institutional reforms would help secure a more sustainable approach to water use without such heavy reliance on the 'boom-and-bust'

Table 4.6 Canadian water transfers existing or under construction, 1980

Province	No. of transfers	Average annual flows in m^3/s	Major use
Newfoundland	5	725	Hydro
Nova Scotia	4	18	Hydro
New Brunswick	2	2	Municipal
Quebec	6	1,854	Hydro
Ontario	9	564	Hydro
Manitoba	5	775	Hydro
Saskatchewan	5	30	Hydro
Alberta	9	67	Irrigation
British Columbia	9	367	Hydro
Canada	54	4,402	Hydro

Source: Day (1985)

construction industry. The French Canadian perspective in Quebec has implications for such projects as the James Bay scheme. As Hamley (1990) reminds us, Quebec has pretentions to autonomy but little oil, coal or gas; the province has worries about becoming industrially unattractive. To be regarded as 'electricity Arabs' in the north-east corner of the continent would be more useful to Quebec than the beaver pelt exports of the Cree Indians, now reduced to 1.3 per cent of their national lands because of the 'dated gigantism' of the James Bay project, 'Quebec's monument to a faded vision of modernism'. More recently, although the Cree settled for $100 million for their resource sacrifices related to the La Grande phase of James Bay they used the James Bay and Northern Quebec Agreement (Peters, 1992) to hold out against the Great Whale phase, a $13 billion extension to the scheme; the project was shelved by the Quebec government in November 1995, although the continuing Environmental Assessment of Great Whale signals the long-term intention of the 'electricity Arabs' to complete.

Canada's relationship with the USA, soured by the acid rain issue and with the remaining threat of NAWAPA (Section 4.2.6), is of considerable day-to-day importance when managing the Great Lakes along their frontier. The International Joint Commission dates back to 1909; the IJC acts as neutral adviser and factfinder while many of the problematic issues of boundary cooperation are retained by individual states and provinces (Colborn et al., 1990). The Great Lakes fluvial system comprises over 80,000 inland lakes, 750,000 kilometres of rivers and a groundwater system providing drinking water for the 7.5 million residents of the basin. However, there is a desire to treat this problem in ecosystem terms and to include the wider environment of air and land resources and hazards. The Colborn et al. (1990) volume presents environmental indicators – of both stress and ecosystem health, together with a future vision of the Great Lakes.

Improvements to municipal waste treatment have considerably reduced the phosphate content of the Great Lakes but their waters are still said to contain 800 toxic solutes. During the 1980s high lake levels caused hundreds of millions of dollars' damage to both sides of the border. Proposals were made to curtail Canadian transfer schemes in Ontario and to augment the Lake Michigan transfer out to the Mississippi. The issue remains unresolved. However, boundary issues are critical to integrated water management, not only with regard to the USA but also because no Canadian province or territory has complete control of its water (excepting Prince Edward Island).

The Water Act 1970 has had little impact on such boundary (territorial and agency) problems except where specific missions such as those covered by the Flood Damage Reduction Program or Heritage River System have provided inspiration. By far the biggest stimulus for basin-scale integrated management has come from provincial governments (see 4.3.3).

4.3.2   Land-use issues in basin development

The European settler/pioneer ethos in North America has, until recently, politically contained an environmentalist backlash against the political supremacy of primary resource exploitation. However, as with the tactics south of the border which led to the Wild and Scenic Rivers Act, activists in Canada (especially in the forested lands of the west) have recently begun an orchestrated campaign in favour of the cultural values of their wild river ecosystems (see e.g. Hume, 1992). Hume says of his volume, 'This book is about the abuse of one of British Columbia's most valuable resources – its rivers'; one of his major concerns, now a prominent political 'hot potato', is the logging activity in the temperate rainforests of British Columbia.

O'Riordan (1986), addressing land-use issues in British Columbia, is yet another Canadian author bemoaning lack of integration and the dangers of overlapping Federal and provincial functions: 'Although water and land issues are closely inter-related, few administrations have managed to coordinate them effectively into a single planning process' (p. 193).

The Cabinet of British Columbia, however, has established an Environment and Land Use Committee which encourages all Resource Ministries to develop strategic plans prior to trading-off and integration of goals. O'Riordan cites three examples of land-use pressure on basin management: water shortage for irrigation (Nicola Basin), hydrological effects of urbanisation (Serpentine–Nicomekl) and logging effects on runoff quantity and quality (Slocan Valley). In each case the land-use/management decisions are taken at a much larger regional scale than that of the basin whose hydrology is affected. He stresses the need for comprehensive information flows up and down a hierarchy of decision and implementation levels. He says:

> In the age of the computer, the massive amounts of data that must be sorted and sieved in such an approach are now manageable. The missing ingredient continues to be man's imagination, innovation and the will of institutions – too often vertically oriented – to make it work.
>
> (O'Riordan, 1986, p. 210)

British Columbia is now a pioneer in the Canadian government's attempt to manage all its natural resources sustainably. The Lower Fraser River basin has been chosen as a unit for State of the Environment Reporting because of its mixture of air, water and land resource pressures (Environment Canada, 1992), making it one of the first river basins in the world to become an eco-region for environmental management purposes.

Developing the state of Canadian knowledge on land/water interactions, Hare (1984) concentrates attention on forestry activity since most of those basins coming under management attention are forested; 43 per cent of Canada's surface is in fact forested and in Hare's words 'forest hydrology is becoming a well developed art in Canada'. In Canada forest hydrology is often snow hydrology since the felling of timber exposes snowcover to direct solar radiation, accelerating spring melting; tightly managed forests offer an opportunity to manipulate streamflow directly in Canada.

Hume (1992) lays out the river-based case against uncontrolled felling of the native temperate rainforest in the Pacific coastlands of British Columbia; 40,000 hectares of forest is logged each year and only one-third of the original 6 million hectares of this land cover remains. Hume puts his case severely:

> The forest industry, which spends millions of dollars a year on campaigns portraying loggers as harvesters of the land, has not yet been able to figure out how to clearcut a watershed without severely affecting the entire hydrology of a valley .... Gravel spawning beds fill with suffocating silt .... The clear water runs brown whenever it rains.
>
> (p. 189)

> And, yes, some areas will have to be cut to preserve jobs, but in deciding what to take and what to leave, there should be no delusions about what it means.
>
> (p. 201)

The prairie provinces of Canada also have land management problems in relation to river flows; the prairies lose many millions of tonnes of soil by both wind and water erosion. Seepage of salinised groundwater to hollows is also a problem: in some areas 15 per cent of irrigated land is salinised. Lake Manitoba is alkaline and brackish from prairie runoff and the sedimentation rate has increased in the last 100 years.

Hare concludes that Canada, despite its abundance of water, faces severe local problems of management. As a British expatriate he writes:

> Having grown up in a densely-populated European country where denser population and more intensive industry are well-served by supply less abundant than that of Southern Ontario I remain unrepentantly of the opinion that Canada's water problems arise from human misuse, not from poverty in the resource.
>
> (Hare, 1984, p. 4)

4.3.3   Bringing about improved integration of policies

Two problems are identified by Shrubsole (1990) which may explain why Canada has avoided management of the water resource around the 'obvious' unit of the river basin: the 'myth of superabundance' (of water resources) and the fact that nearly every province shares its river and lake resources with a neighbour.

The provincial government of Ontario has, however, long favoured river basin authorities to develop local initiative, to promote cooperation with municipalities and to encourage conservation and recreation. The Conservation Authorities Act developed from a conference in Guelph in 1941 and has operated since 1946. The Conservation Authorities themselves (thirty-eight in number) are formed by purely local initiative, are largely municipal, and have the power to purchase land in order to promote rational management of the river resources. Unfortunately, the addition of a third layer of organisation to national and provincial interests has proved problematic; Conservation Authorities (Figure 4.10) have tended to become active in recreation but not in the 'big issues' of basin management. However, the role of Conservation Authorities is important in flood control and this has allowed them to take on a planning role. They have mapped flood hazard

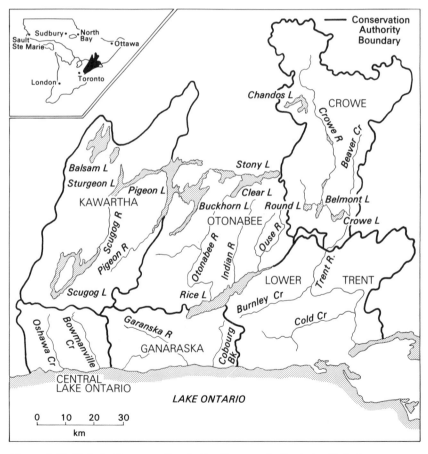

Figure 4.10 Ontario Conservation Authority areas in the Lake Ontario area (Ministry of Natural Resources, Ontario, 1986)

areas, they provide flood warnings and they have carried out structural works. Their success in two fields of water management, recreation and flooding, illustrates that these schemes are not duplicated by other interests and both attract the considerable local interest which the Authorities are able to mobilise.

The most recent extension of this policy of 'intervention with care' in the planning system states that one of the aims of development planning in the river basins of Ontario is 'to encourage a coordinated approach to the use of the land and the management of water'. The Conservation Authorities, where they exist, are given the hub role of providing information and reviewing proposals. For example, the Grand River Conservation Authority has a Master Watershed Planning Committee (Gardiner *et al.*, 1994) which sees as its role 'the promotion of watershed management of small tributary watersheds in an effort to facilitate orderly development and the most suitable types of development whilst not compromising environmental quality'. The possible reasons for Ontario's success in such an holistic vision are the small/moderate size of the basins involved, the fact that river basin boundaries do not cross prominent administrative boundaries and a traditionally high level of support from provincial finances for municipal efforts. In addition, the main river-based issues (flooding, urban drainage, erosion) are powerful educators in the derivation and implementation of integrated policies (see UK, New Zealand). Ontario has also developed a keen approach, through provincial environmental legislation, to reducing the discharge of toxic chemicals from industry both to rivers direct and to sewer systems. 'MISA', the Municipal–Industrial Strategy for Abatement, forces industry to curtail emissions at source according to a 'Best Available Technology' clause and to monitor for its own compliance; in some cases monitoring may be for well over 150 chemical determinands (Environment Ontario, 1988).

In 1985 Canada reviewed its Federal water policy once more; the Pearse Inquiry (Pearse *et al.*, 1985) announced two goals for policy:

(a) To protect and enhance the quality of the water resource.
(b) To promote the wise and efficient management and use of water.

These goals effectively mark the end of a public policy of cheap water, begun as early as 1889 with the first mains supply to Vancouver from dams on the Fraser, and the beginning of 'the polluter pays' principle. The review led on to five strategies and made twenty-five specific policy statements (Table 4.7). The strategies are:

(a) Water pricing.
(b) Science leadership.
(c) Integrated planning.
(d) Legislation to span jurisdictions, e.g. on water quality.

Table 4.7 Canadian Federal water policy

Specific policy statements

1 Management of toxic chemicals
2 Water quality management
3 Groundwater contamination
4 Fish habitat management
5 Water and sewer infrastructure
6 Safe drinking water
7 Water use conflicts
8 Interbasin transfers
9 Water use in irrigation
10 Wetlands preservation
11 Hydro-electric energy development
12 Navigation
13 Heritage river preservation
14 Management of Northern water resources
15 Native water rights
16 Transboundary water management
17 Interjurisdictional water conflicts
18 International water relations
19 Droughts
20 Flooding
21 Shoreline erosion
22 Climatic change
23 Water data and information needs
24 Research leadership
25 Technological needs

Source: Environmental Canada (1989)

(e) Public awareness and consultation.
(Canada has long-established environmental impact assessment procedures.)

However, Shrubsole (1990) is not optimistic that the strategies can be achieved because they rely on the nebulous concept of 'cooperation' (subtitle: bargaining) between Federal desires and provincial expertise and opinion. There is clearly a 'Canadian way' in the integrated management of river basins but until its processes can be unpicked with the somewhat wounding honesty of those south of the border, the resolution of the inevitable conflicts caused by 'bargaining' is a very slow process. Neither the integrated plans of the Canada Water Act nor an Environmental Assessment Review Process appear capable of dealing with the scale of Canadian water resource manipulation or the panoply of agency and regional scales of interest.

Shrubsole quotes the example of the expensively prepared plan for the St John River, whose recommendations were merely incorporated as 'wish lists' by the contributory agencies after completion. Gardiner *et al.* (1994) report that:

> By the close of the 1980s, Canada Water Act studies had become an artifact of the past. Effective river basin management was mainly being accomplished at the provincial and the local level.
>
> (p. 56)

These authors maintain that the Ontario Master Watershed Plans are 'one of the few examples of watershed plans in Canada that are being implemented to their full extent'.

The Canadian approach to water quality control is, however, more optimistic and may well extend the integration of interests around river basin units in a way which has been impossible in a nation obsessed with water transfers *between* basins.

4.4   AUSTRALIA: A LESSON LEARNED LATE

'Drought is not an occasional aberration in Australia but a repeated feature of the climate' (Smith and Finlayson, 1988). These authors go on to demonstrate how both the small amounts and high variability of rainfall (Figure 4.11) make almost all of Australia's vast territory poorly suited to rainfed agriculture; the fact that the average runoff coefficient is 10 per cent also means that irrigated agriculture based on water storage is a poor bet to cope with the aridity of the interior. Smith and Finlayson also say that, 'The aim of Australian farming should be to manage land and water to preserve the soil.' They contrast this with the settler ethos of 'if it moves shoot it; if it doesn't cut it down' and add the heroic phrase 'if it flows dam it' to describe the reality of the European approach to this huge dryland problem for most of the last 200 years. We therefore include Australia in this edition because, even more than the West of the USA, it demonstrates the problems of dryland development (in contrast to its neighbour New Zealand, which has different land and water problems – see below) and because, since a major review ('Water 2000') published in 1983, Australia has made rapid moves to planned joint use of land and water resources.

Australia's population congregates in its coastal strip (79 per cent) and in cities (85 per cent) and it is not without the 'normal' developed-world problems of water quality deterioration (until recently very poorly monitored); however, 82 per cent of its water is used by rural areas. The most important river basin to consider in this light is the Murray–Darling system (Figure 4.11). According to Maheshwari *et al.* (1995) the Murray–Darling river system is 'among the longest in the world' but because, once it has left the snow-fed headwaters it is an exotic stream without runoff contributions, 'its mean discharge has no global significance'. Yet the basin is the key foodstore to the occupation of Australia as a developed economy: a quarter

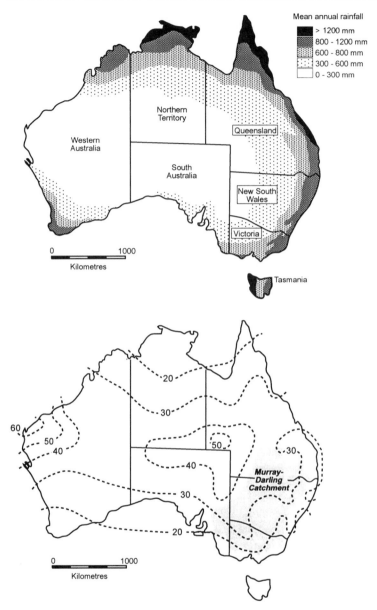

Figure 4.11 Problems of water resource management in Australia (after Smith and Finlayson, 1988):
(a) Mean annual rainfall
(b) Rainfall variability (per cent)

of Australia's cattle and dairy farms are in the basin, half the sheep and cropland and three-quarters of the irrigated land (occupying just 1 per cent of the basin area but using 90 per cent of its water – in fact 75 per cent of all Australia's irrigation water).

The basin nicely bears out a statement by Heathcote (1969) when describing the lack of respect given to droughts by the colonisers, with government finance always given to restoring agricultural communities after drought (until recent years):

> Excessive patriotism sometimes proved costly since officials occasionally promoted intensive land use in areas where it could be only marginally successful because of drought risks.
>
> (p. 191)

The Australian interior is therefore a slightly varied version of the dilemmas of developing the hydraulic civilisation of the US West.

In the late nineteenth century the first diversion weirs were constructed on the Murray–Darling system; however, the main phase of irrigation development occurred between 1920 and 1940. The basin occupies some 14 per cent of the continent; the critical fact is, however, that it covers parts of four states: Queensland, New South Wales, Victoria and South Australia. Because of the devolved nature of water management there has been inter-jurisdictional conflict over the resources of the Murray–Darling. A River Murray Waters Agreement was signed in 1914 but not until the Act was amended in 1987 did a more thorough and holistic Murray–Darling Basin Commission (MDBC) begin work 'to promote and co-ordinate effective planning and management for the equitable, efficient and sustainable use of the water, land and environmental resources of the Murray–Darling Basin' (Crabb, 1991). At the time of writing Queensland was not a signatory to the Agreement.

One of the major successes of the MDBC has been its Community Advisory Council which has used extension techniques to persuade farmers to fight environmental degradation, principally salinisation of soils and nutrient pollution of rivers. It has also had to deal with Aboriginal rights issues. It has been realised for many years that a corporate approach was required to joint land and water management in the basin (Langford-Smith and Rutherford, 1966). Much of the Australian soil cover has large stores of salt accumulated by natural processes over a long period of time; the salt may be released by poor irrigation practices (as in the Murray–Darling) or by the rise of water tables following forest clearance (common in Western Australia). Figure 4.12 separates these types of salinisation and indicates the importance of the forest cover to groundwater quality in dryland Australia. Hydrologists and land managers have combined in two prominent ways to bring about indirect 'cures' for the salinity problem. Conacher et al. (1983a and b) describe the use of throughflow (soil water) interceptors – shallow, contoured 'scrapes'

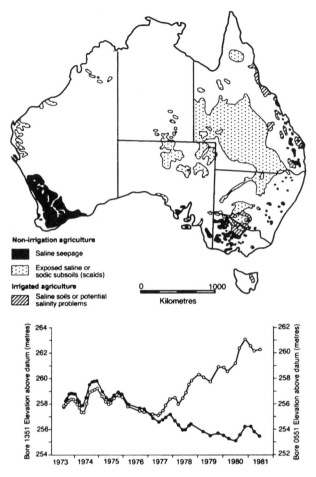

Non-irrigation agriculture

■ Saline seepage

▦ Exposed saline or sodic subsoils (scalds)

Irrigated agriculture

▨ Saline soils or potential salinity problems

Figure 4.12 Soil salinity and its relationship with land use in Australia (after Smith and Finlayson, 1988):
(a) Distribution of human-induced soil salinity in Australia
(b) Hydrographs from two boreholes in Western Australia; both were forested until 1976 when the record marked with open circles was clear cut; the resulting groundwater rise brought saline conditions to the surface

through the soil; George (1990) describes the use of Eucalyptus trees to deprive the groundwater recharge and thus lower the water table below the surface. Annual potential evapotranspiration for the Eucalypts is 2,500–2,900 mm, whereas for the pasture with which the forest is replaced it may be as little as 360 mm. In the Murray–Darling basin a much more direct method is used, whereby groundwater is pumped to evaporation basins to avoid salinity problems in the river (Ghassemi *et al.*, 1988).

Flooding has, paradoxically, been another problem of the settled regions of Australia (Smith and Finlayson, 1988). Floodplains are extensive and dam construction (often by farmers) raises problems of safety and of providing warnings. Smith (1981, 1990) has given examples of the careful analysis of flood damage studies and the ways in which warning systems, particularly after dam failure, can reduce loss of life. In some settlements, such as Lismore (New South Wales) there appears to be a high level of adjustment to the frequent flooding. There are well-established programmes of flood disaster relief and insurance in Australia but as yet no coordinated approach to planning floodplain occupancy, something detrimental to domestic occupiers and small businesses who are not covered by insurance (Smith and Handmer, 1989). In a few locations, however, drastic land-use planning in the form of floodplain acquisition and the relocation of homes is likely and Handmer describes an objective methodology for selecting those properties affected. Hooper and Duggin (in press) have made a pioneering use of ecological floodplain zoning to rural parts of the Murray–Darling basin but it may require the development of basin institutions to apply this concept to other Australian rivers.

The 'Water 2000' initiative of 1983 stressed the need for state water plans (Sewell et al., 1985). Sewell et al. (1985) offer lists of characteristics needed by such plans and stress the need to involve land-use planners and communities (Table 4.8). Progress in implementing planning is variable, with some states needing to be restrained by the Federal Government (e.g. the case of the environmentally disastrous proposal for a dam in a world heritage forest on the Franklin River in Tasmania) and others such as the Capital Territory itself using such aims as public participation and coordination of land-use plans and water development. Mitchell and Pigram (1989) use New South Wales as an example to illustrate how good intentions (the State Premier promised total catchment management during the 1984 election) can be frustrated by slow institutional progress. In the Hunter Valley of New South Wales an existing, mainly rural, Conservation Trust was authorised to make recommendations to the State regarding the 'nature, location, form and extent of any work or proposed work for the purpose of soil conservation, afforestation, reforestation, flood mitigation, water conservation, irrigation or river improvement within the Trust District'. The Trust has been successful in flood damage reduction but has been reactive and supportive rather than forceful and regulatory. The authors conclude that, despite the promise of 'Water 2000', the lack of progress on institutional and budgetary reforms has meant that catchment plans have no 'teeth':

> The policy of integrated resource management has received wide endorsement at the conceptual level in Australia. Progress towards effective implementation, however, has been tentative and hesitant.
>
> (p. 210)

Table 4.8 Essential features of Australian water planning

(a) Broad outlines

Essential features	Key functions	Mainline strategies
Comprehensiveness	Communications between planners Expand purposes Link water to other resources Employ large range of specialists	Foster continuous interaction between planning and other institutions
Resilience	Generate increasing range of options Monitor shifts in preferences Minimise adverse effects on environment Provide for paradigm change Design for future	Facilitate continuous dialogue between planners and the public

(b) Detailed considerations

Scope and content	Planning process
Comprehensive viewpoint (may involve institutional change)	Continuous evolution Interdisciplinary teams Widespread consultation
Maximising options (shift from structural approaches) (shift to demand management)	Interdisciplinary search for options
Water as an economic good (price reflects costs) (transfer rights)	Involvement of economists Use of cost-benefit analysis
River basin as an areal unit (land and water managed)	Identification by public and professionals
Public involvement	Involve in strategic and operational decisions Increased accountability
Environmental considerations	Ecologists in planning team Environment in cost-benefit analysis
Social impacts	Assessed by public involvement

We now turn to Australia's Pacific neighbour New Zealand to examine the long and apparently successful history of catchment management planning there, while bearing in mind that in New Zealand the European settlers found a different set of environmental hazards.

4.5   NEW ZEALAND: WISE MANAGEMENT DETERMINED BY HAZARD

New Zealand is recent both geologically and as a developed economy; there is no national shortage of water. Johnston (1985) has summarised the major phases of the country's development and the environmental policy response; development has included forest clearance and mining, both of which have combined with the hazards of a steep landscape, cloaked in weak sediments, and a wet climate to force considerable government attention to river management (see Moore, 1982 for a broad review). Over 12 per cent of New Zealand's land area suffers from fluvial erosion (7 per cent gullied; 2 per cent tunnelled; 3 per cent bank erosion); erosion surveys are one part of a very large national effort put into land survey.

It is not, therefore, surprising to find New Zealand's geomorphologists reporting some of the world's largest rates of fluvial erosion, for example in the Western Southern Alps, even without the intensive land exploitation found elsewhere in the islands. Soons (1986) reports source-area rates of between $96 \text{ m}^3.\text{km}^{-2}.\text{yr}$ and $8,380 \text{ m}^3.\text{km}^{-2}.\text{yr}$ during the post-glacial period.

During the period since forest clearance, landslides have cost large areas of New Zealand's 'hill country' almost all of their original 2.5 metres of soil; the remaining 10–30 centimetre soils are much less able to store storm rainfall and one result is that flooding is a further hazard to both islands. Plate 4.2 identifies the problem in the Taranaki Hill country and Figure 4.13 shows how an equilibrium soil depth under today's pasture farming is controlled by the landslip hazard.

The *flood hazard* in New Zealand is described in detail by Ericksen (1986). Table 4.9 indicates the options available for adjustment to floods, assuming that upstream mitigation such as *erosion control* and reafforestation are already in the brief of Catchment Authorities (see below). Historically, New Zealand has attempted to modify the floods but continuing disasters led to a programme of modifying the loss burden. However, now that town and country planning and soil and water legislation have begun to line up in New Zealand it is becoming more common to modify the losses. The use of popular devices such as cartoons is very prevalent in New Zealand public life; Figure 4.14 illustrates the desirable features of regulated floodplain development; in order to implement such regulation Catchment Authorities are using a computer model of flood losses developed by the Australian National University in Canberra (ANUFLOOD; Ericksen *et al.*, 1988). ANUFLOOD calculates damage at specific flood heights in relation to known land use and the flood record; it then maps damage under various scenarios of modified floodplain usage.

Table 4.9 A typology of human adjustments to floods

I Modify the flood
  1 Weather modification
    (a) Cloud seeding
  2 Land treatment
    (a) Afforestation
    (b) Conservation farming
    (c) On-site ponding
  3 River control
    (a) Dams
    (b) Levees
    (c) Channel improvements

II Modify damage susceptibility
  1 Land-use management
    (a) Encroachment lines
    (b) Zoning lines
    (c) Subdivision regulations
    (d) Building codes
    (e) Land acquisition
    (f) Floodplain development policies and plans
    (g) Urban renewal policies and plans
  2 Flood-proofing buildings
    (a) Permanent
    (b) Contingent
    (c) Emergency
  3 Community preparedness
    (A) Evacuations
    (b) Flood-fighting
    (c) Flood-forecasting and warning systems
    (d) Rescheduling activities

III Modify the flood-loss burden
  1 Insurance
  2 Tax deductions
  3 Loans
  4 Relief funds
  5 Feeding and sheltering victims
  6 Rehabilitation services: public and private agencies

Source: Ericksen (1986)

4.5.1   Catchment authorities: the New Zealand approach

New Zealand began to organise its river management along catchment lines as early as 1868 with River Boards; Acts were passed in 1884, 1893 and 1908 to strengthen these bodies. However, it was the floods and erosion problems of the 1930s which led to a truly formative piece of legislation, the Soil Conservation and Rivers Control Act 1941, subtitled as follows: 'An act to make provision for the conservation of soil resources and for the

Plate 4.2 Landslide resulting from cyclone Bola, Napier, North Island, New Zealand (photo J. C. Bathurst)

prevention of damage by erosion, and to make better provision with respect to the protection of property from damage by floods'. The Act made New Zealand one of the first countries in the world to recognise land–water links in legislation.

By 1967 the speed of development and intensification of land use in New Zealand forced the incorporation of wider aims in the Water and Soil Conservation Act, subtitled:

> An act to promote a national policy in respect of natural water and to make better provision for the conservation, allocation, use and quality of natural water, and for promoting soil conservation and preventing damage by flood and erosion and for promoting and controlling multiple uses of natural water and drainage of land, and for ensuring that adequate account is taken of the needs of primary and secondary industry, community water supplies, all forms of water based recreation, wildlife habitats and of the preservation and protection of the wild, scenic and other natural characteristics of rivers, streams and lakes.

The regional authorities for water and soil resource management are the twenty Catchment Authorities (see Figure 4.15) whose functions are set out in Table 4.10; they include the function of Regional Water Boards. A vivid pictorial account of the work of the Catchment Authorities is provided by Poole (1983). Until the 1941 Act was strengthened in 1988 coverage was not complete.

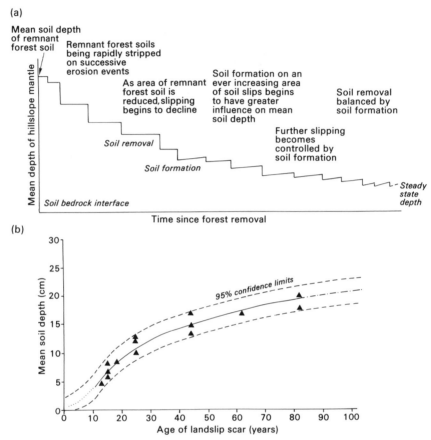

Figure 4.13 Soil slipping – a major hazard in New Zealand river basins:
(a) Development through time
(b) Relationship between time since last slip and soil depth
(NWASCA, 1987)

The 1967 Water and Soil Conservation Act delegated responsibility to the National Water and Soil Conservation Authority (NWASCA) for coordinating the work of Regional Water Boards and Catchment Authorities. By 1976 NWASCA anticipated a move to proactive catchment planning at a regional scale as an alternative to awaiting development proposals before seeking to coordinate activities. Land capability surveys were requested for catchments and farm conservation plans were coordinated into the catchment picture (with legal notices served on those persisting with unwise use of land). As early as 1986 some Catchment Boards and Regional Water Boards were combining to derive catchment plans. For example the Otago Boards (1986) produced a plan with the aim of: 'The wise management of water and soil

(a) Non-regulated floodplain development

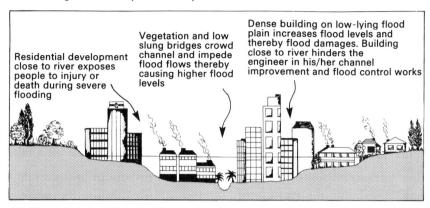

Residential development close to river exposes people to injury or death during severe flooding

Vegetation and low slung bridges crowd channel and impede flood flows thereby causing higher flood levels

Dense building on low-lying flood plain increases flood levels and thereby flood damages. Building close to river hinders the engineer in his/her channel improvement and flood control works

(b) Regulated floodplain development

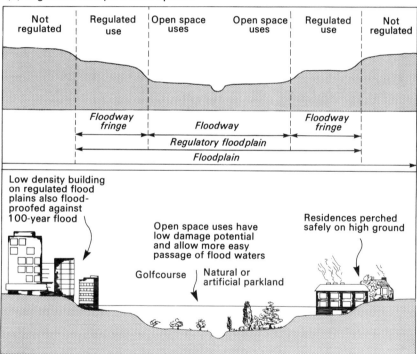

Not regulated | Regulated use | Open space uses | Open space uses | Regulated use | Not regulated

Floodway fringe   *Floodway*   *Floodway fringe*

Regulatory floodplain

Floodplain

Low density building on regulated flood plains also flood-proofed against 100-year flood

Open space uses have low damage potential and allow more easy passage of flood waters

Golfcourse | Natural or artificial parkland

Residences perched safely on high ground

Figure 4.14 New Zealand floodplains:
    (a) Non-regulated development of an urban community
    (b) Means of regulation and regulated development of an urban community (Ericksen, 1986)

Table 4.10 New Zealand Catchment Authority functions

I Soil conservation
  1 Preventing/mitigating soil erosion
  2 Promoting soil conservation
     (advice from Catchment Authority Soil Conservators; financial assistance
       for protection planting, destocking, water control, etc.)

II River control and drainage works
  1 Investigation, survey, design and construction
  2 Encouraging development in areas of lesser flood risk

III Water management
   (Catchment Authorities also function as Regional Water Boards)
   1 Water rights
   2 Water monitoring

IV Resource planning
   (including water and soil management plans)
   1 Water and soil resource surveys
   2 Collection of public submissions to management plans

V Associated environmental activities
  1 Issue of Safeguards Notices
  2 Burning control (grass fires)
  3 Windbreak schemes
  4 Willow control (phreatophyte growth)
  5 Gravel extraction (permits)
  6 Protection forestry

resources to provide maximum benefits from multiple, efficient, equitable and sustainable uses for present and future generations' (p. 4).

However, the preparation of Water and Soil Management Plans has not become a statutory process and, being expensive to carry out, few are complete. One of them, the Waitara River Catchment Management Plan, has attempted to preclude future construction of dams on the river; it exemplifies the fact that planning only occurs where existing conflict has drawn funds. New Zealand is apparently on the verge of the need to integrate town and country planning with soil and water management.

In 1981 an amendment to the Water and Soil Conservation Act was resurrected to permit the conservation of the natural characteristics of New Zealand's rivers and lakes. Popularly labelled the 'Wild and Scenic Rivers Legislation', this legislation can be used to maintain the flow regime of, and curtail development in the catchment of, both nationally and locally important reaches. Interpretations of the amendment give less cause for optimism by calling for multiple use rather than solely conservation of instream values. However, further support for river conservation has come from the interpretation by some authorities of the 1977 Town and Country

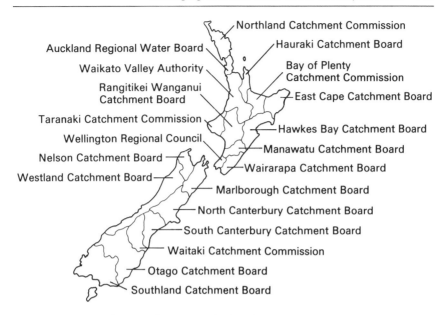

Figure 4.15 New Zealand Catchment Authorities

Planning Act which allows specific attention to river and lake margins. Guest (1987) quotes from the Waimarino County Council's plan which includes the Wanganui River Scenic Protection Zone:

> Although some of the land in this zone was in the past considered to be suitable for farming and exotic forestry, other values such as the quality of the landscape, historic and scenic features and water and soil considerations are now regarded to be of more importance.
>
> (Guest, 1987, p. 11)

The protection of all New Zealand's rivers is a developing issue, involving elements of simple pollution control as well as a growing recreation and conservation interest, fisheries and Maori rights. To the Maori, the essence of life is water and the river or lake of the home area is effectively the address as well as the identity of each individual (Stokes, 1992). Maori culture is threaded with quasi-religious resource management and anti-pollution concepts which are easily offended by European technocratic approaches to wealth and purity (Taylor and Patrick, 1987). A survey of over half of New Zealand's twenty Regional Water Boards (Quinn and Hickey, 1987) has assessed the degree to which the basic water quality criteria laid down in the Water and Soil Conservation Act 1967 are adhered to. Approximately 50 per cent of New Zealand's rivers are classified waters under the Act and the most common discharges are from dairy farms and domestic sewage plants. Quinn and Hickey report disparities in, for example, the application of the suspended solids criteria by some Regional Water Boards to emissions and by others to receiving waters.

Despite eutrophication problems there have not, as yet, been any controls applied to organic discharges or to nitrate or phosphate runoff. A recent upsurge in horticulture as the result of difficulties in the livestock sector has led to problems with pesticides and herbicides. There has been insufficient research to establish the extent of pollution and to develop guidelines or remedies; this is surprising in a nation so dependent on agriculture (see Table 4.11) but reflects the previous domination of erosion control and flood protection. Of the available research on, for example, fertiliser pollution, that by McColl and Gibson (1979) concludes that for hill-country improved grazing lands very little nutrient runoff reaches streams. In certain 'source areas', particularly of surface runoff, there is, however, a clear need for management of agricultural practices (cf. UK – p. 344). Quinn and

Table 4.11 Principal toxicants discharged to New Zealand rivers

Source	Principal toxicants	Number to river
Point sources		
Meatworks	Ammonia, sulphide, organics	18
Pulp and paper	Various organics, heavy metals	6
Tanneries	Chromium, organics	1
Woolscours	Detergents, sulphide	11
Timber preservers	Arsenic, chromium, copper, PCBs, boron	?
Tip leachates	Zinc, other heavy metals, organics	numerous
Mining	Evanide, various heavy metals	?
Petrochemical	Solvents, biocides, heavy metals	1
Pharmaceutical by-products	Chromium, organics	1
Bore drilling	Chromium, lignasulphonate	?
Geothermal (power)	Mercury, arsenic, lead, boron, biocides sulphide, lithium	4
Municipal sewage	Organics, ammonia, (others depending on industrial output)	96
Dairy shed	Organics, ammonia	7,850
Piggeries	Copper, zinc, organics, ammonia	220
Diffuse sources		
Geothermal (natural)	Mercury, arsenic, lead, boron, sulphide, lithium	?
Agricultural and horticultural runoff	Pesticides, herbicides, fungicides	widespread
Urban stormwater	Lead, zinc, copper, cadmium, pesticides, herbicides, petrochemicals, organics	widespread
Weathering and erosion	Heavy metals	?

Hickey identify a channelisation problem in conserving stream habitat, emphasising that Catchment Boards have good control in headwaters but less over riparian activity downstream. The eighty-two hydropower dams throughout New Zealand have also posed problems for fish migration. Detention dams on farmland are now replacing some of the traditional 'hard engineering' solutions to rural flooding.

In an effort to improve water quality control in New Zealand's rivers recent research has attempted to define water quality indices and criteria by expert consultation and by contacting river users. Four groups of quality index, depending on river use (general, bathing, supply and fish), were put to consultation. Replies were compiled on a range of determinands, physical, chemical and biological. From there, numerical values were assessed with a view to weighting scores in simple integrative indices for the four river functions (Table 4.12).

By 1988 the New Zealand Government had recognised the need for 'Resource Management Law Reform' because of the proliferation of new laws and institutions superimposed on some earlier approaches such as Soil Conservation and Rivers Control. The situation, they maintained, had become cumbersome, hard to understand by the public, had led to costly delays and did not respect Maori rights. Interestingly, because of the prominence of land and water issues, the public consultation undertaken in New Zealand focused upon:

(a) The scale of bodies best able to manage resources.
(b) The role of Government in control of ownership.
(c) The voice of local communities.
(d) The role of indigenous culture and values.

Ericksen (1990) also refers to a problem of bureaucratic confusion at the regional level. The proposed reforms would unify the regional councils

Table 4.12 Relative importance and final determinand weightings for the four water quality indices

Determinand	General (G)		Bathing (B)		Supply (W)		Fish (F)	
	Rel. imp.	Wt	Rel. imp.	Wt	Rel. imp.	Wt	Rel. imp.	Wt
Dissolved oxygen	1.33	0.30	2.15	0.15	2.13	0.18	1.00	0.34
pH	3.13	0.13	3.10	0.10	2.79	0.15	2.81	0.12
Suspended material	2.57	0.15	1.73	0.19	2.82	0.15	2.41	0.14
Temperature	3.15	0.12	3.78	0.09	3.59	0.12	1.35	0.26
BODs	2.20	0.18	2.23	0.15			2.48	0.14
Faecal coliforms	3.18	0.12	1.00	0.32	1.78	0.24		
Ammonia					2.59	0.16		

Source: Smith (1987)

(i.e. local government) and Catchment Boards to produce fourteen new authorities. NWASCA has already been disbanded, representing a sad loss of central information, education and synthesis of policy. Nevertheless, the Ministry for the Environment (1989) is convinced of the benefits of devolution and of grouping agencies by legal function rather than by resource.

Scott (1993) reviews the latest piece of New Zealand legislation with relevance to catchment management, the 1991 Resource Management Act. It attempts to develop the concept of sustainable, integrated resource management – of air, water, soil and ecosystems. The concept of the life-supporting capacity of rivers and lakes is introduced and Scott debates whether this means that there will be an 'ecological bottom line' to future land-use developments in an era when economic restructuring is the main political driving force.

In conclusion we may suggest that the scale of the erosion and flood problems in the 1930s, clearly exacerbated by poor land management, led to heavy intervention at a national scale, for instance, riparian rights were handed over to national ownership of water resources. However, while land inventories have remained a national operation, the Catchment Board policy has ensured good local representation and decision-making. Ericksen (1990) has the last word: 'New Zealand has a reputation for throwing hastily-prepared legislation at its problems then making corrective amendments as experience highlights the shortcomings' (p. 83).

4.6   CONCLUSIONS: NATIONAL PRIORITIES IN THE DEVELOPED WORLD

To characterise the nations whose approaches to river basin management have been outlined in this chapter one would say that:

(a) The USA was the victim of the rapidity of its colonisation and of the ability of the financial resources thereby created to 'hold' hazardous environments, particularly drylands, in settlement and in productive use. Superimposed on this resource picture is the cultural and legal battle between an enormous national technological skill to operate rational planning and the private enterprise culture of a Gold Rush. Politically, there are continuing tensions between the national view and that of the individual states.

(b) In Canada a more sympathetic view to resource management might well prevail for a number of physical and cultural reasons but the plentiful water is badly distributed and the issue of clean energy generation also contributes to the 'gigantism' of the many water-transfer schemes. The resource problems of the neighbour to the south are intricately bound up with those of Canada and it is possible that the Canadian policy, replete with centralist mission statements on the environment, will not settle down until the future of US use of Canadian water has taken shape, possibly following climate change. Canada also has its own political battle between nation and provinces, notably Quebec.

(c) Australia has begun to emerge from its settler ethos of heroic exploitation of the interior and support for its farmers after frequent droughts and floods. The Federal Government has had to request the states to develop integrated management but progress is slow except where it is easy. The Murray–Darling basin is, however, perhaps the most critical exotic river system on a world scale for which integrated and interactive planning is being attempted.

(d) New Zealand, uniquely, appears to have experienced a relatively stable period of careful water management policies, made necessary by earlier misdeeds and by a very unstable natural background. The ability of the New Zealand system to combine central strategy with local planning and implementation seems, however, to be crumbling in the face of devolved policies and economic uncertainty.

To the reader in the UK, many of the aspects and approaches described above may be familiar during various stages of the evolution of UK policies (see Chapters 7 and 8). The UK may be characterised by: poor distribution of water in time and space, a drastic neglect during early industrialisation of water quality, and over-simplified 'free resource' approaches to water consumption.

The current direction of UK policy shares with the USA, Canada and New Zealand a number of other cultural and political controls, for example the development of a national position on all aspects of environmental management, the enforced cutback on public expenditure and the degree to which the population is consulted about policy development. These work against each other and produce tensions which may or may not be creative in all four developed nations we have considered. For example, the dismemberment of centralised strategic planning may have the advantage of encouraging more local action, but not if that action is financially driven or constrained; in that case the increasing feeling in the developed world that environmental matters require a strategic overview swings the argument back towards central action – or possibly towards international action.

The really important lessons to be learned from this review are:

(a) That national and regional uniqueness (climatic, hydrological and cultural/political) 'sets up' the pattern of land and water development to date.

(b) However, developed nations are keen to 'export' models of development for river basins which pay little attention to the growing evidence of lack of sustainability in, for example, dryland irrigation supported by federal handouts during a settler economy.

A recent retirement from heading the US Bureau of Reclamation said 'I think it is a serious mistake for any region in the world to use what we did on rivers as examples to be duplicated.' Sadly, the message is too late for much of the developing world.

Gangstad (1990) has another relevant message for developed world basin management:

> Engineers and planners are at a nexus between the divergent faith in democracy and low institutional esteem. Traditionally they provide critical services for a civilization's ability to survive and adapt. But this tradition quickly erodes when a society's engineering capability diverges from its changing social values.
>
> (p. 77)

The problem is when a ready market exists among politicians abroad for those outmoded social values, that is, in newly settled lands. Politicians should learn that water-based development must proceed in two phases – provision of essential infrastructure followed by processes by which the economic and environmental costs are amortised by all parties.

In view of the importance attached to water-based development in drylands in both this chapter and the next, perhaps the best piece of caution comes from the conservationist sage Aldo Leopold who wrote of the US experience:

> That what we call 'development' is not a uni-directional process, especially in a semi-arid country. To develop this land we have used engines that we could not control, and have started actions and reactions far different from those intended. Some of these are proving beneficial; most of them harmful. This land is too complex for the simple processes of 'the mass mind' armed with modern tools. To live in real harmony with such a country seems to require a degree of public regulation we will not tolerate, or a degree of private enlightenment we do not possess.
>
> (Bates *et al.*, 1993, p. 150)

Chapter 5

River basins and development

5.1 GENERAL CHARACTERISTICS AND NEW PHILOSOPHIES

There are many characteristics of developing world water projects which are reminiscent of the distribution philosophies of the early hydraulic civilisations; a principal difference may be summarised by the word 'identity'. Modern schemes have tended to lack the identity of communal national effort which seemingly characterised earlier eras, partly because of marked regional disparities in technological expertise but mainly because of lack of popular national involvement. The strong social structures which upheld prehistoric irrigation economies are imitated by bureaucratic resettlement schemes and innovative, but rarely successful, land tenure arrangements.

At the end of the previous chapter we alluded to the habit of the developed world of transferring technologies which its own citizens had come to realise were outmoded physically or culturally, part of a global trade pattern which complicates all development issues. In addition we must face the fact that the power élites of the developing world are unlikely to be generous to the ecosystem concept of sustainable riverbasin development at a time when intensive exploitation of basin resources appears to be the most rapid route to a national trade surplus, wealth and political success. The spontaneous regulatory functions of the basin are understated without a proper investigation, which the urgency of development precludes for the élites. The indigenous peoples of the developing world have practised extensive exploitation for centuries, largely leaving the spontaneous regulators intact.

Rural communities in an urbanising nation have little incentive to bring forward their own water development schemes at a scale which could improve national economic performance; some form of imposition, if only of development alternatives by extension workers, is clearly inescapable. Equally clearly, *indigenous people* have both human rights and huge reserves of adaptive strategies involving local environmental management (see Section 5.7.2).

Table 5.1 Guidelines for river basin management

1 Preservation or improvement of the spontaneous functions fulfilled by the river, by:
  (a) Restoring erosion/sedimentation processes, through countering increased silt loads caused by upstream erosion (improvement of watershed management!).
  (b) Preserving genetic diversity, through conserving natural areas and threatened species.
  (c) Preserving the self-purifying capacity of the river, through combating pollution (water-treatment plants, at-source anti-pollution measures).
2 Conservation of the natural values of the river basin, by:
  (a) Preventing deterioration/destruction of natural resources, by means of legislation (including compulsory environmental impact assessments) directed towards industrial development, impoldering schemes and drainage activities.
  (b) Establishing reserves in the most vulnerable ecosystems, with surrounding buffer areas.
  (c) Establishing environmental education programmes.
  (d) Initiating programmes to promote sound, durable exploitation of eco-systems (particularly fisheries, herding and forestry).
3 Conservation of the river basin's extensive exploitation functions, by:
  (a) Guaranteeing the protection of productive zones, such as floodplains, estuaries and lakes, by allocating appropriate quantities of water (in relation to para. 4) and by means of the programmes mentioned in para. 2).
  (b) Implementing reafforestation schemes for supply of firewood, in relation to sound watershed management (cf. para. 1).
4 Development of sustainable intensive exploitation functions, by:
  (a) Drawing up a water allocation plan for the entire river basin, to achieve a better match between water demand and supply; this should give due consideration to the water requirements of the spontaneous functions (para. 1), natural values (para. 2) and extensive exploitation functions (para. 3).
  (b) Developing small-scale projects, e.g. irrigation, fishponds, forestry.
  (c) Improving product processing, sales and marketing, e.g. by making better use of the river as a transport route.
  (d) Ensuring that detailed plans for the above objectives are thoroughly checked against the other criteria (paras 1–3 and 5) within the framework of an environmental impact assessment procedure.
5 Improvement of the overall health situation in the river basin, by:
  (a) Combating water-borne diseases.
  (b) Improving the food situation, both quantitatively and qualitatively.
  (c) Establishing a drinking-water programme for rural areas, with the objective of making clean, healthy water available for the whole population.
  (d) Ensuring that detailed plans for the above objectives are thoroughly checked against the other criteria (paras 1–4) within the framework of an environmental impact assessment procedure.

Table 5.1 (*continued*)

6 Guiding principles for regional planning:
   (a) Work with, not against the environment.
   (b) Start work from the existing situation, i.e. existing infrastructure, technical know-how, perceptions of subsistence security, cultural needs, etc.
   (c) Protect the authentic evolution of local culture, institutions and know-how.
   (d) When undertaking action or introducing social change, prefer those actions involving decision-making at the lowest possible level.
   (e) Assess the carrying capacity of extensive agricultural and water-use systems, as well as their present value.
   (f) Assess the required inputs for intensive systems of land and water use and their value if a growing number of people are to be fed.
   (g) Intensify or introduce intensive land- and water-use systems at locations with the best soil and superior climatological and market conditions.
   (h) Preserve, develop and utilise nature's spontaneous functions.
   (i) When planning reserves for species or ecosystems, endeavour to make them as large, as varied and as interconnected as possible.
   (j) Preserve rare species and ecosystems in their authentic ecological setting, giving due consideration to the long-term effects of isolation.
   (k) Avoid land- and water-use systems exhibiting irreversible dependence on a single crop or market (especially narrow or foreign markets), also on the input side.

Source: Marchand and Toornstra (1986)

5.1.1 Pre-empting exploitation

Is there a rationale with which the technology of water development could have a more acceptable social focus when viewed both from the centre and periphery of a developing nation? Marchand and Toornstra (1986) first demand the treatment of the entire river basin as an ecosystem and develop this theme through to a set of impact matrices for specific intervention by development agencies, one of several such matrices now available (see Chapter 8, plus Horberry, 1983; Hellawell, 1986). Table 5.1 summarises Marchand and Toornstra's guidelines.

The two major current problem areas in terms of rapid water-based development are the world's drylands and the tropical forest lands; they are very different hydrologically but the sensitivity of their river basin eco-systems (in terms of the lack of resilience and sensitivity of their natural regulators) means that great damage is already occurring. We focus most of our attention on drylands because at present there is a greater wealth of historical and scientific knowledge about their development; for obvious practical reasons tropical forest basins have been harder to study and monitor – the petrochemical wealth of Middle East drylands has also sponsored long-running studies of semi-arid and arid lands.

5.2   PROBLEMS OF FOOD, POWER AND TRADE IN DRYLANDS

The forty-two LDCs (least developed countries) show a high correlation with the tropical arid and semi-arid lands; 340 million people live in these countries (Figure 5.1). Many of the major river basins of the world impinge on the arid and semi-arid lands and the exploitation of their basins has a disproportionately vital effect on indigenous peoples. Most of the important basins are also international – they cross the boundaries of sovereign nations (Figure 5.1). While the population of most of the LDCs is rising fast, much of the land within their boundaries is limited by drought (Table 5.2a, b) and it is very difficult to increase food production for the home population as well as creating exportable surpluses of cash crops; Africa has the most formidable problem of all continents in this respect (Figure 5.2a, b). Drought years place an unmanageable burden upon food production systems by both delaying the crop cycle and curtailing the period of growth (Figure 5.2b). While acknowledging along with Brandt and Bruntland (Independent Commission on International Development Issues, 1980; World Commission on Environment and Development, 1987) that the 'world order' of trade and militarism are responsible for the dilemma of development, the management of individual river basin projects and the cross-boundary allocation of water resources and hazard mitigation is a necessary focus of the hydrologist.

A much wider group of world population is, in fact, threatened by low levels of water resource development. The UN's International Drinking Water Supply and Sanitation Decade (1981–90; see Figure 5.3) set out to

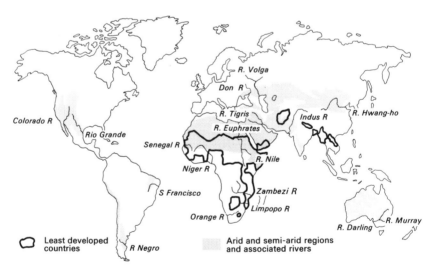

Figure 5.1 Major river basins, least developed countries and the world's arid and semi-arid lands

Table 5.2(a) Areas of land limited by drought

Region	% of land limited by drought
North America	20
Central America	32
South America	17
Europe (mainly Spain)	8
Africa	44
South Asia	43
North and Central Asia	17
South East Asia	2
Australia	28

Source: Finkel (1977)

Table 5.2(b) Distribution of drylands by country

Group	Description	No.	% of nation arid/ semi-arid	Countries
1	Arid	11	100	Bahrain, Djibouti, Egypt, Kuwait, Mauritania, Oman, Qatar, United Arab Emirates, Saudi Arabia, Somalia, South Yemen
2	Predominantly arid	23	75–99	Afghanistan, Algeria, Australia, Botswana, Burkina Faso, Cape Verde, Chad, Iran, Iraq, Israel, Jordan, Kenya, Libya, Mali, Morocco, Namibia, Niger, North Yemen, Pakistan, Senegal, Sudan, Syria, Tunisia
3	Substantially arid	5	50–74	Argentina, Ethiopia, Mongolia, South Africa, Turkey
4	Semi-arid	9	25–49	Angola, Bolivia, Chile, China, India, Mexico, Tanzania, Togo, USA
5	Peripherally arid	18	<25	Benin, Brazil, Canada, Central African Republic, Ecuador, Ghana, Lebanon, Lesotho, Madagascar, Mozambique, Nigeria, Paraguay, Peru, Sri Lanka, USSR, Venezuela, Zambia, Zimbabwe

Source: Paylore and Greenwell (1979)

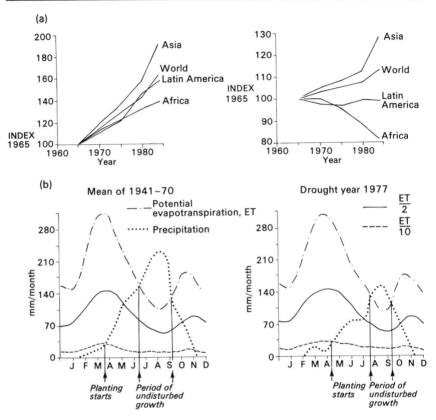

Figure 5.2 Problems of the developing semi-arid world:
  (a) The growth of population (*left*) and food supplies (*right*) (Higgins *et al.*, 1988)
  (b) The sensitivity of the growing season to the water balance, both depths and timings, 'normal' and drought (Falkenmark, 1986)

reduce the appalling statistics of inadequate or polluted water supplies for human use (Agarwal *et al.*, 1980): 30,000 children die each year because of inadequate water supply or sanitation and half of the world's hospital beds are occupied by patients with water-related diseases. At the close of the Decade, 700 million extra people had been supplied with clean water but nearly 2 billion people remained in danger – a vanishing perspective.

Despite the urgency of providing basic human facilities the internal and external demands of burgeoning developing-world cities pose special problems for maintaining sustainable ideals (i.e. of protecting basin ecosystems). The natural water environment of developing-world cities exhibits many signs of stress, particularly to the important water and sediment transfer systems (problems of pollution may, in the longer term, be surmountable). In the humid tropics, for example, urbanisation around

Plate 5.1(a) Encroachment by dryland development on to hazardous alluvial fan environments, Eilat, Israel (photo M. D. Newson)

Plate 5.1(b) Nature asserts itself: flood-damaged highway north of Eilat, Israel (photo M. D. Newson)

Harare, Zimbabwe, has destroyed the water storage capacity of the dambos (African valley wetlands), channelised the natural watercourses and created downstream channel erosion problems (Whitlow and Gregory, 1989). Within city limits, as was the case with rapidly urbanising cities in Britain in the Victorian period, open channels are used as sewers and refuse dumps. In south-western Nigeria, Ebisemiju (1989) points to a decrease in channel capacity brought about by this depositional activity, with an increased flood hazard resulting. Elsewhere in Nigeria, Odemerho and Sada (1984) describe the erosional gulleys which result from 'the increase in impervious land

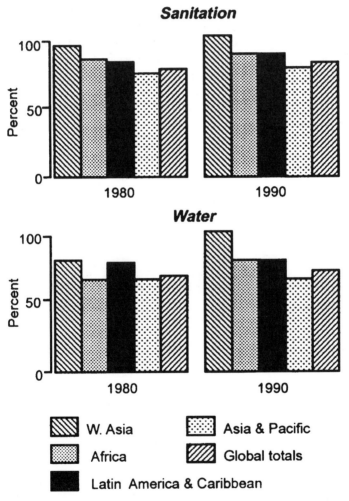

Figure 5.3 Progress towards vanishing targets? Sanitation and water supply provision levels at the start and close of the UN's International Drinking Water Supply and Sanitation Decade (after Rogers, 1992)

uses, the discontinuous urban sewerage and generally unplanned layout of buildings'.

Urban wastewater management is particularly problematic in the humid tropics because of the very intense rainfall (Gladwell, 1993); five-minute intensities of 160–250 mm.h^{-1} are common, overwhelming runoff collection systems and spreading the copious surface wastes (especially of squatter cities) into the nearest watercourse.

In rapidly urbanising areas of the world's drylands a major hazard is that of extreme floods in what, otherwise, is a semi-arid climate. Schick (1995) records the case of Eilat, on the Red Sea coast of Israel, a city undergoing rapid tourism development (see Plate 5.1a) but situated on an area of coalesced alluvial fans. Forty per cent of the rainfall falls in storms reaching an intensity of 20 mm. h^{-1} and although the streets of the town are designed to carry flood waters there is considerable disruption, damage (Plate 5.1b) and deposition as the result of choosing such a hazardous site. Eilat has many a badly sited urban example to follow, e.g. Las Vegas in the USA. It is clear that urban areas in the developing world are worthy of geomorphological study ('fluvial audit' – Chapter 2) to avoid such damage and Douglas (1985) makes a plea for the new topic of 'urban sedimentology'.

Urbanisation around dryland rivers inevitably leads to problems of river and groundwater pollution; in fact there are strong potential linkages between environmental degradation in urban and rural areas of the developing world. A study of a region in south-east India where the leather tanning industry is a notorious polluter (Bowonder and Ramana, 1987) has demonstrated how a combination of rural land pressures and urban/

Table 5.3 Major urban dryland centres with more than 2 million people

Continent	Estimate date	City proper	Urban
Africa			
Cairo (Egypt)	1974	5,715,000	—
Alexandria (Egypt)	1974	2,259,000	—
Asia			
Peking (China)	1970	7,570,000	—
Tientsin (China)	1970	4,280,000	—
Teheran (Iran)	1973	4,002,000	—
Delhi (India)	1971	3,287,900	3,647,000
Karachi (Pakistan)	1972	3,498,634	—
Lahore (Pakistan)	1972	2,165,372	—
America			
Los Angeles (USA)	1975	2,727,399	7,032,075
Santiago (Chile)	1970	3,273,600	3,350,680
Lima (Peru)	1972	2,833,609	3,303,523

Source: Cooke *et al.* (1982)

industrial developments leads to health problems, community stress and environmental decline.

The special problems of the world's drylands (comprising one-third of the earth) have recently been highlighted by Beaumont (1989). His comprehensive survey of all aspects of development demonstrates that this is not a rural problem; rural communities in drylands have shown superb adaptations including 'primitive', yet sophisticated, technologies based upon folk knowledge of environmental variables (e.g. Australian Aborigines), but today's use of drylands is more intensive and urbanised, as Table 5.3 shows. There is but one water lifeline for such an intensive use of drylands – irrigation, but only 15 per cent of the world's land area is irrigated, producing 40 per cent of the food crop. Despite this impressive productivity

Table 5.4(a) Continental distribution of irrigated area, 1984

	Irrigated area (10⁶ ha)	% of world total
Asia	136.865	62.30
North America	20.461	9.31
Soviet Union	19.485	8.87
Europe	15.710	7.15
Africa	10.390	4.73
South America	7.979	3.63
Central America	6.914	3.15
Oceania	1.869	0.85
Developing countries	157.198	71.56
Industrial countries	62.475	28.44
World	219.673	100.00

Source: FAO (1986)

Table 5.4(b) Countries with major involvement in irrigation: irrigated areas in 1984 (10⁶ ha)

China	45.42	Spain	3.14	Sudan	1.70
India	39.70	Italy	2.97	Argentina	1.66
United States	19.83	Afghanistan	2.66	Australia	1.63
Soviet Union	19.48	Romania	2.61	Philippines	1.43
Pakistan	15.32	Egypt	2.47	Chile	1.26
Iran	5.73	Brazil	2.20	Bulgaria	1.21
Indonesia	5.42	Turkey	2.14	Nigeria	1.20
Mexico	5.10	Bangladesh	1.92	Peru	1.20
Thailand	3.55	Iraq	1.75	South Korea	1.20
Japan	3.25	Vietnam	1.75	France	1.16

Source: FAO (1986)

ratio the irrigated area is a relatively small proportion, especially in areas which are in serious need of food (e.g. Africa at 4.7 per cent; see Table 5.4).

There are thus two components of the dryland river development problem – rural irrigation which in theory 'feeds' urbanisation (surplus labour, food exports) and lack of urban infrastructure (as revealed by the Water and Sanitation Decade); both damage the fragile river ecosystems of dryland basin; we concentrate here on the land/water linkages emerging in rural drylands.

5.2.1   Desertification

Paradoxically, despite this need, the drylands of the world are mismanaged to the extent that in the Sahel zone of Africa as much land is lost to food production by the processes known as desertification as is gained by new irrigation schemes. The term 'desertification' and the extent of the problem are both hotly debated by hydrologists, agriculturalists and development specialists. Promoted vigorously by the 1977 UN Commission on Desertification meeting in Nairobi, the phenomenon is said to affect 21 m. ha. yr^{-1} at a cost of \$26 b. yr^{-1}. Desertification is defined as: the diminution or destruction of the biological potential of land that can lead ultimately to desert-like conditions (UNCOD, 1977).

Despite such specificity Thomas and Middleton (1995) examine the hypothesis that desertification is a myth and claim (p. 9) that 'desertification is a shorthand term rather than a specific process with a specific cure'. They examine four dimensions of the myth – the size and progress of the problem, the fragility of dryland ecosystems, the physical effect on human communities and the role of the United Nations; in each case they find that we are guilty of applying a spatial definition to a general problem of human occupation – that of degradation.

Spectacular schemes to abate desertification, such as tree-planting in Mali and Algeria to 'halt' the Sahara have forced a re-think. It is possible that UNCOD's estimates of the extent of desertification risk were based on a poor database, mainly comprising early remote sensing images.

Desertification as a theme to discourage profligate rural development is still intensively pursued. For example, Grainger (1990) considers the causes as overcultivation, overgrazing, poor irrigation management, and deforestation.

Among the outcomes are wasted water, disease, erosion, salinisation, pollution by pesticides. As such UNCOD's recommendations (Table 5.5) remain valid.

Most authorities, however, now see desertification as an issue utterly complicated by climatic change. Mensching (1986) emphasises the need, therefore, for the population of the Sahel to have a flexible response and for exploitation potentials to be reviewed downwards to reflect the inherent variability of environmental controls.

Table 5.5 UNCOD plan of action: checklist of priority measures to combat desertification

A Land use and rehabilitation
  1 Introduce methods of planning land use in ecologically sound ways.
  2 Improve livestock raising by means of new breeds of livestock and better range management.
  3 Improve rainfed cropping by introducing more sustainable techniques.
  4 Rehabilitate irrigated cropping schemes that have failed owing to waterlogging, salinisation and alkalinisation.
  5 Manage water resources in environmentally sound ways.
  6 Protect existing trees, woodlands, and other vegetative cover and restore tree cover and vegetation to denuded lands.
  7 Establish woodlots as sustainable sources of fuelwood and encourage the development of alternative energy sources.
  8 Conserve flora and fauna.
  9 Ensure the fullest possible public participation in measures to combat desertification.

B Socio-economic and institutional measures
  1 Investigate the social, economic and political factors connected with desertification.
  2 Introduce measures to control population growth, as appropriate.
  3 Improve health services.
  4 Improve scientific capabilities.
  5 Expand local awareness of desertification and skills with which to combat it by training and education, both by means of mass media and courses at various educational institutions.
  6 Assess the impact of settlements and industries on desertification, and keep desertification in mind when planning or expanding new settlements and industries.

Source: Summarised from Grainger (1990)

The main requirement is to maintain the natural regenerative capacity of the vegetation. This calls for far-sighted settlement and development plans which, in the last analysis, can be prepared only by the desert countries themselves.

(Mensching, 1986, p. 17)

Binns (1990) concludes that:

The evidence seems to suggest that the idea of the Sahara desert sweeping southwards on a broad front, as Stebbing postulated in 1935 and others have argued more recently, is no longer tenable. However, *land degradation* of varying degrees does exist and the causes and possible remedies of this problem must be understood. The search for better systems for indigenous participation and co-operative management of resources is also important.

(Binns, 1990, p. 112)

The emphasis on indigenous participation is supported by Thomas and Middleton (1995), who stress that desertification has been conceptualised mainly by Western minds with little or no cultural experience of adaptation to dryland conditions; the Western perception also dominates the international response to a social corollary of degradation – poverty and starvation. They are bold in stating that:

> The life cycles of plant and animal populations in drylands are characterised by 'boom and bust' cycles in tune with a variable environment, and human populations may have adjusted in a similar manner under environmental stress in the past. When this occurs today, it is viewed from the West as unacceptable.
>
> (p. 154)

Bearing in mind the dangers of myth and cultural prejudice, it is important that we now look at the problem of dryland river basins, the management of which is complicated by the international nature of many of the world's most prominent dryland rivers; we therefore begin with the experience of Iran in managing basins entirely within its national boundaries.

5.3   RIVER BASIN MANAGEMENT IN IRAN: THE ZAYANDEH RUD EXAMPLE

It would appear that river-basin management to integrate land and water resources should be an overriding preoccupation for drylands if sustainable development is a realistic ideal – such is the pre-eminence of the relationship. The Zayandeh Rud (River) basin in Iran offers a suitable case study without the complication of international boundaries. It covers 41,503 km^2, approximately half lying within the mountain (snowmelt) source of most of the flow and half in a semi-arid environment (Figure 5.4). There is rapid urbanisation in and around Isfahan (population now 1.3 million), an ancient Persian capital city but now home to many refugees from the Iran/Iraq war.

Most of the eastern half of the basin relies on irrigated farming in what Beaumont (1989) calls the Isfahan Oasis. Only 20 per cent of the natural flow of the river remains in-channel at the Gavkhuni marsh – the 'end' of the river, which does not reach the sea. Thus the physiography, climate and layout of the Zayandeh basin is typical of the dryland basin, but there are three extra reasons for a special study: the very long (c. 2,000-year) history of irrigation, its links to the Islamic culture of resource management and the impacts of the post-revolutionary Islamic state since 1979.

The flow of the Zayandeh to Isfahan and its 'oasis' has long been augmented by engineering schemes to store spring snowmelt from the Zagros Mountains (there being virtually no summer runoff and little in the

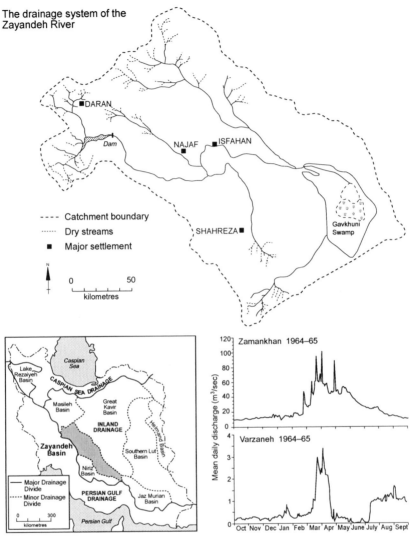

Figure 5.4 The Zayandeh Rud basin, Iran: the basin, its setting and two annual
        hydrographs, one from upstream and one from downstream of the
        Isfahan centre of population (note different flow scales)

lower basin) and also to transfer runoff from neighbouring basins with
headwaters in the Zagros. The modern 'hinge' of the basin is the Zayandeh
Rud Dam (Figure 5.4). Adding to the inflow to the dam is water transferred
from a neighbouring catchment to the south through artificial tunnels with a
total flow capacity of 95 cubic metres per second. The transfer system also
requires minor dams and the least sustainable aspect of the entire scheme is
the lack of compensation water allowed down the Karun to sustain that river

on its path to the Persian Gulf (Newson and Ghazi, 1995). There is also 'clear-water erosion' of the Zayandeh's upper tributaries where the transfer flows enter; twenty-two check weirs are required to control erosion on the 35 kilometre reach to the Zayandeh Rud Reservoir (which took four years to fill after completion in 1970).

Other aspects of the management of the basin demonstrate, however, the importance of a blend of good engineering, environmental concerns and a strong cultural inheritance in terms of equity and flexibility.

The origin of Isfahan as a settlement is an irrigation system based on canals ('Madi') from the river and subterranean 'quanats' from the alluvial fan aquifers of the mountains to the north. The use of these supplies has, according to Islamic law and tradition, been based upon multiple use (e.g. a sequence through villages of drinking water, washing water, water to bathe the dead, animal water and irrigation water) and common, community property rights (see Chapter 1).

Irrigation water has traditionally been used four times in rotation in the villages; nitrogenous fertiliser is collected from traditional pigeon roosts. While the quanats have now fallen into disrepair in places the legacy of such cultures can be seen, for example, in the way in which urban neighbour-hoods in Isfahan regularly clean out and maintain the Madi system near their homes. In the modern irrigation economy east of the city there has been regular attention to the problems of salinisation and other soil degradation; new, concerete-lined canals reduce leakage waste and other drainage canals remove polluted seepage.

Overall management of the Zayandeh basin is now in the hands of the Isfahan Water Company; choice of the latter name is recent, a change from 'Organisation' perhaps reflecting the opportunity seized by the Company to regulate water usage by a new pricing system (which would be the envy of parts of the US West!). The post-revolutionary religious society of Iran is as surprisingly open to economic instruments and to town planning as it is closed to the import of foreign experts and dam builders (though before the Revolution the French and British provided much dam-building expertise in Iran). Pollution control legislation is slow to keep pace with the rapid industrialisation of towns like Zarrin-shahr and there are fears that continued growth will place stress on the ability of the Zayandeh to dilute and disperse contamination from the industrial suburbs before the river passes the holy sites along the Isfahan river front, already a growing focus for the tourist ambitions of the city. Local people believe, however, that such is the importance of the cultural and religious meaning of the river that the catchment managers will bring about controls before this capacity is exhausted. The 'jury' is, however, still 'out' on whether a moderately successful example of dryland river basin development will be truly sustainable as growth becomes exponential.

Within a single culture, tradition and management institution and with the focus of one major city, the Zayandeh has advantages not enjoyed by the next dryland river we feature – the Nile.

5.4 THE NILE: A DEFINITIVE CASE OF HYDROPOLITICS

The Nile (Figure 5.5) is the world's longest river (6,825 km); its catchment area (3M km^2) covers one-tenth of Africa and its annual flow is measured in cubic kilometres but the scale of the Nile basin is not the cause of its river management problems, except in so far as its scale stretches its boundaries across several climatic zones, nine sovereign states and a number of racial and religious boundaries. Moorehead (1983) describes just one of the boundaries, on the Blue Nile, in this way:

> No one crosses this border with impunity. When the Arab invades Ethiopia his camels die in the mountains and he himself loses heart, in the fearful cold. When the Ethiopian comes down into the desert his mules collapse in the appalling heat, and he is soon driven back to the hills for the lack of water. It is the conflict between two absolutely different forms of life, and even religion seems unable to make a bridge since Christianity falters as soon as it reaches the desert and Islam has never been really powerful in the mountains. Only the river binds these two conflicting worlds together.
>
> (Moorehead, 1983, p. 16)

Moorehead (1973; 1983) has given us a very graphic insight into the physiography and exploration of this river's two main branches – the Blue Nile which carries the majority of the flow and the White Nile whose source proved so elusive to men like Burton and Speke. Moorehead's books also weave the themes of exploration and exploitation closely; once European nations had discovered the wealth of Egypt (notably from Bonaparte's invasion in 1798) they became launched inevitably on controlling the rest of the basin. This destiny has never proved attainable and the cost in lives, reputations and failures was enormous in the nineteenth century alone.

5.4.1 Egypt: product of the river

Modern Egypt has a severe population problem; 48 million people live there, sustained only by the waters of the Nile. Unlike some Middle Eastern states Egypt's oil resource is small by comparison with its size; furthermore its population continues to grow by over 2.5 per cent per annum. Egypt's situation is dire as Table 5.6 indicates. Biswas (1993) points out that Egypt needs very careful land-use planning in connection with its joint needs for food and for water to produce that food; urbanisation is a double blow, covering much-needed agricultural land and raising the water demand.

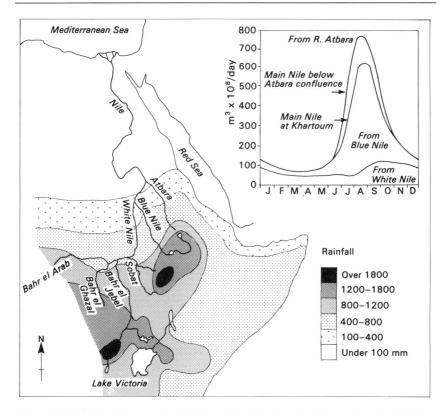

Figure 5.5 Nile basin annual rainfall and the Nile flood hydrograph, subdivided by contributing catchment

Egypt's need for water has had violent political repercussions already in modern times; in 1956 when the USA refused finance for the construction of the Aswan High Dam, Egypt nationalised the Suez Canal to raise the capital herself but then accepted Soviet backing for the venture, forcing the USA to consider damming the Blue Nile in Ethiopia (a move cut short by the Communist revolution in that country!).

The Egyptian leader President Nasser used words which evoke the complex political motives behind water-based development: 'Here are joined the political, social, national and military battles of the Egyptian people, welded together like the gigantic mass of rock that has blocked the course of the ancient Nile.'

Aswan is now complete; it is a controversial dam (see below) but its existence has both saved Egypt from famine and constantly posed the problem of providing guaranteed inflows from upstream.

Another historian with a fascination about the Nile is Robert Collins, who uses the word 'hydropolitics' to connote the sequence by which the basin

Table 5.6(a) Estimated present water use in Egypt

	Volume in billion cubic metres
INFLOW	
Aswan release	55.5
OUTFLOW	
Edfina to sea	3.5
Canal tails to seas	0.1
Drainage to sea	13.2
Drainage to Fayoum	0.7
Sub-total	17.5
WATER USE	
Municipal and industrial	2.4
Evapotranspiration (irrigation)	33.6
Evaporation from water surfaces	2.0
Sub-total	38.0

Source: Chesworth (1990)

Table 5.6(b) Growth of Egyptian population and cultivated and cropped land

	Estimated population (million)	Cultivable area (000 feddans)[a]	Cropped area (000 feddans)[a]	Cropping intensity (%)
1821	2.51–4.23	3,053	3,053	100
1846	4.50–5.29	3,746	3,746	100
1882	7.93	4,758	5,754	121
1897	9.72	4,943	6,725	136
1907	11.19	5,374	7,595	141
1917	12.72	5,309	7,729	146
1927	14.18	5,544	8,522	154
1937	15.92	5,312	8,302	156
1947	18.97	5,761	9,133	159
1960	26.09	5,900	10,200	173
1966	30.08	6,000	10,400	173
1970	33.20	5,900	10,900	185
1975	37.00	5,700	10,700	188
1986	49.70	6,000	11,400	190

Source: Chesworth (1990)
[a] One feddan = 4,200 m².

has reached its present level of development. Collins (1990) traces history through the lives of those who harnessed the rivers water in both Egypt and Sudan. For more than six thousand years the benign flood of the Nile (compared with the flashy snowmelt regime of Tigris and Euphrates) provided simple gravity irrigation to the cereal-growing flood basins of Egypt. However, rulers and occupiers of the nineteenth century needed to extend the area of cultivation and to diversify into cash crops such as cotton; from 1843 barrages were built on the delta to raise water throughout the year for perennial irrigation. The size of this task is demonstrated by the river's markedly seasonal regime.

5.4.2   Control of the river

A new phase of development on the basis of dam-building began with the first Aswan Dam in 1902. Collins (1990) records the history of the long British engineering/hydrological involvement with the river; Garstin's hydrological surveys between 1899 and 1903 heralded the modern phase, although flow records at the ancient 'nilometers' date back to AD 672. Throughout the centuries Egypt had established its claim to Nile waters by historic use. However, by 1929 it proved essential to reconcile the Egyptian needs with the growing use of irrigation in the Sudan; the Egyptians were suspicious of the ruling British in that country. The Nile Waters Agreement of 1929 merely partitioned the annual flow between the two countries, it did nothing to develop the basin's resources as a whole. After the resignation of the most famous engineer/hydrologist, MacDonald, in 1921 the basin entered what Collins calls 'the years of indecision' in which a huge extra effort went into river gauging and the analysis of records but very little into proactive planning.

> They were terrified of not having sufficient data but few were able to forge the enormous amount of information they gathered into a rational, coherent, holistic view of the Nile. They were also dull, rigid and unimaginative. Reports they could write; dreams they suppressed.
>
> (Collins, 1990, p. 162)

H. E. Hurst was the first hydrologist to combine data with dreams and he labelled the product 'century storage', a proposal to smooth out the regular variability in Nile flow by increasing storage (principally on the White Nile) to the point where 100-year mean discharges could be guaranteed. Figure 5.6 compares Hurst's design with the existing management and utilisation pattern in the Nile Basin.

One major obstruction has, however, always thwarted the notion of century storage – the Sudd, a huge area of floating vegetation and wetlands which physically blocks the river in southern Sudan, detaining and evaporating half of the inflow. The area of the Sudd fluctuates according to climatic conditions in the headwaters of the White Nile as Table 5.7 shows.

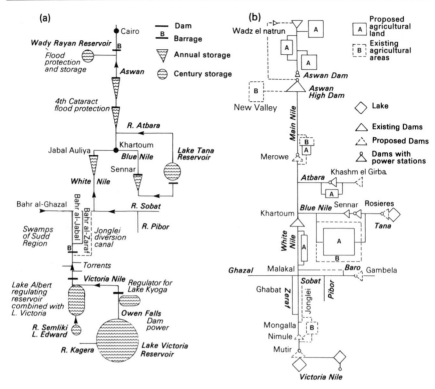

Figure 5.6 Nile water resources – dreams and reality:
  (a) Hurst's plan for 'century storage', indicating the value of the upper White Nile (Collins, 1990)
  (b) Actual developments in resources, showing the concentration on Egypt/Sudan (Fahim, 1981)

Table 5.7 Losses in the Sudd

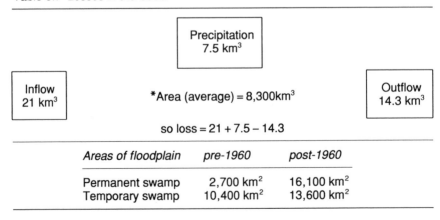

Areas of floodplain	pre-1960	post-1960
Permanent swamp	2,700 km²	16,100 km²
Temporary swamp	10,400 km²	13,600 km²

In 1925 the Egyptian government approved a scheme to cut through the Sudd in order to reduce detention and evaporation. The Egyptian irrigation service told the Sudan and the Governor of the Upper Nile nothing of their proposed Bor-Zaraf Cut and the scheme became politically impossible. Subsequently a large number of ways of breaking through the Sudd were evaluated and by 1938 the Jonglei Canal Diversion Scheme was selected by Egypt, though once again the British in the Sudan were not consulted and no surveys were made of the impact the Canal might have on the large indigenous population (Collins, 1990).

5.4.3   Jonglei: long promised, incomplete through war

After the Second World War, Sudanese plans to develop the South and H. E. Hurst's proposals for century storage and a canal from Jonglei to the Sobat converged; British officials in Khartoum assembled a Jonglei Investigation Team which was to survey the desirability of reducing evaporative loss with minimum impact on the living and economy of the people of the Upper Nile, an early but unlabelled example of environmental impact assessment. The Jonglei Investigation Team's four-volume report was published in 1954 but its findings have been refined and updated (Howell *et al.*, 1988). As part of the work undertaken by the Jonglei Investigation Team, Sutcliffe (1974) surveyed the floodplain regions of the Sudd in relation to inundation patterns and the resulting spatial distribution of vegetation, the basis of the existing rural ecology. Later Sutcliffe and Parks (1987) used a simple mathematical model to predict the areas of inundation under various natural and modified flow regimes. Howell *et al.* (1988), in their volume devoted to the Jonglei Investigation Team's work, conclude:

> The fact of the matter is that water evaporated in the Sudd region is not a total loss; it has its vital local value in the subsistence economy and has done from time immemorial (Laki, 1994). It also has potential for future development in the Jonglei area. In this context there are fundamentally different perceptions of riparian rights. For downstream users, water saved from evaporation by major drainage or diversion works financed by them is 'new' or 'additional' water; it is water saved and theirs by right [conveniently forgetting that the 2m of water lost annually from the surface of Lake Nasser represents twice the losses from the Sudd - Pearce, 1994]. For those who live in the Jonglei area ... in the process of seasonal inundation of the floodplain valuable economic assets in pasture and fisheries are created.
>
> (Howell *et al.*, 1988, p. 468)

At the time of writing their book, Howell and his colleagues reported that, as a result of the outbreak of civil war in Sudan in 1983, all work on the Jonglei Canal had ceased. They conclude that a 260-kilometre 'vast trench

Table 5.8 The Aswan High Dam debated: a summary sheet

Logistics	Benefits to date (1980)	Major side effects	General assessment and prospects
Land development was a necessity to cope with the incredible imbalance between the country's population growth and agricultural production. Agriculture is basic to the Egyptian economy.	Controlled high floods and supplemented low ones; saved Egypt from the monetary cost to cover the damage from both high and low floods.	Water loss, through seepage and evaporation, is likely to affect the water supply needed for development plans; studies show that the water loss is within the predicted volume.	The Aswan High Dam is solid engineering work; more importantly it is fulfilling a vital need for 40 million people.
Egypt found no option but to increase the water supply for land development policies of both horizontal and vertical extension.	Allowed for increase in cultivated land area through reclamation, and increased the crop production of the existing land through conversion from basin to perennial irrigation.	Loss of the Nile silt would require costly use of fertilisers; it has also caused riverbed degradation and coastal erosion of the northern delta.	All dams have problems; some are recognised while others are unforeseen at the time of planning.

The lesson learned from the Aswan project, however, is that dams may be built with missionary zeal but little careful planning and monitoring of side effects. |
| Building a dam and forming a water reservoir in an Egyptian territory minimises the risks of water control politics on the part of riparian countries. | Improved Nile navigation and changed it from seasonal to year-round.

Electric power generated by the dam supplies 50% of | Soil salinity is increasing and land in most areas is becoming waterlogged, due to delays in implementing drainage schemes. | As a result of the new semi capitalist policies and also the technical and monetary |

Egypt perceived the Aswan High Dam as a multipurpose project, basic to national development plans.

Egypt's current consumption, although the dam was built primarily not for power generation but for water conservation.

The new lake resources are potentially economic, including land cultivation and settlement, fishing, and tourist industries.

Increased contact with water through irrigation extension schemes is expected to affect schistosomiasis rates adversely; evidence exists that rates have not increased due to use of protected water supplies.

The Nile water quality has been deteriorating; studies indicate the occurrence of change in the water quality parameters; they do not consitute a health hazard at present.

aid that Egypt has lately been receiving from several Western countries, it is expected that several of the dam's problems will be efficiently controlled and monitored.

Dam-related studies have given recognition to present problems and have provided possible solutions. The dam would meet its expectations on several aspects providing that research findings are utilised.

The development potential and economic returns of this water project are expected to be very rewarding in the long run, if the project's developments are systematically studied and monitored.

Source: Fahim (1981)

which catches water during the rains' will need repairs and crossing points for people and wildlife whatever the outcome of the Project, once peace is re-established.

5.4.4   Aswan and Lake Nasser: plus and minus

For the foreseeable future Egypt and Sudan will need to maximise the benefits of the Aswan High Dam. In order to carry through this project a second Nile Waters Agreement was signed in 1959, dividing 'the spoils'. It is fortunate that the Aswan scheme has been documented (e.g. by Fahim, 1981) in terms of construction and early impacts, both positive and negative. Fahim writes:

> As the controversy over the Aswan High Dam began to intensify during the early years of the 1970s and was especially aggravated by increasing conflicts between facts and fiction, science and politics, in both domestic and international circles, it became essential to tackle the dams dilemma on a scientific basis and in a way that would realistically account for the technical and human issues combined.
>
> (Fahim, 1981, p. xiii)

Of the human issues, Fahim notes how closely the Dam came to President Nasser's ideals of a new social democracy and the abolition of feudalism; its scale (5 kilometres long and 1 kilometre thick at its base) was also important, Nasser remarking that, 'In antiquity we built pyramids for the dead. Now we will build pyramids for the living.'

How well has the pyramid performed? Four major problems are commonly identified:

(a) Water loss from Lake Nasser by seepage and evaporation.
(b) Sedimentation in the Lake and degradation below the Dam and in the Delta.
(c) Waterlogging and soil salinity from year-round irrigation.
(d) Increase in disease, especially schistosomiasis.

There is no sign of an end to controversy, rival sets of views and 'facts' offered freely on every topic (see Table 5.8). Egyptians are sure of the impact of the dam in preventing famine in 1972 and 1984 – for this they are prepared to accept some costs. Uptake of commercial farming is not yet complete and Fahim is critical of the unsystematic way in which human, community resources have remained uncoupled from the technical aspects of the Dam; he proposes Figure 5.7 as an integrating system for the proper consideration of such projects. For example, the Nubians displaced by the project (100,000 in both Egypt and Sudan) were resettled and have received the social benefits of modern living but without any infrastructure for community development, as if planners feared this. Many people chose to

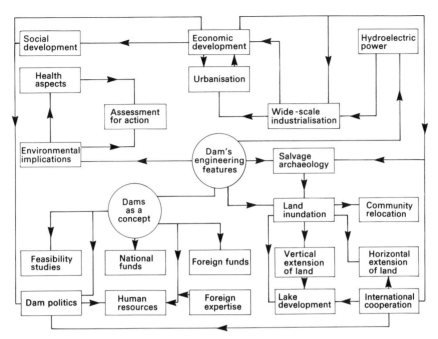

Figure 5.7 The development of a major dam scheme (e.g. Aswan) seen as an interconnected socio-economic system (Fahim, 1981)

stay on or around Lake Nasser and those Nubians who moved to the New Halfa agricultural settlement in Sudan have encountered shortage of irrigation water, weed infestation, poor yields, absenteeism and houses which were too small. The site of the settlement was chosen against the wishes of those to be moved 'in the national interest' (Davies, 1986; El Arifi, 1988). One of the systematic problems with such re-location schemes, indicated by a study of the White Nile Province of Sudan (Hulme, 1986), is that migration is an essential component of drought response by communities outside such schemes. Government policies on water supply should reflect this by encouraging flexibility of sources.

Efforts continue to bring about political consensus on the planning of the Nile as a basin resource. It is astounding that a river for which we have so much data cannot utilise a fraction of that information, although more rational irrigation modelling is promised if the Jonglei Canal is completed (Stoner, 1990). The major hidden agenda on the Nile is development of the major source of the river's flow – the Blue Nile. During negotiations over the Nile Waters Agreement of 1959 Ethiopia launched a major study of the river's potential through the agency of the US Bureau of Reclamation. This study proposed four major dams on the Blue Nile, producing three times more electricity than the Aswan High Dam, virtually eliminating the seasonal fluctuations of flow into Sudan and reducing the total flow by 8.5 per cent. The losses are accountable to evaporation but they are, of course, much less on the Blue Nile in the mountains and these are optimal dam sites – the challenge of political consensus is therefore to throw much more of the Nile's control on to the Blue Nile. Post-revolutionary Ethiopia is unlikely to enter negotiations easily and has no capital (nor the desire for provincial development) to develop the Blue Nile unilaterally, preferring, it seems, to hold the potential to do so as a threat (see Table 5.9).

As for multinational forums in which use of the Nile can be discussed, the Nile Basin Commission (set up by the 1959 Agreement) is mainly a technical and hydrometric organisation but it has spawned a grouping of riparian nations (Egypt, Sudan, Uganda, Zaire and the Central African Republic) as the ENDUGU ('Brotherhood') group. Rwanda and Burundi have also joined; the aim is to encourage Kenya, Tanzania and Ethiopia to join, moving towards a Nile Basin Economic Community (Samir, 1990).

External interest in the Nile is represented by a UNDP Commission which has collated information on population growth and the expected loss of food and power security in the basin (as population doubles every twenty years). In fact a joint electricity grid may be a more tangible political expediency than an agreement on flow allocation.

Further organic progress in cooperation is, however, likely to need changes of international law governing river basins; the Nile needs a substitution of the principles of the International Law Association's *Helsinki Rules* for the myriad of *ad hoc* treaties currently in use (or abuse) in the basin (Okidi, 1990).

The most important Helsinki Rules comprise:

(a) Equity of distribution is the governing factor among riparians.
(b) Equity does not mean distribution by equal share, but by fair shares which can be decided by the following factors:
   —The topography of the basin; in particular, the size of the river's drainage area in each riparian state;
   —The climatic conditions affecting the basin in general;
   —The precedents about past utilisation of the waters of the basin, up to present-day usages;
   —The economic and social needs of each basin state;

Table 5.9 Divisive economic problems of major Nile basin partners

	Egypt	Sudan	Ethiopia
GNP/Cap($)	680.0	330.0	130.0
Annual growth (%)	3.5	−0.5	0.1
Debt bill (bn$)	40.2	11.1	2.6

—The population factor;
—The comparative costs of alternative means of satisfying the economic and social needs of each basin state;
—The availability of other water resources to each basin state;
—The avoidance of undue waste and unnecessary damage to other riparian states.

Such is the vital importance of the international dimension in any future vision for sustainable development in many of the world's humid tropical and dryland basins that we return to it at the end of this chapter and in Chapter 7.

5.5 RIVER BASIN DEVELOPMENT AUTHORITIES: EXPERIENCE ELSEWHERE IN AFRICA

Having considered the seemingly impossible task of integrated river basin management across nine nations' boundaries in the case of the Nile, we briefly investigate two 'model' river basin authorities wholly within national boundaries elsewhere in Africa: the Awash Valley Authority in Ethiopia (Winid, 1981) and the Tana and Athi Rivers Development Authority in Kenya (Rowntree, 1990). Both authorities are responsible for basin areas of approximately 120,000 km^2 but, while some of the lessons they teach us about development of power and irrigated agriculture are similar, their political context and structure illustrate the importance of individual national and regional development pathways in Africa. Nigeria is also a nation with successes and failures in river basin planning (Adams, 1985).

5.5.1 The Awash Valley Authority, Ethiopia

The Awash Valley Authority was established in the pre-revolutionary Ethiopia of Emperor Haile Selassie in 1954. Ninety per cent of Ethiopia's 30 million people live in the central highland area where rainfed agriculture is feasible most years but where population pressure has led to huge soil erosion losses. While 6 million hectares are currently cultivated, half as much again could yield food and cash crops with irrigation. Rivers radiate from the highlands ('the water tower of Africa'). Situated in convenient proximity to the nation's capital, Addis Ababa, the Awash Valley stretches 700 kilometres towards the Djibouti border across the Rift Valley, its flow sustained by fourteen tributaries but its course succumbing to aridity in the shallow Lake Abe (Figure 5.7). The highland rainfall of 1,000 mm/yr decreases to 200 mm/yr in the Rift and rainfall is very unreliable; nevertheless irrigable soils constitute 24 per cent of the valley area. The indigenous agriculture consists of livestock, with cattle, sheep, goats and camels migrating in search of grass and water.

The Awash Valley Authority was set up to coordinate the activities of government ministries in the Valley, to charge for the use of water, to conduct surveys and to administer water rights.

The Development Plan followed by the Authority was highly ambitious prior to the Marxist revolution of 1974, itself a result of Sahelian drought. Winid (1981) describes the plan's objectives as including policy change to promote development, social and health surveys, feasibility studies for irrigation, flood control and hydropower, establishment of agro-industries and infrastructure improvements to permit tourism. Winid describes seven run-of-river irrigation sites in the upper and middle Awash, together with some smaller areas of the lower Awash (Figure 5.8); however, dam sites to increase the scope of irrigation and to generate power for industry are continually prospected.

The calculation of benefit/cost ratios for developments in the Awash is, according to Winid, markedly influenced by the balance between the cultivation of food and cash/industrial crops. To date the latter dominate; this in turn determines the fate, in development, of native nomadic

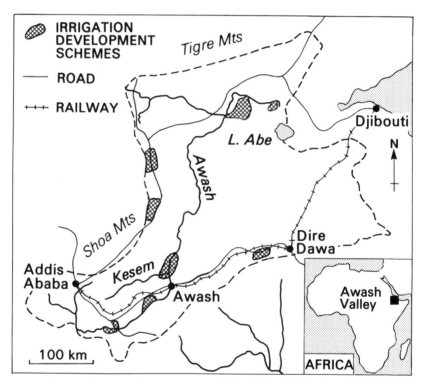

Figure 5.8 The Awash valley, Ethiopia and schemes developed by the former Awash Valley Authority (after Winid, 1981)

pastoralists who must be settled to form a labour force, joined by other migrants from troubled regions of Ethiopia. However, the Awash schemes for resettlement have been failures, with the true costs of resettlement soaring. No attempt was made to integrate an improved traditional livestock sector with the plantation irrigated agriculture.

Winid offers twelve reasons for the failures of the Awash Valley Authority (Table 5. 10), which was wound up by Ethiopia's centralist Marxist government in 1981 in favour of stronger ministerial roles in water developments.

In 1986 the present author was a member of the British team which investigated the feasibility of a dam on the Kesem tributary of the Awash, one of four dam sites suggested by the Awash Valley Authority in the 1960s. A Kesem Dam would irrigate 20,000 hectares of land on the left bank of the Awash. Sir M. MacDonald and Partners, who carried out the Kesem study for FAO and UNDP, set out to use all resources to maximise agricultural potential under a balanced environmental and ecological system, producing cash and industrial crops for export and/or import substitution, plus food crops at least for local self-sufficiency. These objectives are much more broadly based than those of the Valley Authority's consultants in the 1960s, reflecting a new social and environmental conscience in river basin development. Referring to the demise of the Authority in 1981 the MacDonald report concluded that, while centrally the roles of ministries were now more defined, at local level there was 'frequent lack of communication, coordination and co-operation'. The pace of the proposals for a Kesem scheme was made compatible with the changing attitude of the Afar tribespeople; the Afar would be granted their own lands, avoiding a traditional plantation culture. Components of the new proposals include roads, power supplies, housing, domestic water supply, offices and workshops, clinics and health services, schools and education, police and civil administration, and recreational facilities. Preferred crops are tobacco and citrus, with smaller areas of cotton, wheat and maize.

One problem with the proposed Kesem scheme is the high sediment yields of the river (largely the result of erosion in the highlands – see Plate 6.1) which would fill the reservoir in less than 100 years. Finally the recommendations concluded that the internal rate of return for the original scheme would be between 2 per cent and 4 per cent. A smaller scale scheme was recommended.

Abate (1994) has recently updated the water resource development picture for Ethiopia. From 1977 the Awash Valley Authority had its powers limited to the development of water resources as the Valleys Agricultural Development Authority; in 1987 it became the Ethiopian Valleys Development Studies Authority. The EVDSA has national coverage, collects hydrometric and other basic information, identifies development projects and ensures environmental protection; its eventual aim is an Ethiopian Water

Table 5.10 Reasons why the Awash Valley Development was not a success

1 Knowledge on the part of the French firm of consultants (SOGREAH) of the political, economic, social, cultural, scientific and technical aspects of the Awash valley region was lacking as a result of:
   (a) The short period during which the consultants were employed, namely, four years.
   (b) Their limited contact with Awash valley peoples which was mainly restricted to a small proportion of the population, the egoistic 'élite'.
   (c) The consultants living very much within the foreign UN community.
   (d) Frequent changes of advisers from different schools of thought.
   (e) Ineffective supervision policies (local counterparts were too young).
   (f) Relations between the home country and the international UNO bureaucracy.
   (g) The introduction of extensive techniques and excessive investments.
   (h) No interest in continental African 'integration' policy.
2 There were great differences in the Ethiopian technical levels between sectors, branches of the economy, industry, and infrastructural elements (services, education, health, science).
3 Small consideration was given to planning the future AVA situation in relation to Ethiopia, the African continent or the world (especially the Horn of Africa).
4 Foreign advisers were unaware of the true political and governmental obstacles to development activities.
5 There was little cooperation among those concerned with AVA development or between them and the countries adjacent to Ethiopia.
6 Advisers' preparation of plans was:
   (a) Not nationally integrated and divided up between specialised UN agencies.
   (b) Focused on the mining and consumptive industries.
   (c) Export-oriented in agricultural production.
   (d) Not related to existing locally trained manpower.
   (e) Minus much large-scale investment.
7 There were difficulties in the implementation stages, namely:
   (a) The existence of other advisory bodies/institutions (e.g. Australians were active in the AVA region).
   (b) The difficult financial situation – no domestic means were available.
   (c) There was instability, political and economic (both internally and externally to Ethiopia).
8 The creation and enlargement of the 'dual economy':
   (a) Weakened the links between the 'new economy' and the 'old' which had engaged 80–90% of the population.
   (b) Promoted a pattern of demand which required importation from industrialised countries.
9 New investments are very often owned or controlled by foreigners who view profit goals as the deciding factor.
10 Locational policies for industry were mostly drawn up in Addis Ababa, or the major Ethiopian ports which are in general the crystallisation or evolution of the colonial city.
11 The towns in AVA were only administrative centres for the exploitation of the countryside with the exclusion of the natives from government.
12 Production was in the form of goods and in quantities best suited to the ruling (minority) group.

Source: after Winid (1981)

Plan defining the long-term use, development, protection and control of water resources (surface and groundwater). The need is dire; given the annual population growth rate of 2.9 per cent, the annual growth rate of per capita food demand of 6.3 per cent and the proclivity of the region to severe drought, EVDSA faces every temptation to act unsustainably. Ethiopia's irrigated area must multiply one hundred-fold by the year 2000. Abate is critical of the fact that 'there is no thorough attempt to base organisational structure on the requirements of development programmes. What should have been considered is the functional compatibility of the organisation (EVDSA) with the requirements of the development programme' (one which Abate considers should have been more 'bottom up' and much less based, by the former Communist government, on social settlement models). The present government of Ethiopia is encouraging regional resource development and, as yet, the implications of this decentralisation for water resources have not become firm.

5.5.2   The Tana and Athi Rivers Development Authority, Kenya

Like Ethiopia, Kenya has a burgeoning population highly concentrated in its humid zones with an acute problem of peripheral aridity. However, it has a more pronounced and more recent colonial European past and, while currently a one-party state, has fewer problems of drought or famine than has Ethiopia.

The Tana and Athi basins, whose joint Authority, TARDA, is reviewed by Rowntree (1990), total 132,700 km^2, containing a significant proportion of Kenya's population (62 per cent) including the capital, Nairobi (Figure 5.9). The principal water yield for both basins comes from the slopes of Mount Kenya and the Aberdares which receive 2,000 mm of precipitation each year; out on the plains a potential evaporation of 1,200 mm exceeds rainfall and *irrigation* forms the basis of most agriculture. An immediate problem of this twofold hydrological division is that, as in the case of the Kesem, the desire to exploit *rainfed agriculture* in the uplands has led to deforestation and soil erosion. Sediment yields of between 109 t. km^{-2}. yr^{-1} and 433 t. km^{-2}. yr^{-1} are quoted by Rowntree. (Kenya is fortunate at least to have had hydrometric surveys which include measurement of sediment yields.) As well as threatening irrigation schemes downstream by excessive sedimentation, the populous uplands also have a high demand for hydroelectric power (coal and oil must be imported and exploitation of wood fuel leads to further erosion) and for domestic water supply/sanitation.

Like many other emerging nations, Kenya introduced strongly centralised planning in 1974 and set up a Ministry of Water Development; at the same time the Tana River Development Authority (the Athi was added in 1982) was instigated with aims as shown in Table 5.11. The Authority, like that of the Awash in Ethiopia, was designed to coordinate the work of ministries

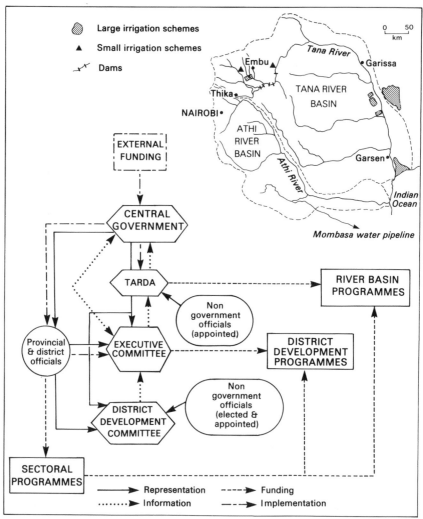

Figure 5.9 The area of the Tana and Athi Rivers Authority and its decision-
network diagram (after Rowntree, 1990)

and to balance the competing demands of domestic water supply,
hydropower and irrigation in the basin. Such coordination was vital in view
of three major problems of development:

(a) Hydropower dams in the middle course appeared incompatible with
irrigation in the lower reaches.
(b) Dry-season water supply abstraction in the upper catchment could
reduce the hydropower potential.
(c) Siltation could only be controlled by a vast conservation programme.

Table 5.11 Functions of the Tana River Development Authority

1 To advise the Government generally and the scheduled Ministries in particular on all matters affecting the development of the Area including the apportionment of water resources.
2 To draw up, and keep up to date, a long-range development plan for the Area.
3 To initiate such studies, and to carry out such surveys of the Area as it may consider necessary, and to assess alternative demands within the Area on the resources thereof, including electric power generation, irrigation, wildlife, land and other resources, and to recommend economic priorities.
4 To coordinate the various studies of, and schemes within, the Area so that human, water, animal, land and other resources are utilised to the best advantage, and to monitor the design and execution of planned projects within the Area.
5 To effect a programme of monitoring of the performance of projects within the Area so as to improve such performance and establish responsibility therefor, and to improve future planning.
6 To ensure close cooperation between all agencies concerned with the abstraction and use of water within the Area in the setting up of effective monitoring of such abstraction and use.
7 To collect, assemble and correlate all such data related to the use of water and other resources within the Area as may be necessary for the efficient planning of the Area.
8 To maintain a liaison between the Government, the private sector, and foreign agencies in the matter of the development of the Area with a view to limiting the duplication of effort and assuring the best use of technical resources.
9 To render assistance to operating agencies in their applications for loan funds if required.

Source: After Rowntree (1990)

The first major project of the Authority was the commissioning of the Masinga Dam in 1979. It was designed both to generate power itself and to improve the regime for existing downstream dams and irrigation schemes; as with the Awash settlement, plans for 40,000 people are part of the irrigation projects and once again social problems and climatic contrasts with the over-populated uplands are anticipated to give trouble. Nairobi is continuing to expand and to require power and water; its City Commission appears to Rowntree to act independently of TARDA.

Rowntree's sad conclusion about TARDA is as follows:

We are left with the conclusion that TARDA does not represent an effective framework for regional planning, neither on its own nor through integration with the district focus policy. Both are controlled by top-down planning, by political allegiance to the power élite and by the interests of foreign aid agencies. It may represent a forum through which technocratic solutions to resource development can be promulgated but it is unlikely to achieve the type of grass-roots development that is so essential to effective and lasting development programmes.

(Rowntree, 1990, p. 39)

5.5.3 Basin Authorities elsewhere in Africa: Nigeria's failings and the South African challenge

River Basin Development Authorities (RBDAs) in Nigeria date from 1960; in 1976 an extraordinarily wide brief was laid down by government to include water resource development and flood protection but also watershed management (including afforestation), control of pollution, resettlement, land clearance, agricultural research, crop processing and rural water supply. Adams (1985) discussing the performance of RBDAs (he cites eleven), reveals the following shortcomings:

(a) Over-reliance on large projects: dam construction and irrigation development.
(b) Inadequate economic, environmental and social appraisal.
(c) Ineffective population resettlement.
(d) Almost total lack of attention to watershed management and pollution control.

He quotes state/federal rivalry as a (now familiar) obstruction to progress and advocates both development from below and a concentration on improved operation of existing projects as priorities. A worrying paradox is that to strengthen the role of the Authorities would seem to require the collapse of the river basin unit into a state or regional bureaucratic unit, merely to improve the chances of collaboration and adequate financial resources. Salau (1990) demonstrates the mismatch between administrative and river basin boundaries and points to the need for RBDAs to have properly trained, properly remunerated personnel, as well as more power relative to the state corporations.

We can draw out certain common constraints on institutional river basin management in those developing countries with a drylands settlement problem:

(a) In the last analysis the control is in the hands of those creating the new settlements and wealth; technocratic speed in the early stages of development (e.g. dam-building) is always liable to be forsaken once farming begins.
(b) If not overly centralised, multi-agency authorities can potentially deliver a unified picture to the people on the ground; however if they have been forcibly moved from the humid zones, or settled from a dry-land nomadic lifestyle, there are still problems without much expensive attention to patterns of tenure, infrastructure and timing.
(c) Newly emerged nations, particularly facing problems of infant democracy or totalitarian control, often are troubled by regional rebellion verging on civil war. The present author has been held at gunpoint by rebels while sediment sampling in the Kesem. Local participation is therefore particularly difficult to 'release', there are cultural clashes within the local population and, within government, ministries are keener to compete than to collaborate.

(d) The use of foreign finance, foreign expertise and foreign personnel, often with the aim of growing exotic crops for foreign consumption or supplying electricity to distant cities, further threatens the local element of development schemes so widely seen as desirable.

In view of these problems a particularly interesting challenge is that now faced by the people of South Africa; their colonial past, engineering skills and relative wealth have led to a highly structural set of solutions to massive water shortage. The ending of apartheid in 1993 and the election of a government committed to the rights of the black population effectively means that the clock begins ticking afresh on all issues of water resources, rights, distribution systems and planning within a programme of broadly based economic development within the world trading community. In an atmosphere of political release, democratisation and popular empowerment it will be extremely interesting to see how the previously technocratic management of rivers will 'open up' and how the needs of non-human biota can be incorporated against a background of huge human needs. The picture is, at the time of writing, only just beginning to emerge.

A recent White Paper (Department of Water Affairs and Forestry, 1994) puts the South African problem succinctly:

> In a country with nuclear power, cellular telephones and vast inter-catchment water transfer schemes, more than 12 million people do not have access to an adequate supply of potable water; nearly 21 million lack basic sanitation.
>
> (p. 1)

As part of the Reconstruction and Development Programme of the Government of National Unity a strong central Strategic Management Team has been set up in a new Department of Water Affairs and Forestry. Its Chief Engineer has put the geographical and management problems succinctly:

> The mean annual rainfall of 500mm is only 60% of the world average. It is poorly distributed relative to areas of need. On average only 9% of rainfall reaches the rivers. The country is vulnerable to droughts and floods. The groundwater is sparse and often saline. South Africa's abundant natural riches are scattered across the country, while most of the rain falls near the south-eastern seaboard.
>
> (Conley, 1989)

Despite these problems there is optimism and resolve to utilise the principle of sustainability:

> A river basin is often the most appropriate unit for managing water and land resources to gain the full benefits of multipurpose use and to coordinate the activities of various agencies and other bodies interested in resource utilisation for their own purposes. Therefore, integrated water

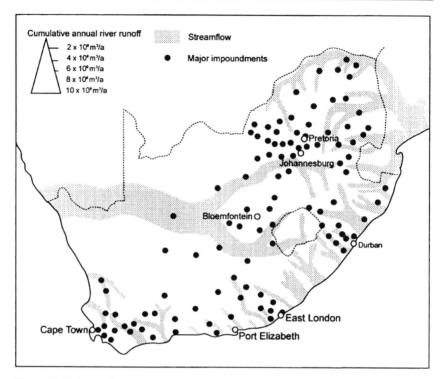

Figure 5.10 River flow and major impoundments in South Africa. Note how the major centres of demand are remote from the major sources (after Department of Water Affairs, 1986)

management requires the simultaneous assessment of economic, social and environmental consequences in local, adjacent and possibly remote catchments (a reference to water transfers). A vital component of the water supply mechanism is the watershed itself, which should be managed according to sound principles of environmental management and resource allocation as these come to be understood better.

Policy principles listed by the 1994 White Paper are:

- Development should be demand driven and community based.
- Basic services are a human right.
- 'Some for all', rather than 'all for some'.
- Equitable regional allocation of development resources.
- Water has economic value.
- The user pays.
- Integrated development.
- Environmental integrity.

Information and decision support systems are seen as vital to the programme; fortunately (and despite its political harshness) the wealth and sophistication of the water management system imposed before the birth of the new South Africa has already given a large database of both hydrometric and land-use statistics, together with modelling and decision-support capabilities. The former Department of Water Affairs published in 1986 'the Red Book', an almost totally comprehensive compilation of information about the national and provincial water situation, forming an ideal basis for decision-making and in marked contrast to the situation facing river basin development authorities in the rest of Africa – in fact in most drylands. Conley (1989) has emphasised the role of high-technology solutions to sustainable river basin development – not as construction technology, but information technology to plan, interact with communities and resolve conflicts.

Among the land-use conflicts facing the new South Africa in relation to water management are:

- Afforestation (partly the reason for the joint Department).
- Thermal power generation and use of cooling water.
- Irrigation – the biggest water user.
- Construction of farm dams on headwaters.
- Informal settlements ('townships') peripheral to cities.
- The needs of high-profile sites (e.g. Kruger National Park) and species.

(Conley, 1989)

Eco-tourism has the potential to become a major employer and earner of foreign exchange in South Africa and so protection of biodiversity is crucial, both of terrestrial species against loss of habitat to irrigated lands or the reservoirs which feed them and of freshwater species needing ecologically acceptable flows in rivers beneath abstraction points. Issues such as the proposed construction of a dam on the Palmiet River in the Kogelberg Nature Reserve in the Cape 'floral kingdom' and the development of irrigation, forestry and public water supply upstream of the Kruger National Park (Gore et al., 1992; van Niekerk and Heritage 1994) indicate that conflicts will be enormous. Dam construction (both on a large scale and at the farm level) has been the widespread solution to supply problems; demand management has hitherto been politically unthinkable. However, progress is being made on the assessment of the environmental resources and sensitivities of South African freshwater ecosystems (King et al., 1992; Wadeson, 1994) and decison-making on ecologically acceptable flows to sustain instream biota is made during workshops which deliberately integrate all 'sides', rather than by the use of technocratic mathematical models.

There are rapid moves towards integrated catchment management in South Africa (Stoffberg et al., 1994). Basin studies began in 1985, often carried out by consultants but directed by Steering Committees representing all the water-use sectors, plus local, regional and state authorities. As these authors state: 'Possibly the greatest advantage of the catchment study is that it enables the planners or managers to see the whole picture and to gain a holistic perspective of the problems and the pressures that are being placed on the limited water resource' (p. 460).

Conley (1995) states that 'A vital component of the water supply mechanism is the watershed itself, which should be managed according to sound principles of environmental management and resource allocation as these come to be understood better.' He goes on to describe the use of CRAM (Catchment Resource Assessment Model) as an aid to decision-making on basin strategies. As yet vehicles for public consultation and participation are but slowly developing; the problem is particularly hard for the majority black population to whom empowerment has come suddenly and late, a feature of the new South Africa perhaps best demonstrated by the national 'voter education' programmes before local elections.

Finally, South Africa is necessarily concerned with transfers across international boundaries and has 'eyed' border rivers jealously for decades. The Lesotho Highlands Water Project, on the drawing board since the mid-1950s, has now been financed by the World Bank, following the end of sanctions against South Africa. The scheme seeks to dam the headwaters of the Orange River (South Africa's major surface water source – see Figure 5.10) and to divert the regulated flows northwards towards Johannesburg and Pretoria. In return Lesotho (which is landlocked by and totally dependent on South Africa) gains hydroelectric power generation. Local people who have lost land have been compensated with corn or with jobs on the project but protest is rife (according to Horta, 1995) and the prediction of 'no major environmental obstacles to the project' made in 1986 is being reconsidered as biologists catch up with surveys of the biota impacted and the impact of soil erosion in Lesotho is considered in relation to siltation. The downstream impacts on the Orange River are also now thought likely to be much more damaging than before – one symptom of the long delay implicit in many developing world schemes. At the time of writing the Katse Dam, the Project's first, is well under construction and South Africa's promising attitude to its water resource allocation problems may well suffer as criticism of its collection policy mounts. There is no doubt that in future the water needs of all the southern African nations need to be considered jointly as part of a trade grouping in which 'virtual water' (in the form of foodstuffs) is moved, rather than the long-distance transport and inefficient use of the raw material itself (Conley, 1996 and Chapter 7).

5.6   THE LAND-USE DIMENSION: HIMALAYAN HEADWATERS AND THE INDIAN SUBCONTINENT

In Africa (outside South Africa) there is, as yet, only the thinnest body of empirical knowledge on the hydrological effects of land use and management. Of more than fifty papers presented at the Harare conference on African water resources (Walling *et al.*, 1984) only four reported field measurements and all of these were in the area of erosion, which completely dominated the Harare Symposium. The books by Pereira (1973; 1989), while full of qualitative and extrapolated wisdom on the subject, are obviously short of experimental information and much reference is made to North American and European experimental material in both volumes.

This gap is particularly significant because in the typical situation of a dryland river, 'exotic' because it draws runoff only (or mainly) from mountains in its headwaters, mountain development is clearly one of the main threats to downstream sustainability. Fifteen years ago the United Nations University and International Mountain Society, realising the critical importance of the world's mountains, initiated a journal entitled *Mountain Research and Development* which focuses on the dilemma of rapid development in the face of uncertainty. Books by Stone (1992) and Gerrard (1990) have further highlighted the mountain development problem and Chapter 13 of Agenda 21 makes specific reference to the importance of mountains as catchment headwaters.

The region to which we next turn has no waiting to do in terms of seeking a solution to the downstream impacts of headwater, mountain development. Like the Nile, the Ganges is a truly remarkable international river basin, 900,000 km^2 in area, with a dominant exploiting nation in the form of India for whom the river comprises a quarter of the available water resources, whose basin comprises 26 per cent of the nation's land and 43 per cent of its irrigated land. Additionally 'Ganga' has constituted a basis for successive dynasties in India's history and a firm religious theme of life-giving significance; the Hindu god Shiva collected Ganga's waters from heaven in his matted hair that the ashes of human dead might be purified for their return to heaven (Darian, 1978).

Indians bathe, water cattle, wash clothes and utensils, defecate, cremate their dead and conduct the other waste activities of twenty-seven cities of over 100,000 people in the River Ganges; it is sacred in what we might call a 'pre-hygienic' way. While the Indian government can control the activities of industrial polluters via the 1974 Water Pollution Control Act and, through the 1985 Ganges (Ganga) Action Plan, is improving urban sewerage and sewage purification, it is clear that a popular change in perception of the river's true role in an evolving society is required. Waterborne diseases in India account for 80 per cent of the health problems; 73 million man-days are lost each year through them at a cost of £250 million. The Ganga Action Plan targets sewerage improvements for twenty-seven cities larger than 100,000 population along its banks, notably Varanasi, where Hindus believe they will achieve liberation for the soul by the river (Ahmed, 1990).

5.6.1   Nepal: a rush to judgement

However, pollution of the Ganges is not the major focus of international attention in the basin. This is divided between the international tension which exists as the result of India's prime role in exploiting the river (by comparison with its poorer riparian neighbours Nepal and Bangladesh) and the apparent recent increase in sedimentation and flooding in these countries, phenomena widely attributed to headwater land-use changes. When one considers that the headwaters of the Ganges are the Himalayas one might anticipate causes other than development pressures alone; one might also anticipate that 'solutions' to the 'problem' might be taken at a regional (Himalaya/Gangetic plain) level rather than locally or nationally. All of these latter themes have been strongly developed in writings on the Ganges basin in the last five years.

The link between the Ganges and the Nile is strengthened in a paper by Hellen and Bonn (1981) demonstrating that both Egypt and Nepal, central problem nations to the management of both basins, are experiencing population growth of 2–5 per cent per year. However, these authors express the dangers of assuming that the step from population growth to environmental degradation is inevitable and a one-dimensional problem.

In a recent book devoted entirely to the Ganges and the development of the mountain zone in Nepal (Ives and Pitt, 1988) Ives considers the elements of *'the perceived crisis'* in the Himalaya–Ganges region. They are:

(a) That a population explosion was initiated shortly after the Second World War due to the introduction of modern health care and medicine and the suppression of malaria and other diseases.
(b) That increased population in subsistence mountain societies has led to:
     —Reduced amount of land per family;
     —Deepening poverty;
     —Massive deforestation.
(c) That mountain deforestation, on such a scale, will result in total loss of all accessible forest cover by AD 2000 and is the cause of accelerating soil erosion and incidence of landsliding.
(d) That destabilised mountain slopes resulting from points (a)–(c) above cause:
     —Increased flooding on the Ganges and Brahmaputra plains;
     —Extension of the delta and formation of islands in the Bay of Bengal;
     —Massive siltation and drastic reduction in the useful life of highly expensive water resource projects;
     —Drying up of wells and springs in the hills and lower dry-season river levels downstream.
(e) That deforestation also leads to climatic change in general and reduced rainfall amounts in particular.

Newspaper and TV journalists, together with the World Bank, have keenly fallen upon the deforestation perception of cause with such statements as: 'at the present rate of cutting the Himalayas will be bald in 25 years, topsoil will have disappeared and the climatic effects threaten to turn the fertile plain into a new Sahel' *(Sunday Times,* 11 September 1988). Gilmour (1988) cites this as an example of the *'uncertainty principle'* (Thompson and Warburton, 1985) in Himalayan development, since, when conducting actual ground surveys, the Nepalese government and the hill farmers appear jointly to have reversed the loss of forests in the mountain zones – the problem is actually worse in the 'terai' or piedmont zone.

Ives and Pitt quote further direct survey evidence of the erosion problem from Nelson (1979):

> The blanket indictment of Nepal for high erosion rates is obviously unwarranted. It is probable that, of the 14 Ecological Land Unit Associations, 3 or 4 will be the site of most of the watershed problems. This complex picture means that watershed management in Nepal must be a mix of educational, remedial and protection activities which must vary by geographic location.
>
> (Ives and Pitt, 1988, p. 66)

Thus, while logic foretells a huge downstream impact for careless deforestation of mountain lands, the inherent local instability of a very extreme physiographic and climatic setting has much to offer as an explanation for the observed effects (Figure 5.11). If one considers the heavy, seasonally concentrated rainfalls, the very high altitudes, huge floods and the 6 mm/yr rate of geological uplift of the region, the list of causative factors becomes much broader than just deforestation (Sharma, 1987):

(a) Steep topography.
(b) Weak geology.
(c) Heavy and intensive rainfall.
(d) Glacial lake outburst floods.
(e) Deforestation.
(f) Seismicity.
(g) Human factors.

Nepal has suffered nine major landslide events since 1819 and fifteen major earthquakes since 1866.

Nevertheless, land-use planning and management is worthwhile both for its intrinsic benefit to Nepalese society and as an ameliorative policy in river basin management. Gilmour *et al.* (1987), undertaking the all-too-rare factual field investigation in Nepal, point out that, while soil does become compacted after deforestation and grazing, only 17 per cent of rainfall exceeds the infiltration capacity of such soils to become flood-producing, erosive surface runoff; they also point to the fact that deep-seated landslides

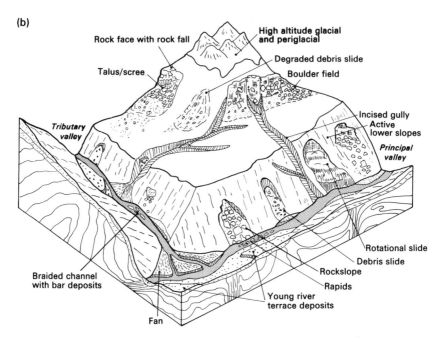

Figure 5.11 The physiographic instability problem of the Himalaya/Ganges
region: montane geomorphology and hazard (modified from Fookes
and Vaughan, 1986)

are promoted by efficient infiltration, not surface flows (though undercutting of slopes by stream incision must also be a factor). Hamilton (1988)
stresses the importance of maintaining a litter layer in forests (Table 5.12).
A recent hydrological analysis of trends in Himalayan rainfalls and river
flows finds no significant increase in flood flows (Hofer, 1993).

Elsewhere in the Himalayas, successful land management schemes in
India (Dhruva Narayana, 1987) have reduced the river impacts of deforestation (Figure 5.12) and the Chipko movement has popularised the
environmental and local political benefits of forest covers.

Water is an important resource within Nepal's own boundaries; 25 per
cent of cultivated land is irrigated and hydro-electric power (as yet only 0.15
per cent developed) perhaps offers the best chance to halt the woodfuel
crisis which threatens the forests of the terai. Sharma (1987) stresses that
within Nepal aggradation has rendered river structures useless, ruling out the
feasibility of dam-building, and has produced a substantial flood threat
which extends, on the Kosi, into India. A review of the scanty sediment
transport data for Nepal by Ramsay (1987) concluded that total erosion
keeps pace with orogenic uplife (1–5 mm/yr), implying natural causes, but
locally deforestation or poor grazing leads to surface erosion, gullying and
shallow landsliding. Efficient delivery to channels sets up positive feedbacks

Table 5.12 Erosion in various tropical moist forest and tree crop systems (ton/ha/year)

Forest system	Erosion		
	Minimal	Median	Maximal
Multistoried tree gardens (4/4)[a]	0.01	0.06	0.14
Natural forests (18/27)	0.03	0.30	6.16
Shifting cultivation, fallow period (6/14)	0.05	0.15	7.40
Forest plantations, undisturbed (14/20)	0.02	0.58	6.20
Tree crops with cover crop/mulch (9/17)	0.10	0.75	5.60
Shifting cultivation, cropping period (7/22)	0.40	2.78	70.50
Taungya cultivation (2/6)	0.63	5.23	17.37
Tree crops, clean-weeded (10/17)	1.20	47.60	182.90
Forest plantations, burned/litter removed (7/7)	5.92	53.40	104.80

Source: From Wiersum (1984)
[a] (x/y) x = number of locations; y = number of treatments or observations.

but Ramsay concludes that the loss to farmers is more serious than the 'gain' to rivers.

The work of Ives and his colleagues in the region has had considerable benefits in the setting of new research agendas in watershed management, particularly relevant to developing mountainous zones, but equally applicable in principle to any large river basin. Ives *et al.* (1987) conclude that the Himalayan region 'is one of extreme complexity from both physical and human points of view, and panacea-type solutions of perceived and illusory problems will likely exacerbate growing human and environmental crises'. Further:

> Problems, though they do involve all sorts of real physical processes, are not defined by those processes. They are defined by people – people, moreover, who are embedded in very different social and cultural contexts and who naturally define their problems very differently. The system, in other words, is a system of mountains and people and we must do everything we can to avoid treating it as if it were two separate slices.
>
> (Ives *et al.*, 1987, p. 335)

Figure 5.13 emphasises this multi-faceted system; Ives and his colleagues demand a scaled, geographical approach to the search for solutions in which the people, the village, the region and international policies are reconciled through the flow of viewpoints and information. A more recent, local, argument for a scaled, downstream approach has been put by Bandyopad-hyay and Gyawali (1994), who list twenty-two large dams proposed for the Himalayan region in an environment which has no basin organisations and, as a result, no integrated information base from which to make the projects sustainable.

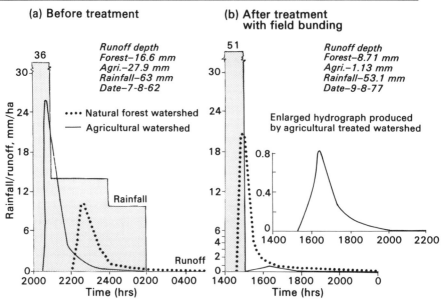

Figure 5.12 Sucessful management of land in the Indian Himalayan region has controlled runoff (Dhruva Narayana, 1987)

5.6.2   The Ganges as an international basin: India dries out, Bangladesh floods

The problem of managing the River Ganges long predates perceptions of environmental degradation in the basin (Crow, 1995). As far as international activity is concerned, this is currently remote from basin-scale environmental management, focusing instead on India's demands for water from the barrages erected across the river at Farakka and Gandak. Following the Farakka diversion of water to benefit navigation to Calcutta, India and Bangladesh set up a Joint Rivers Commission to study flows (Robinson, 1987); a treaty, resembling the Nile Waters Agreement, on 'Sharing of the Ganga Waters at Farakka and on augmenting its flows' was signed in 1977. Crow points out that the navigation (siltation) problems at Calcutta may have originated when the Damodar Valley Corporation (modelled on the Tennessee Valley Authority) closed its dams in 1955. The Joint Rivers Commission was set up to maintain liaison between the participating countries and study flood control and irrigation projects. The barrage has had a number of interconnected and unforeseen impacts on the Ganges downstream, including saline intrusion (Crow, 1995) but, as might be expected, India and Bangladesh disagree on the areas impacted.

The treaty with Nepal has little to do with that country's environmental and population problems, though it may come to have: the Gandak Barrage Treaty allows Nepal rights to withdraw water upstream until India's

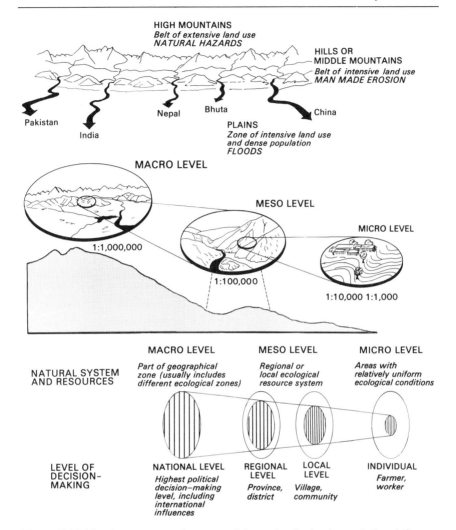

Figure 5.13 Himalayan research: appropriate scales for implementation (after Ives *et al.*, 1987)

irrigation schemes downstream are threatened by shortage; the shortage is then shared! Robinson concludes that the knowledge base for a Ganges environment convention is not available, making the research agenda set by the Ives team yet more urgent if further tensions and damage are to be avoided. Crow's suggestion is that commodity exchanges between India, Nepal, Bangladesh, Bhutan and the international community would be a more hopeful means of reaching agreement on the Ganges. For example, Nepal could exchange hydro-power and water storage with India for finance, engineering expertise and navigation rights.

The situation is undoubtedly complicated by India's apparent unwillingness to confront the shortcomings in its water policy, particularly its approach to irrigation and to drought. Newly independent India in 1947 set about catchment management along the lines of the Tennessee Valley Authority with corporations such as the Damodar Valley Corporation. Dam building and large-scale irrigation have been central planks in national and DVC policy despite the threats posed by floods, seismic activity and, principally, siltation resulting from poor catchment management. India now has 1,554 large dams (Bandyopadhyay, 1987); very little pre-scheme monitoring was carried out and, for example, silt loads were 'fixed' at 1/5,000 of the inflow stream discharge. Chettri and Bowonder (1983) itemise the sorry state of the Nizamsagar Reservoir which has lost 60 per cent of its gross water storage capacity in forty-three years and will be filled by sediment after seventy-two years! The Indian government appears to the authors to be unconcerned about remedial action or monitoring, instead coping with drought emergencies in a way which makes Mathur and Jayal (1993) detect political cunning: the provision of relief measures shows the government to be the true guardian of public welfare. Mathur and Jayal make a claim for catchment management as part of dealing with drought (Bandyopadhyay, 1987, suggests 'The crisis of surface water is mainly due to the collapse of water conservation in the upper catchments'): 'watershed development should ideally be seen as a first step towards the more important objective of an equitable sharing of scarce resources so developed' (p. 51).

The Indian government initiated a National Watershed Development Programme for Rainfed Agriculture in 1986 with the following features (Mathur and Jayal, 1993):

1 Vegetative soil and moisture conservation.
2 Fodder, fuelwood and horticulture development.
3 Forest production.
4 Stabilisation of drainage lines, village water supply.
5 Land use planning; institutional capacity for same.
6 Training.
7 Interagency coordination.
8 Introduction of innovative, simple, low-cost, replicable technologies for soil/water conservation.

The emphasis is on agricultural development and water supply rather than river basin management and it must be said that one of the benefits of such small-scale efforts has been erosion protection, notably in the Himalayan foothills of India where community-based schemes involving earth dams, water conveyance, forest management and improved agricultural productivity are increasing (Figures 5.12 and 5.14; Grewal et al., 1990; Arya et al., 1994). Sivanappan (1995) illustrates the good sense of soil and water management strategies based upon the catchment outline.

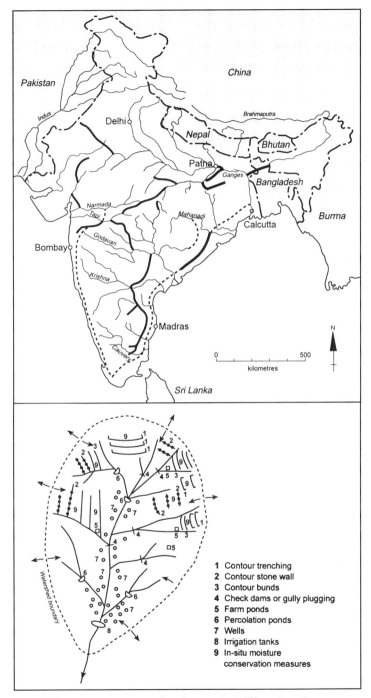

Figure 5.14 Indian water resource development at different scales:
(a) The proposed water grid
(b) Catchment scale soil and water management (after Sivanappan, 1995)

Part of the watershed management debate in 'dry India' will inevitably centre on forests (they are important to the culture and economy of rural India) and in particular on the hydrological and erosional impacts of eucalyptus plantations. Recent intensive research in Karnataka, southern India, suggests that the prodigious rates of evaporation reported for eucalypts in Australia (see Chapter 4) are unlikely in dry regions where soil moisture is limiting (Calder, 1994). Eucalypts use similar amounts of water to indigenous trees, but twice as much as most traditional agricultural crops. In the interests of conserving natural forests, says Calder, the productivity of eucalypts should be modelled so that irrigated patches or rotation cropping could be used to integrate their production with agriculture and other water users. Calder *et al.* (1992) report in detail the research behind these recommendations, which were made to a symposium in Bangalore in 1991. Nevertheless, such penetrating scientific outputs will never reach policy implementation without a greater willingness on the part of government to adopt watershed management. Because the Indian government appears unable to cope with the crisis of drought, however much it affects national development, it will need to keep the resources of the Ganges 'up its sleeve' for the foreseeable future, making international agreement on the river unlikely. As Figure 5.14 shows, there are plans for a 'water grid' in India, a scheme which will need yet more abstraction from the Ganges, supported perhaps by dams built in Nepal.

Meanwhile Bangladesh is reacting to disastrous flooding in 1987 and 1988 (Brammer, 1990a). Floods are an integral part of Bangladeshi life but these were the worst on record, killing 1,200 people and inundating 60 per cent of the nation's land. Up to 15 per cent of the grain production was lost but certain flood-loving rice varieties (Bangladesh has 10,000 varieties, each with specialist water needs – Custers, 1992) performed better. Custers points out that flood management does not mean flood prevention. However, as one journalist put it, 'the world rallied round. It was open season for hydrologists and technologists and what emerged was a dog's breakfast of massive flood defence proposals.' The World Bank has now selected a coordinated scheme of the best proposals, avoiding large dams, but nevertheless having a large impact on people adapted to riparian or island livelihoods. This is clearly important in the light of early conclusions from socially based investigations (e.g. Thompson and Sultana, 1996) which reveal that flood losses can be greater within the flood embankments, there is no stabilisation of livelihoods, and that fishermen and boatmen are disadvantaged by the schemes; only nine of sixty-three flood projects justified themselves economically. The UNDP has stipulated that the flood control strategies adopted should research river morphology and justifies a 'small is beautiful' approach to irrigated food production in Bangladesh (Brammer, 1990b).

The Lower Atrai basin, a tributary of the Ganges, represents a much less structural approach to flood control and has involved a broader approach to the needs of local farmers (Franks, 1994).

5.7 DAMS, ALTERNATIVES AND THE NEED FOR A NEW INTERNATIONAL ORDER

It is likely that the most commonly occurring term in this chapter and the preceding one is 'dam', such is the almost monotonous linearity of the river basin development model evolved in the Western world and exported, like soft drinks and burger bars, across all cultures, economies and stages of development. Perhaps the precursor to further investigation of sustainable use of a river basin ecosystem model in development should therefore be the rather rhetorical question, 'Do we need dams?'

It is rhetorical partly because so many are already there (e.g. 5,609 in the USA and 1,554 in India) but also because many would claim (as do the builders of the Aswan High Dam in Egypt) that they prevent the starvation of millions of humans. We are put in mind of Joseph Krutch's words: 'Technology made large populations possible; large populations make technology indispensable' (Conley, 1989). Thus, even those who have evidence or feelings antagonistic to dams and dam projects must consider that some form of alternative technology is needed; 'do nothing' is only an option when there is a proven, negotiated case for conservation or preservation and even conservation normally needs technology (see Chapter 6).

Clearly, one of the most compelling needs in the technological reappraisal of river basin development is a new look at large dams. Often the central, triumphal feature of a project, the large dam has a growing popular and political protest movement set against it. It is of note that Pakistan, for example, now has a Small Dams Directorate in the North West Frontier Province.

Since the early 1980s the International Dams Newsletter (now International Rivers Network) has coordinated a campaign against such projects, not least because of the large number of people relocated as a result of reservoir filling or of changes to local agriculture and infrastructures.

5.7.1 The case against large dams and associated developments

There are many technical problems in dam-building and dam operation, some of which are covered in Chapter 6, but in relation to development processes it is essential to remember the message of Fahim's study of Aswan.

> Since dams are often associated with development, which I conceive as an ultimate task involving a complex and longitudinal process of concept-making, strategy-building and implementation, people should present a central theme and a basic element in this process. Although capital and technology are important pre-requisites for development, people, in my view, represent a far more significant and non-depleting resource. A

basic premise in this volume is that economic benefits and human welfare should constitute part and parcel of the development process of water projects. Otherwise, dams will result in situations of 'growth without development'.

(Fahim, 1981, p. xv)

Goldsmith and Hildyard (1984) begin their litany of criticism of large dams with a poem by Kenneth Boulding, reproduced here instead of a table of environmental impacts!

A BALLAD OF ECOLOGICAL AWARENESS

The cost of building dams is always underestimated
There's erosion of the delta that the river has created,
There's fertile soil below the dam that's likely to be looted,
And the tangled mat of forest that has got to be uprooted.

There's the breaking up of cultures with old haunts and habits loss,
There's the education program that just doesn't come across,
And the wasted fruits of progress that are seldom much enjoyed
By expelled subsistence farmers who are urban unemployed.

There's disappointing yield of fish, beyond the first explosion;
There's silting up, and drawing down, and watershed erosion.
Above the dam the water's lost by sheer evaporation;
Below, the river scours, and suffers dangerous alteration.

For engineers, however good, are likely to be guilty
Of quietly forgetting that a river can be silty,
While the irrigation people too are frequently forgetting
That water poured upon the land is likely to be wetting.

Then the water in the lake, and what the lake releases,
Is crawling with infected snails and water-borne diseases.
There's a hideous locust breeding ground when water level's low,
And a million ecologic facts we really do not know.

There are benefits, of course, which may be countable, but which
Have a tendency to fall into the pockets of the rich,
While the costs are apt to fall upon the shoulders of the poor.
So cost-benefit analysis is nearly always sure
To justify the building of a solid concrete fact,
While the Ecologic Truth is left behind in the Abstract.

Kenneth E. Boulding
From T. Farvar and J. Milton [1973] *The Careless Technology*
(Tom Stacey, London)

After their extensive, and as yet incomplete, review Goldsmith and Hildyard claim to demonstrate:

(a) How little of the extra food grown through irrigation schemes ever reaches those who need it most; how, in the long run, those irrigation schemes are turning vast areas of fertile land into salt-encrusted deserts; and how, too, the industry powered by dams is further undermining food supplies through pollution and the destruction of agricultural land.

(b) How millions of people have been uprooted from their homes to make way for the reservoirs of large dams; how their social lives have been shattered and their cultures destroyed; and how, also, their health has been jeopardised by the water-borne diseases introduced by those reservoirs and their associated irrigation works.

(c) How dams are now suspected of triggering earthquakes; how they have failed to control floods and have actually served to increase the severity of flood damage; and how, in many instances, they have reduced the quality of drinking water for hundreds of millions of people.

(d) How the real beneficiaries of large-scale dams and water development schemes have invariably been large multi-national companies, the urban elites of the Third World, and the politicians who commissioned the projects in the first place.

Goldsmith and Hildyard's (1986) case studies involve thirty-one contributions from North America, Europe, the Soviet Union, Africa, South America, India and other Asian countries. One of these, the Narmada Valley Project (Kalpavriksh and the Hindu College Nature Club, 1986), involving two major reservoirs and up to 3,000 smaller structures, has now been the subject of an individual book of criticism (Alvares and Billorey, 1988). The key to the Narmada controversy, according to Alvares and Billorey, was that a 'majestic and sacred' river would be equipped with a plethora of dams which

> would, in addition to generating irreversible environmental changes, also uproot over a million people, including a large number of tribals, and submerge a total of 350,000 hectares of forest lands and 200,000 hectares of cultivated land.
>
> (Alvares and Billorey, 1988, p. 10)

Gita Mehta (1993) poetically explains the cultural significance of the Narmada:

> You grace the Earth with your presence
> The devout call you Kripa: Grace itself.
> You cleanse the Earth of its impurities
> The devout call you Surasa: The holy soul.
> You leap through the Earth like a dancing deer
> The devout call you Dewa: The leaping one.
> But Shiva called you Delight and Laughing
> Named you Narmada.

Alvares and Billorey provide detailed scrutinies of 'the human tragedy' as well as of the adverse environmental impacts (notably sedimentation) and (on their calculation) minimal benefit/cost ratios of the project. The alternatives they propose are:

(a) Reducing the height of dams.
(b) Lift irrigation – direct from rivers.
(c) Small-scale, community-run dams.

In March 1993 the World Bank announced that it was withdrawing its funding from the Sardar Sarovar Dam, centrepiece of the scheme, but the Indian government immediately said that it would proceed alone or find other funding agencies after another review of options. This followed a threat by villagers to drown themselves in the rising waters behind the dam and a hunger strike by two members of the Save the Narmada movement.

Figure 5.15 and Table 5.13 point up the huge scale of the Narmada, a project in the tradition of the Tennessee Valley Authority but without its justification and having, if anything, the reverse impact.

Dislocations of traditional societies on such a scale have been relatively common in development schemes involving dam construction. Barrow (1987) tabulates more than twenty such schemes with resettlement costs ranging from $3.6 million to $100.2 million. The most recent concern of anti-dams activists in terms of dislocation has been the '2010' Plan of the Brazilian government, involving thirty-one large dams on tributaries of the Amazon (Cummings, 1990). The Indian tribal people lose land both to permanent flooding above the dams and by lack of seasonal flooding below them.

Beauclerk et al. (1988) have listed the following characteristics of indigenous peoples:

(a) They make sustainable use of resources.
(b) Their land is held in common.
(c) They have relatively unstratified economies; wealth is generally even.

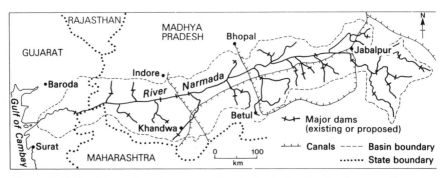

Figure 5.15 The Narmada scheme, western India, indicating the heavy reliance on dam construction (after Kalpavriksh, 1986)

Table 5.13 Villages and population to be submerged by Narmada Dams

Project	State	District	Villages	Population
Sardar Sarovar	Gujarat	Bharuch	19	
		Baroda		30,000
	Maharashtra	Dhulia	33	
	Madhya Pradesh	Dhar	80	
		Jhabua	26	70,000
		West Nimar	76	
Narmada Sagar	Madhya Pradesh	Khandwa	167	
		Dewas	39	170,000
		Hoshangabad	48	
Omkareshwar			27	13,000
Maheshwar			58	14,000
TOTAL			573	297,000
			Nearly	300,000

Source: Alvares and Billorey (1988)

(d) Their societies are rooted in kinship.

(e) They are highly vulnerable.

In the volume edited by Carruthers (1983), officials from major financing agencies such as the World Bank (which makes 38 per cent of agricultural sector loans to irrigation projects and has loaned over $10 billion to schemes since 1948) stress the need for all the improvements advocated in this chapter: participation, environmental appraisal, realistic benefit/cost accounts, long planning cycles, river basin scales, etc. In addition, Carruthers himself pleads for better assessment of whether irrigated or improved rainfed agriculture is more sustainable and for much more investment in operations and maintenance of existing schemes. The latter statement tacitly admits that benefit/costs have been poorly evaluated in the longer term: development 'lift-off' has been illusory. The former needs little technological elaboration – it simply builds on the lessons of traditional land and water use (Gilbertson, 1986). The long-term and broader scale considerations needed for investment in irrigation schemes are also stressed by Welbank (1978), who prefixes his reflections of two major schemes in the Sudan with the words, 'on the day of first water, dramatic changes take place in the project area'. Little thought and no investment, he claims, is available for *settlement* lay-out and planning, *social welfare* and the *multiplier effects* of the anticipated regional growth.

5.7.2 Respect for tradition: 'bottom-up' water development in drylands

Much of the fascination for dam building programmes in the developing world derives from the need for irrigated agriculture and power generation as part of general national development; we deal with small-scale irrigation techniques in Chapter 6 but here consider briefly the techniques employed in drylands to make optimum use of what precipitation falls, techniques employed in antiquity and by indigenous peoples today. Gilbertson (1986) provides a general introduction to 'floodwater farming'; traditional forms of 'dry farming' cultivation are also worthy of encouragement by NGO aid or government extension services. The 'Khadin' system of cultivation in India uses basins of internal drainage in undulating semi-arid land to focus both surface and groundwater supplies; the ratio of catchment to the 'plain of accumulation' (Figure 5.16) is 11:1 (Tewari, 1988). Wheat and chickpeas are grown without irrigation after the summer rains have infiltrated the basin (also recharging wells for domestic supply and for livestock which manure the basin). Khadin farming may date back to 3000 BC but currently the Indian government has revived 500 Khadin farms under the Drought Prone Areas Programme.

Perhaps surprisingly a similar venture is being attempted around Tucson, Arizona, where groundwater irrigated agriculture is being 'retired' to be replaced by *water harvesting* via catchment basins (Karpiscak *et al.*, 1984). Runoff from the non-agricultural area is increased by removing weeds, compacting soil and adding salt to decrease infiltration. The resulting 'agrisystem' is at its most efficient in small units of 0.2 hectares. Runoff from catchment strips is concentrated in channels and flows into a sump. Annual average rainfall does not exceed 250 mm but evaporation reaches 2,860 mm; consequently losses from storage are potentially ruinous but are curtailed by ingenious techniques such as the use of black film canisters

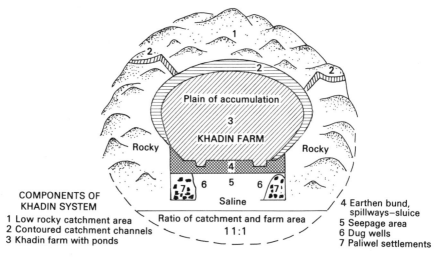

Figure 5.16 Traditional Indian dryland farming strategy: the Khadin irrigation system (after Tewari, 1988)

which float on the reservoirs. High-yield, high-value and drought-tolerant crops such as grapes, jojoba, olives and pines are grown.

The aid agencies are coming to appreciate the lessons learned, mainly by NGOs operating on small budgets, that gaining acceptance for techniques is an innately socio-political process and that without it failure is likely however much capital is poured in. Conroy and Litvinoff (1988) describe, with case studies, 'sustainable rural livelihoods', including catchment protection in Indonesia and river basin development in Ecuador. These authors emphasise that sustainability entails 'staying with' projects well into the execution phase to avoid 'unravelling' of integration – it is at such a stage that regional development authorities can perform best.

Putting yet another slant on the sustainable water development conundrum, Dankelman and Davidson (1988) advocate a special role for women, 'the invisible water managers'. Women were mentioned specifically by the UN's Drinking Water Supply and Sanitation Decade. States and agencies were urged to 'promote full participation of women in the planning, implementation and application of technology for water supply projects'. Women in villages know where to find water, how to judge its quality and how to cope with shortages; however, much of their day needed for food production may be occupied in fetching water. Examples of women's schemes for local water development quoted by Dankelman and Davidson include a women's dam in Burkina Faso and two Kenyan schemes to improve water supplies. Chauhan et al. (1983) also develop the crucial role of women in community-based water supply and sanitation schemes.

There are supreme challenges in the management of development through the vehicle of improved water use. As Greenwell (1978) pointed out to us, our species originated on arid plains and has prospered, developing culturally through the hardships of dryland life. 'In fact, our very existence may be due to the now much-feared process of desertification' (p. 10). Greenwell sees many of the features of human life – bipedalism, binocular vision, birth of poorly developed infants, and communication – as being related to resource hardship.

Modern resource hardship has, however, other complications, of nation, politics, war and above all human numbers. It also has 'developmentalism'. Robert Lee sounds a warning about our perceptions of the development process:

> the notion of reaching a universal model of economic, social and political change has failed. This failure, partially acknowledged in the West, has not yet gained full recognition in the Third World, which limps along in pursuit of Western goals to which the West no longer subscribes.
>
> (quoted in Fahim, 1981, p. 1)

The truth of his statement for the West is brought home by Eckerberg's (1990) study of Swedish forestry in which environmental protection is achieved mainly because non-professionals want it.

Table 5.14 International river basin cooperation

Basin (and/or project)	Organisation	Sign	Countries concerned	Date of agreement
Senegal	Organisation pour la mise en valeur du fleuve Senegal	OMVS	Mali, Mauretania, Senegal	11.5.72
Gambia	Organisation pour la mise en valeur du fleuve Gambia	OMVS	Gambia, Senegal	19.4.67
Niger	Commission du fleuve Niger		Benin, Cameroon, Chad, Guinea, Ivory Coast, Mali, Niger, Nigeria, Burkina Faso	20.10.63
Lake Chad	Commission du Bassin du lac Chad		Cameroon, Niger, Nigeria, Chad	22.5.64
Kagera	Organisation for the Management and Development of the Kagera River Basin		Tanzania, Rwanda, Burundi	24.8.77
Mano	Mano River Union		Liberia, Sierra Leone	3.10.73
Nile	Permanent Joint Commission for Nile Waters		Egypt, Sudan	8.11.59
Plata	Comité Intergubernamental Coordinador de los Paises de la Cuenca del Plata	CIC	Argentina, Bolivia, Brazil, Paraguay, Uruguay	23.4.69
Plata–Paraná (Itaipu)	Itaipu		Brazil, Paraguay	26.4.73
Plata–Paraná	Comisión Técnica Mixta Argentino–Paraguaya del Río Paraná	COMIP	Argentina, Paraguay	16.6.71
Plata–Paraná	Ente Binaciónal Yacireta	EBY	Argentina, Paraguay	23.1.58
Plata–Uruguay (Salto Grande)	Comisión Técnica Mixta del Salto Grande	CTM	Argentina Paraguay	30.12.46
Plata–Uruguay	Comisión Administradora del Río Uruguay	CARU	Argentina, Uruguay	7.4.61
Puyango–Tumbes	Comisión Mixta Peruano–Equatoriana para el aprovechamiento de las cuencas hidrográficas binacionales Puyango–Tumbes y Catamayo–Chira		Equador, Peru	27.9.71

All water courses near border	Comisión internacional de Límites y Aguas		Guatemala, Mexico	21.12.61
All water courses near border	Comisión internacional de Límites y Aguas	IBWC	Mexico, USA	1.3.1889
All water courses near border	International Joint Commission	IJC	Canada, USA	11.1.09
Lower Mekong	Comité pour la Coordination des Etudes sur le Bassin inférieur du Mekong		Laos, Thailand, Vietnam	31.10.57
Indus	Permanent Indus Commission		India, Pakistan	19.9.60
All water courses near border	Indo-Bangladesh Joint Rivers Commission		India, Bangladesh	24.11.72
Ganges	Mixed Commission for the Ganges downstream of Farakka		India, Bangladesh	5.11.77
Rhine	Commission centrale pour la navigation du Rhin		Belgium, France, Netherlands, Germany, Switzerland	17.10.1868
Rhine	Commission internationale pour la protection du Rhin contre la pollution		Germany, France, Luxembourg, Netherlands, EC	29.4.63
Danube	Commission du Danube		Austria, Bulgaria, Hungary, Romania, Czechoslovakia, Ukraine, USSR, Yugoslavia	18.8.48
All water courses near border	Finnish−Swedish Frontier Rivers Commission		Finland, Sweden	16.9.71
All water courses near border	Comisión Hispano−Portuguesa para la Reglamentación del Uso y Desarrollo de las Aguas Fronterizas sobre los Tramos Comunes a los dos países		Spain, Portugal	11.8.27
Vardar−Axios	Joint Greek−Yugoslav Commission for the development ot the Vardar−Axios		Greece, Yugoslavia	18.6.59

Source: Translated by Marchand and Toornstra (1986)

5.7.3   Avoiding 'water wars' and sustaining ecosystems: international river basin management

Marchand and Toornstra recognise that their ecosystem basis for the technological choices in river-basin development often needs to be applied across national boundaries, particularly in the developing world. Table 5.14 shows their summary of the existing institutional structures for such diplomacy. Carroll (1988) is, however, sceptical about such bodies, citing the failure of the International Joint Commission to control pollution along the USA/Canada border as an example of failure to set and achieve the real objectives.

> I suggest we must start to measure such entities as boundary waters treaties and the institutions they spawn not on the basis of how much they seek to accomplish nor in how much they claim, but, in the final analysis, in the health of the environment they are meant to protect, and that means the long-term economic and political environment as well as the natural ecosystem – for they are all ultimately one.
>
> (Carroll, 1988, p. 276)

There are signs at present that 'political environment' can include the threat of 'water wars', particularly prevalent in the Middle East. During 1990, for example, Turkey closed the Ataturk Dam across the Euphrates, a culmination of the 'Pride of Turkey' project on both the Tigris and Euphrates designed to boost electricity supplies by 70 per cent and to irrigate 1.6 million hectares of land; it is self-funded because of its political sensitivity (Hellier, 1990). Syria, next in line for the river's waters, anticipates flow reductions of 40–70 per cent volume; already electricity generation from its Assad Reservoir is reduced to one-third capacity (a fault also blamed on Russian technology). Iraq, next in line for the waters, fears not only reduced quantity but also reduced quality, since some of the flow reaching its own fields will have been salinised twice by upstream irrigation. With its own irrigation systems damaged by the Gulf War the prospects for resource tensions in the region clearly extend from oil to water in Iraq. Elsewhere in the Middle East, the abundance of energy reserves to power pumping has resulted in an over-reliance on groundwater reserves. The aquifers being 'mined' were recharged during the Pleistocene glaciations further north – this is 'fossil' water. With a current average annual rainfall of 90 mm, Saudi Arabia, the largest arid nation in the region, faces a future in which only demand management and recycling can avoid inevitable decline in agricultural production (Al-Ibrahim, 1991). Rowley (1990) describes how, within the general problems posed by Middle East aridity, competition occurs for groundwater resources on the West Bank. Since Israeli settlers entered the region in 1967 their numbers have expanded to 68,000 – representing a considerable water-supply burden. However, much of the water pumped from the region is destined for Israel proper and there are already signs of over-exploitation. The shallower wells and funnel systems of the Arab population are being depleted by the superior Israeli technology. Abu-Maila (1991) reports on similar problems caused by population growth and the establishment of Israeli settlements in the Gaza Strip; over-exploitation of fossil groundwater is essential because of the negative water balance (306 mm precipitation, 1,800 mm evaporation). Water quality is also declining rapidly and with an anticipated resource deficit of 200 Mm^3 by the year 2000 the only options appear to be desalination, recycled sewage effluent or import of supplies from elsewhere in Israel, a prospect which must be considered in the context of Israel's relationship with the Palestinians and with its neighbours (see Chapter 7). Conflicts of interest between Jewish and Arab rural settlements within Israel reduce the effectiveness of the Israeli Drainage Districts – the nearest approximation to river basin authorities. Kliot (1986) describes how the Drainage Districts expanded their interests from purely drainage and flooding to soil conservation, water supply and some aspects of water quality. However, incomplete spatial coverage and a wide national variation in their functions blunt their influence on the crucial need to conserve water and protect the environment throughout Israel. Nevertheless, Israel has made considerable progress towards irrigation efficiency (Chapter 6) and is anticipating that 80 per cent of all wastewater will be recycled for agricultural use by 2000 AD (Shuval, 1987).

Falkenmark (1989) uses the Turkey–Syria–Iraq grouping as one of three potential water conflicts in Middle East hydropolitics, the others being on the Nile and the Jordan. There are, however, some optimistic signs. Turkey has proposed a 'peace pipeline' for water, supplied from new sites in central Turkey, via Syria, Jordan, Israel and Saudi Arabia to the Gulf, a formidable distance of 3,000 kilometres but one emulated by President Quaddafi's 'Great Man-Made River' in Libya (1,900 km). The peace pipeline could be completed in 8–10 years and could eventually serve 6–9 million people; it would not, of course, remove the potential for conflict – indeed it might strengthen the hand of terrorists. Pipelines once proved to be vulnerable to attack within the UK (e.g. Wales 1969).

Falkenmark (1989) actually doubts whether technology transfer from temperate zones unused to water shortage will ever meet the needs of the two-thirds of Africa now facing serious water scarcity. She develops a risk spiral of water demand now threatening Africa and calls for:

(a) Long-term national water master-plans.
(b) Water strategies for appropriate socio-economic development within supply limits.
(c) Top-level autonomous authorities for integrated land and water management.

Table 5.15 Tropical water resource development: problems, causes and possible relief

Problems	Causes	Possible relief
Data incomplete/uneven, i.e. geographically 'patchy' or lacks depth or continuity or collection has not progressed long enough to establish trends	Research largely in hands of expatriates	Train and encourage indigenous consultants
	Much knowledge/monitoring techniques based on temperate latitudes experience	Promote tropical research/monitoring
	Too few indigenous experts	Improved training/incentives encourage experts to stay in own country
	Useful research 'unfashionable', unattractive to researchers trained in the West	Promote vital research fields to discourage 'brain drain' to high-income nations
	Little exchange of information between (often environmentally similar) developing nations	Encourage exchange of ideas, data, personnel and promote shared research between developing nations
	Difficult environment, communications, political instability	
	Data not widely available (restricted by ministries/armed forces)	Better publicity; encourage flow of information between countries
	Data not standardised; difficult to compare between regions	Adopt internationally agreed standards; compile international registers and data banks

Inadequate use of data	Data not understood by planners, decision-makers, local people	Involve planners and public in research; publish data in simplified form to interest people
	Expedience, political or other pressures overrules sound advice of ecologists/planners	Improve public accountability; establish large powerful authorities to control water resources to avoid intimidation by other institutions
	Research too sectoral	Promote interdisciplinary research
Data unsatisfactory when applied to water resources development	Data inadequate or irrelevant to problem requiring solution	Adopt 'problem-orientated' research
	Data too site-specific; difficult to transfer experience between projects	Establish models, checklists, suitable environmental impact assessment procedures

Source: Barrow (1987)

(d) Possible use of local rain by integrating soil and water conservation on a catchment basis.

(e) Population control in relation to water resources.

The most pressing need is for developing indigenous expertise by prioritising national programmes of research, education and training.

Further exploitation of developed-world philosophies and technologies must clearly await a more introspective review of these, a process only recently begun (Gunnerson and Kalbermatten, 1978; Kruse *et al.*, 1982; Green and Eiker, 1983). Reforms in the education and training of engineers are also necessary and this is a theme returned to in Chapter 9.

5.8   DEVELOPMENT AND RIVERS: BROAD TRENDS

At the beginning of Chapter 4 we addressed two almost rhetorical questions about the size of large water-based development schemes, particularly those involving dams, and about the need to respect local variations in 'appropriate technology'. The two chapters can be viewed as a series of disparate national and regional case studies, hardly helping to answer these big questions. However, there seem to be certain threads in the material which indicate scenarios for better management of river basins.

The sequence of priorities listed by Table 4.1 is broadly accurate, the most challenging aspect being that, while issues such as conservation, recreation and pollution are the 'luxury' issues of a developed world, the care and attention they bespeak to river management practice brings benefits for all forms of water resource management. For example, while the USA, Canada, New Zealand and the UK hastily develop control strategies for non-point pollution sources, such as those for nitrate (not yet proven as a toxic threat), they indirectly incorporate measures which protect against soil erosion and encourage water recycling.

At earlier stages of national development these same nations made the sorts of gross errors for which we now criticise the least developed countries; indeed, in some cases the same technicians have been exported, their failings intact! In an environment of rapidly increasing population and urbanisation the developed nations felled their forests and let generations die from cholera; under the same conditions of growth but under the additional stress of dryland conditions it is likely to remain normal for large water development schemes to dominate, proportional in scale to the management of international finance. Since the monopoly of technical skill at such a scale is retained by expatriate experts we will rely very heavily on their steep learning curve to make such schemes more sensitive.

They appear to need to learn three important lessons:

(a) That the physical sensitivity of river basins in drylands to development pressures is high, yet the resilience of adapted ecosystems and indigenous cultures copes.

(b) That a requirement of strategies based on sustainability makes it necessary to anticipate the multiplier effects of development on the river system, e.g. future pollution patterns.

(c) That the indigenous population, while unlikely to participate in the design stage of a scheme, from then on effectively control the future of that design.

Within these broad limits large schemes, including dam schemes, are likely to continue. The World Bank, responsible for funding, has recently introduced new guidelines for water resources management in developing countries (World Bank, 1993). They include conservation of water and pollution control for industry, efficient and accessible delivery of water services, modernised irrigation practices (especially for small farmers), environmental protection and the alleviation of poverty. The Bank will also 'help countries improve the management of shared international water resources by, for example, supporting the analysis of development opportunities forgone because of international water disputes' (p. 19).

However, a 'second front', mainly operated by non-government organisations, will deliberately exploit a knowledge of, and sympathy for, the local talents of indigenous peoples for environmental management – even when that environment is changing. These organisations can combine physical and social improvements much more efficiently at the local level. Under the title 'Social aspects of sustainable dryland management' Stiles (1995) has brought together those elements which are essential to the social development of those communities upon whom we depend to protect the environment. Evers (1995) lists the guiding principles as 'support, empower and enable' while Hausler criticises the entire Western model of development, the backdrop to large dam schemes and their universal, if inappropriate, application:

At the core of this process lies a universalist economistic framework of the human/nature relation ... The universal introduction of the Western industrial, export-led, economic growth model of development, despite some positive changes, has caused considerable damage to local life styles, cultures and – very visibly – to the environment.

(p. 181)

Chapter 6

Technical issues in river basin management

Despite the power of holistic science to open up issues to interdisciplinary analysis, and despite the impression perhaps created in the book thus far that the role of the engineer should now be minimised in river management (see also Chapter 9), it is likely that *technocentric* solutions will remain influential, if not dominant, in a world of teeming population and widespread famine and disease.

We therefore now return to the technical aspects of the river basin system, having seen from Chapters 4 and 5 the dire consequences of the unsympathetic application of present technologies. Improving 'the sympathy' is one part of the research agenda; we also need to improve the technology of river basin management itself in order to stress the need for attention to the larger-scale and longer-term environmental elements, which has proved inadequate in both developed and developing worlds. Inevitably such improvements are brought about by 'normal science' – the conventional, experimental route championed by Francis Bacon and orientated to 'the relief of man's estate'; also inevitable for the foreseeable future is that research results will be applied by engineers – agricultural, civil and environmental. Nevertheless, at each stage of this chapter the 'people' issues are never far from the technical ones.

We examine in this chapter the technical problems facing management; it has been necessary to be very selective – omitting, for example, coastal impacts of river basin development. We have included restoration of rivers and wetlands to emphasise the overall ecosystem approach and as a warning that those developed nations which have damaged the spontaneous regulators of their river basins are now making large investments in reinstating them.

The first area of technical facilitation required by sustainable development of a basin's resources is that of controlling soil loss by erosion and salinisation. It is the crudest paradox that in order to develop plenty we tend to squander sufficient and, while erosion is a natural process, we must come to understand it well enough to work in balance with it. Rainfed agriculture will, in many circumstances, be more sustainable in poor economies than large-scale irrigation, yet the unreliability and intensity of that rainfall make it a hazard as well as a resource in many countries. Another natural process, when achieved by lakes, floodplains and wetlands, is water storage; however, dam/reservoir construction, done to extend natural storage capacity, produces a hiatus in the river system and associated developments such as land drainage, flood control or irrigation can in fact reduce other parts of that capacity. Furthermore, there is a direct loss of capacity if a dam impoundment slowly fills with sediment, possibly representing the soil grains lost by erosion upstream. Soils are also lost when they are salinised by unsophisticated irrigation systems. It is unlikely that we have witnessed our last major dam scheme despite the protests illustrated in Chapter 5. Neither have we seen our last socially and environmentally damaging large-scale irrigation scheme. However, by treating river basins in the ecosystem framework we have suggested that we can again work with the processes of nature, producing net, long-term gains rather than short-term gains ultimately exceeded by losses. One need not be a transcendental philosopher to hold this view, merely a well-trained engineer (see Chapter 9). The need for vision enters, however, in the field of conserving or restoring Marchand and Toornstra's spontaneous regulation functions of the channel, its corridor, valley and associated wetlands. Definitions are important here and we reflect below on the differences between restoration, rehabilitation and enhancement of these regulators.

Finally, these important elements of any river basin development scheme must be refined to the point where they can be *interactively managed* through time to reflect the changing needs of society and the environment. On the latter count, therefore, the likely impact of climatic changes must be researched. This presents major problems to 'normal science', requiring highly public research ventures conducted urgently, without much prospect of a certain answer. A system which is destabilised physically (one may say geomorphologically) is much more difficult to restore than a chemically polluted system. Once again therefore, as in Chapter 3, we will make relatively limited reference to water pollution as a technical problem in basin management. Restoration of unpolluted conditions is a challenge to law and to economics as much as it is to science; restoration of a physically degraded river system is a daunting task and the requirement to do so should be avoided by anticipation and good science.

6.1   SOIL EROSION

The erosion of soils by running water (we do not here deal with wind erosion) is not in itself a problem; soils are produced from a bedrock or drift mineral base by weathering (and the incorporation of organic material) and are then eroded as part of the long-term evolution of landscapes. Soil erosion only becomes a problem when its rate is accelerated above that of other landscape development processes (see Table 6.1) – notably weathering – because it becomes visible; it becomes a river management problem when it constrains agricultural production and leads to river and reservoir sedimentation. Erosion is not the only way in which exploitation of river basins can damage soil. Mismanagement of this thin, interface system between lithosphere, biosphere and atmosphere takes many forms as Figure 6.1 illustrates.

Soil erosion is often quoted as one of the clearest symptoms of global environmental abuse by humans; as with desertification there are many opportunities for even the well-intended to misunderstand the extent and significance of erosion. Lal (1993) lists some of the key statistics – erosion affects 24 per cent of the planet's inhabited land area; 1,094 Mha of land are impacted by water erosion and 548 Mha by wind erosion. Because soil erosion rates are accelerated most commonly by human intervention in the soil formation/denudation process its identification and control involves the social as well as the physical sciences; Hallsworth (1987) describes the 'anatomy, physiology and psychology' of erosion while a book by Blaikie (1985) explores the 'political economy' of soil erosion in developing countries. Boardman (1990) writes of the 'costs, attitudes and policies' in soil erosion control.

6.1.1   The physical processes of erosion: identifying controls

The need to control soil loss by interventionist measures, both agricultural and hydrological, was first perceived in terms of national crisis by the USA

Table 6.1 Recommended values for maximum permissible soil loss

	$kg\ m^{-2}\ y^{-1}$
Meso-scale (e.g. field level)	
Deep fertile loamy soils, values used in the Mid-West of USA	0.6–1.1
Thin, highly erodible soils	0.2–0.5
Very deep loamy soils derived from volcanic deposits, e.g. in Kenya	1.3–1.5
Soil depths: 0–25 cm	0.2
25–50 cm	0.2–0.5
50–100 cm	0.5–0.7
100–150 cm	0.7–0.9
over 150 cm	1.1
Probable realistic value for very erodible areas, e.g. mountains in the tropics	2.5
Macro-scale (e.g. drainage basins)	0.2
Micro-scale (e.g. construction sites)	2.5

Source: Morgan (1980)

Use of such rates under a 'permissible' heading implies a standardisation of practice; this, however, is misleading and the fixing of indicator rates is highly controversial.

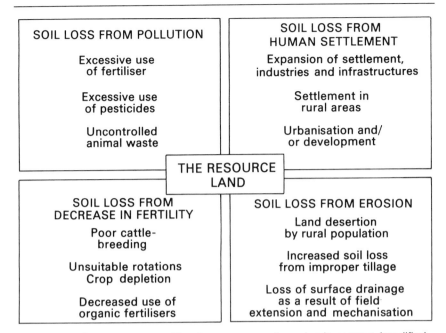

Figure 6.1 Mismanagement of land resources: soil erosion in context (modified after Guerrieri and Vianello, 1990)

in the early 1930s following extremes which had created dust-bowls and badlands from essential pioneer farmland. This is not to deny that other civilisations have faced problems with the soil resource, as Seymour and Girardet (1986) and Hillel (1992) draw out in stimulating historical reviews. Hillel points out that terracing began in the second millennium BC – an important feature of the Iron Age expansion of agriculture into hill country; terracing increased the vulnerability of farming and erosion was still a major problem at the time the *Iliad* included these lines:

> Many a hillside do the torrents furrow deeply, and down to the dark sea they rush headlong from the mountains with a mighty roar and the tilled fields of men are wasted.

The measures adopted in the USA during the modern era have, however, been exported across the globe, rightly or wrongly coming to dominate over the local, traditional measures practised by traditional agrarian communities. Equations to predict soil loss were developed on a rational basis to incorporate soil, slope and rainfall factors during the 1930s and 1940s. However, these tended to be of only local or regional relevance and in the 1950s field research at the plot scale was combined to produce the USLE (Universal Soil Loss Equation), originally described by Wischmeier and Smith (1965).

The USLE is phrased as a simple multiplicative expression predicting a mass of soil removed from a unit area per annum, based on causative factors (see Figure 6.2) identified and measured at thiry-six locations in twenty-one states in the USA (totalling >10,000 plot-years of data).

$$A = RKLSCP$$
where A = soil loss, $kg.m^{-2}$
R = rainfall erosivity
K = soil erodibility
L = slope length
S = slope gradient
C = cropping management factor
P = erosion control practice

The USLE is codified in a simple fashion, with nomographs to ease calculation of the basic factors from minimal data collection in the field. Clearly, care and experience in applications are essential and as shortcomings have been identified in aspects of its operation further research has 'filled in the gaps'. The USLE has become much more 'universal' in a geographical sense than originally intended and is now applied well beyond its original range but has the great advantage of rationality and simplicity; scientists who may arrogantly claim that USLE is unsophisticated and outdated have seldom seen it in practice, supporting essential conservation schemes.

The USLE may be used to:

(a) Predict annual soil loss under specific land uses.
(b) Guide the selection of cropping and management options.
(c) Predict the effects of changing land use or management.

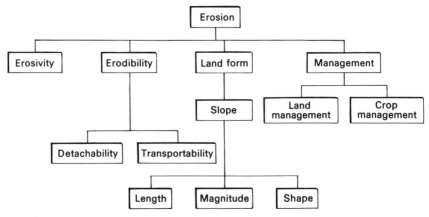

Figure 6.2 Controlling factors built into the calculations of the Universal Soil Loss Equation (USLE)

(d) Determine conservation practice.
(e) Estimate losses from non-agricultural developments (e.g. mine spoil).

Numerous authors have reviewed the refinements produced by the considerable research support that the USLE has achieved (e.g. Mitchell and Bubenzer, 1980). Two main components of research improvements have been made:

(a) Improvements to the rationality of the process mechanisms of the equation or to the way that it agglomerates processes.
(b) Improvements to the applicability of the equation in specific (often tropical) environments.

As it stands the Equation performs best for medium-textured soils on 3–18 per cent slopes less than 122 metres in length. Improvements have, for example, sought to move from plot scale to field scale and thence to catchments; predictions for shorter time periods under variable input conditions are also desirable and have resulted in alternative approaches through dynamic models (see Morgan *et al.*, 1984). Morgan (1995) describes three empirical (data-based) models and four deterministic (physically based) models. Such models seek to reproduce the process mechanisms of the soil detachment and transport processes (Figure 6.3).

Keeping strictly within the USLE structure, major research improvements have been made to extend the range of calibrations for slope form, for soil erodibility and, especially, for rainfall erosivity (to include refined estimates of raindrop size and the role of plant canopies – which tend to increase the kinetic energy of drops). A further problem has been to address *gully erosion* rather than *sheet erosion*. Rational equations, similar to the USLE, are available but tend to predict the rate of growth of gullying rather than initiation of this damaging process.

Renard *et al.* (1994) describe the Revised Universal Soil Loss Equation (RUSLE). It retains the same nomenclature and format of the original but has incorporated the following broad improvements. The 'R' term now incorporates more climatic information, making the equation more 'transportable' to other regions of the world. The 'K' term can be varied seasonally and includes soils containing rock fragments (common in drylands). Renard *et al.* note that, while 'L' has attracted controversy its effect on outcomes is far less significant than 'S', which has been improved to incorporate complex slope profiles and plans. The conservation term 'C' has been extended to include surface soil roughness and ground cover plants as well as crop cover. 'P' is still the least reliable term because of the difficulty of incorporating farmer variability in practice. The US Department of Agriculture is also pioneering WEPP (Water Erosion Prediction Project), a model which incorporates the small-scale geometry of eroding fields, i.e. rills and inter-rill areas; it also extends from the plot and field

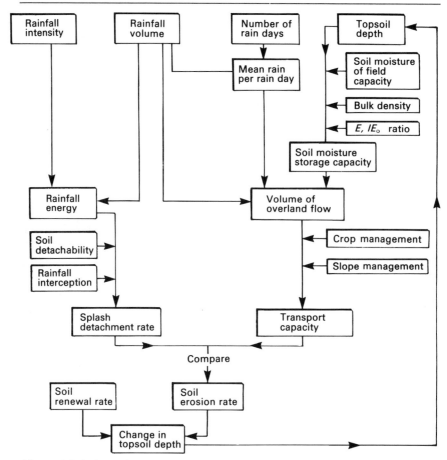

Figure 6.3 A dynamic model of soil erosion (Morgan *et al.*, 1984)

scale to the watershed via hillslope, channel, impoundment (terraces, stock tanks) and irrigation modules. Nearing *et al.* (1994) describe the paradox that the greatest need for improvement in WEPP is now in modelling *deposition* (mirroring the geomorphologist's view that much eroded material becomes colluvium before it becomes alluvium; see below).

As examples of the often dire need to extend and apply the USLE to new conditions we may select the work described by Hurni (1983), who has improved the rainfall erosivity and slope components in Ethiopia, and Harper (1988), who additionally improved the C, K and P components for North Thailand.

Hurni's studies (Figure 6.4) contribute a basic statement of erosion as a long-term resource problem facing rainfed agriculture in the Ethiopian Highlands. Figure 6.4 compares the timescales of soil formation with those of its erosional loss; rates of loss should not exceed those of production – a

Plate 6.1 Sheet (mid-foreground) and gully erosion, Ethiopian Plateau (photo M. D. Newson)

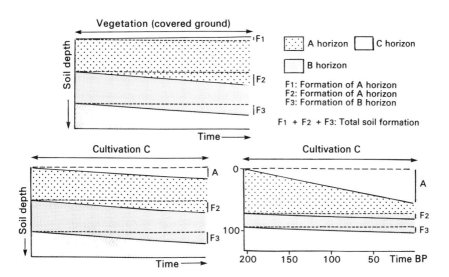

Figure 6.4 Soil erosion and sustainability: comparing the rate of soil production by weathering and loss by careless cultivations (Hurni, 1983)

key test of sustainable development. His project led to the production of a simple field manual, using cartoons to select appropriate estimates of erosion rates for different agroclimatic zones and to design simple protection strategies. Such advice in the hands of capable extension workers touring the farms achieves very rapid conservation results. The statistics of Ethiopian soil erosion and attempts to control it are impressive. Losses of soil from the Ethiopian Highlands exceed 20 tonnes per hectare; since this region comprises 40 per cent of the country but receives 90 per cent of the rainfall and is home to 88 per cent of the population, remedial action is urgent. Since 1978 the Ethiopian government has sponsored a national campaign of Food-for-Work; the 20,000 peasant associations formed after the revolution in 1974 are allocated extension workers and between 1978 and 1984 11 per cent of the land in need of protection was treated, with a total of 590,000 kilometres of contour bunds; 600,000 kilometres of diversion ditches; 1,100 check dams; and 470,000 kilometres of afforestation terraces. These are merely locally practicable examples from a wide range of available techniques for soil protection shown in Figure 6.5.

6.1.2 Sophistication and the role of research

Experience such as that of Hurni and his collaborators, together with that slowly emerging from other problem regions such as the Himalayas (Chapter 5), points to some interesting but daunting problems of effective erosion control.

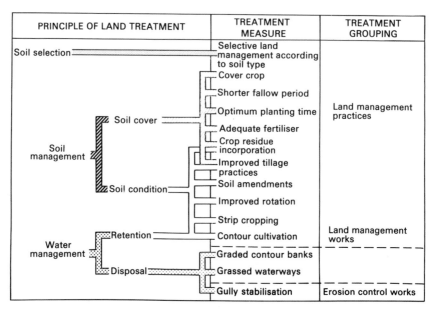

Figure 6.5 Methods of combating soil erosion (FAO, 1977)

In terms of scientific research the paradox becomes one of a need for increasing sophistication yet the virtual impossibility of applying it. For example, multi-parameter hydrological erosion models, either for sheet or gully erosion, require expensive instrumentation in the field. While this is not impossible it is likely that such instrumentation can only be provided by foreign aid and only run by foreign experts, immediately breaking an important bond with local land users (Table 6.2 and Blaikie, 1985).

Boardman *et al.* (1990) reveal two fundamental elements of mismatch in research support for soil conservation. One is that research is often best

Table 6.2 Contradictions between foreign aid and soil conservation programmes

Essential elements in conservation programmes	Objectives and limitations of foreign aid projects/ programmes	Results
Long maturing	Measurable benefits within 3–5 years	Emphasis on short-term and often peripheral objectives to soil conservation
Diverse, timely and highly coordinated inputs	Inability to deal with a large number of line ministries	Either disorganisation or attempts to set up independent foreign-staffed implementation agencies
Outputs diffuse, diverse and difficult to quantify	Quantifiable benefits need to be predicted at proposal stage for purposes of justification	Concentration on physical and often less important objectives of conservation
Implementation deeply involved in sensitive political issues	No overt inter-ference in internal political affairs of recipient country	Acute problems of implementation
In-depth analysis of political economic circumstances	Short-term consultants with the necessity to be tactful on political feasibility	Project documents full of rhetoric and technical details
Sustained political will at central government level	Short-term consult-ancies (1–3 years usual)	Uneven and uncertain back-up of project/ programme after aid finishes

Source: Blaikie (1985)

supported in regions where there are few erosion problems: technology transfer has significant restrictions in land-use and land management fields. Another is that 'problems in erosion studies' are not equivalent to 'erosion problems'; a mismatch can therefore occur between the intellectual satisfaction of process studies and the practical needs of application. Once again the USLE, or techniques with a similarly practical but validated orientation (e.g. Harris and Boardman's simple expert system, 1990), is seen to occupy an essential niche.

Practical field experience in soil conservation reveals the true importance of social and cultural influences on the outcome; we should not consider this to be only a feature of the developing world. Napier (1990), reviewing the successes and failures of erosion control in the USA since the inception of the Soil Conservation Service in the 1930s, stresses the very high levels of financial and infrastructural support needed to achieve the initial, voluntary thrust by farmers. Such support found political approval because of the perception of the initial devastation of erosion and because of the strength of the economy: 13,000 advisers took to the fields on 23 million acres. An 'education-subsidy' approach to erosion broke down in the 1950s when 'soil banking' schemes to reduce production in sensitive zones were often abused; the Food Security Act of 1985 is much stricter in its Conservation Title, a set of clauses which imposes penalties for the cultivation of, or failure to protect, erodible soils. Napier produces a list of twelve conditions to be satisfied by effective soil conservation programmes (Table 6.3).

In the developing world the translation of soil conservation goals into practical agricultural action must bear Napier's assessment and the USA's experiences in mind but with the additional elements of local sensitivities. As Eckholm (1976) puts it: 'Land-use patterns are an expression of deep political, economic and cultural structures; they do not change overnight when an ecologist or forester sounds the alarm that a country is losing its resource base' (p. 54).

6.1.3 Social science and erosion

We are now closer to Hallsworth's 'psychology' and Blaikie's 'political economy' of erosion. Blaikie writes of the need for judgement as to whether an agrarian community can respond to the task of making good land degradation; he is sceptical as to whether 'induced innovations' can cope with the problem.

Land tenure is an especially relevant element in erosion and its control. Whittenmore (1981) concludes that: 'unjust land tenure systems and the political, economic and social policies which enable these systems to prevail are the chief causes of hunger and poverty in the Third World today' (p. 1).

Stocking (1987) states, however, that part of the problem of introducing erosion controls in Africa is the tendency to hyperbole of specialists, mainly

Table 6.3 The elements of a successful soil conservation programme

1 The development of a political constituency which supports action to reduce the social, economic and environmental costs of soil erosion.
2 The allocation of extensive human and economic resources on a long-term basis by national governments to finance soil conservation programmes.
3 The creation of government agencies commissioned to address soil erosion problems with sufficient autonomy to be immune from short-term political influences.
4 The development of well-trained professionals to staff soil-conservation agencies.
5 The development of an informed farm population which is aware of the causes and remedies of soil erosion.
6 The development of a stewardship orientation among land operators to protect soil and water resources.
7 The creation of national policies which place high priority on the protection of soil and water resources.
8 The creation of national development, agricultural and soil conservation policies and programmes which are consistent and complementary.
9 The creation of national environmental policies which are consistent and complementary.
10 The development of physical and social scientists who are committed to the generation of scientific information which will contribute to the creation, implementation and continual modification of soil and water conservation policies and programmes.
11 The creation of an interdisciplinary professional society committed to the maintenance of environmental integrity of soil and water resources.
12 The emergence of political leadership which will be willing to implement policies and programmes which some segments of the agricultural population will find oppressive.

Source: Napier (1990)

from outside the continent, who speak and write of an erosion 'crisis' – he quotes examples. The measurement of erosion rates for Africa reveals a huge variability inherent in the measurement techniques themselves. Furthermore, the essential participation of people on the ground is unlikely if their folk knowledge insists that there is no, or a very slow, acceleration of erosion rates over what is 'natural'. Morgan (1995) says 'Most farmers are aware of the problem and its effects and the notion of the peasant farmer damaging the land through ignorance is severely mistaken' (p. 98).

Stocking's other conclusion is that while Africa may not have an 'erosion crisis' in terms of the depth of physical soil loss, the depletion of soil nutrients in a continent short of food is very serious. Zimbabwe loses 1,635,000 tonnes of nitrogen and 236,000 tonnes of phosphate each year through soil erosion; the cost of equivalent fertilisers would be $1.5 billion.

The return on investment for widespread structural protection against soil erosion (e.g. terracing) has been poor throughout the developing world; while traditional structural methods have a role this is more to do with their

Table 6.4 Planning a soil and water conservation project

Guiding principle	Questions to ask
Participation is the key to a successful project	Respect and cooperation of local people? Are we answering their needs? Are we involving them at every stage?
Training puts skills into the hands of people	Are we using appropriate technology? Are we taking training needs seriously?
Work with existing groups	Are there traditional working groups? Which are the strongest local institutions? Which could help with village planning?
Flexibility is strength	Does the workplan allow flexibility? Are we ready to evaluate progress and make changes?
Don't expect dramatic results too quickly	Have we planned for a long enough period of activity? Is there provision to extend the project?
Choose incentives with care	Do we use food-for-work? How can we avoid creating dependence?
The poorest are often the hardest to reach	Is the programme reaching poor people? Can they afford to participate? How can we channel help to them more effectively?
Keep records – measure success	Can we say exactly how the programme works? Is there a plan to monitor? How do we use the information?
Collaborate to avoid confusion	Have we made sufficient contact? Have we discussed the workplan with official agencies? Do we circulate reports – to the right people?
Build on what people already know	Has anybody made a study of local practices? Are there local traditions? How could such systems be improved?
Suitable systems survive with minimum support from outside	Are we introducing a technique appropriate to the area? Is it the most appropriate system? Has it been tested locally by other projects?
Farmers want benefits now!	Does the technique improve yields? Does it give the crops more moisture?
Think before you mechanise	Do we really need machines? Do the people have the means to maintain them? What are the alternatives?
Conserving fields only the starting point	Is the village committee ready to plan land use? Are there plans for grazing and fuelwood? How can the community be motivated to act collectively?

Source: Critchley (1991)

maintenance by involved farming communities than the inherent value of braking the transport of detached particles downslope. What is now required is joint management of soil and water needs by farmers. As Shaxon *et al.* (1989) put it: 'The new message is this: aim to improve the soil conditions for root growth and crop production and, in so doing achieve better conservation of water and soil' – they call this approach 'land husbandry'. While we might advocate coordination of soil protection efforts at the watershed scale it may be that management units based on village communities are more appropriate if that is the identification of local people. Van Dyke (1995) describes, from a very detailed study, how the Beja nomads of the Sudan adapted and improved their traditional 'teras' for collecting runoff and curtailing erosion; profitability was raised at the household level. Crichley (1991) provides a list of questions (Table 6.4) which overseas NGOs such as OXFAM can follow in promoting land husbandry schemes. Atampugre (1993) describes the very successful village-based scheme which has developed in Burkina Faso since 1988; crop yields were improved by between 10 per cent and 45 per cent through the construction of stone diguettes or bunds, an operation in which local women were more heavily involved than men. This new approach to technical aid for development is spreading very rapidly to, for example, rural water supply schemes (Logan, 1987) and urban drainage (WHO, 1991).

6.1.4   Tropical deforestation: a particular erosion problem

The cover type intervening between the very high rainfall intensities of the tropics (Table 6.5) and the underlying soil layer occupies a critical control on erosivity. In seasonal rainfall regimes such as those of the Ethiopian Highlands the present erosion 'crisis' has been precipitated by a mixture of deforestation and the extension of arable cropping into areas of steep slopes. However, in the remaining areas of tropical 'rain' forest, the rapid current rate of exploitation lays an extremely thin and sensitive soil cover whose fertility and stability depends on the rapid recycling possible with the forest cover intact.

Ross *et al.* (1990) point, however, to the danger of blanket conclusions about the erosional effects of tropical deforestation. In their Brazilian study site, erosion rates following virgin forest clearance varied according to soil texture, the presence of impermeable horizons, slope location and other variables. They further support the view that erosion is the more serious as a form of nutrient loss and that retention of soil organic matter is essential in management regimes.

Rates of soil erosion measured from various tropical cover types are shown in Table 6.6. Clearly, loss of the canopy cover involves a penalty of soil loss. It is generally assumed that traditional tropical farming practice, well adjusted to local conditions, can protect the soil and this has largely

Table 6.5 Some examples of high intensity rainfall from tropical regions

Duration	Rainfall (cm)	Location	Date
15 min.	4.06	Ibadan, Nigeria	16 June 1972
15 min.	11.94	Monrovia, Liberia	1 August 1974
15 min.	19.81	Plumb Point, Jamaica	12 May 1916
90 min.	25.40	Colombo, Sri Lanka	1907
24 hr	50.80	Kalani Valley, Sri Lanka	May 1940
24 hr	116.81	Baguio, Philippines	14–15 July 1911
48 hr	167.10	Funkiko, Formosa	19–24 July 1913
63 hr	200.96	Baguio, Philippines	14–17 July 1911
96 hr	258.68	Cherrapunji, India	12–15 June 1876

Table 6.6 Measurements of soil loss in Peninsular Malaysia

Land use	Rate $(kg\ m^{-2}\ y^{-1})$	Source
Rain forest	0.034	Cameron Highlands
Rain forest	0.004	Headwaters of R. Gombak[a]
Tea plantation	0.673	Cameron Highlands
Vegetables	1.009	Cameron Highlands
Mining	0.495	Ayer Batu, near Kuala Lumpur[a]

Source: Morgan, 1979
[a] Data have been converted using a bulk density value of $1.0g\ cm^{-3}$.

been borne out in plot experiments (e.g. Table 6.7a). However, under field conditions, land pressure, for instance in small-holder cropping regions of southern Africa, leads to over-intensification of cultivation and erosion (Table 6.7b).

As yet there is little written on the social implications of soil loss and soil protection schemes in the humid tropics; this is perhaps related to the remoteness of the regions affected (e.g. Amazon, Congo basins).

There are many similarities between the concepts of soil protection and sustainable river basin management. Protection of soils against degradation implies protection for the quantity and quality of runoff over and through soils. Howard *et al.* (1989) see protection against heavy metals, agrochemicals, compaction, farm waste, acidification and erosion as comprising a 'total package'; they employ the concept of *critical loads* to define the maximum stress which the soil will bear without adverse changes in function.

Table 6.7(a) The impact on soil erosion rates of tropical
deforestation

Treatment	% runoff	Soil eroision (kg ha^{-1})
1 Forest control	0.05	0.32
2 Cassava	3.27	29.49
3 Oilpalm and maize	15.78	170.10
4 Traditional farming	0.06	27.75
5 Alley cropping and rice	7.20	94.19
6 Plantain	8.69	157.34
7 Pasture and coconut	10.36	183.11
8 Improved forest	10.29	172.40

Source: Mahoo (1989)

Table 6.7(b) Seasonal rainfall, runoff and soil loss totals

Catchment	Season	Rainfall (mm)	Runoff (mm)	Soil loss (t ha^{-1})
Bvumbwe (full land plan)	1981/2	957	29.7	0.12
	1982/3	822	19.0	0.21
	1983/4	836	6.7	0.03
	1984/5	1,334	62.2	0.13
Mindawo I (traditional cultivation)	1981/2	890	81.0	10.06
	1982/3	910	156.7	13.70
	1983/4	865	42.0	4.44
	1984/5	1,191	154.5	14.32
Mindawo II (physical conservation)	1982/2	–	–	–
	1982/3	975	54.8	2.31
	1983/4	914	17.4	1.18
	1984/5	1,207	85.6	5.11
Mphezo (plantation)	1981/2	–	–	–
	1982/3	951	7.7	0.09
	1983/4	923	4.2	0.03
	1984/5	1,137	8.4	0.06

Source: Amphlett (1990)

6.2   DAMS: PROBLEMS OF SEDIMENTATION AND RIVER REGULATION

We have observed the popularity of dam construction as a means of
moderating the natural extremes of flow which rivers exhibit in response to
precipitation episodes. Since the first aim of reservoir storage is flow
moderation, this is the first impact of dam construction; river habitat is
adjusted to a natural pattern of flows (or 'regime'). The storage itself is also
vulnerable from sedimentation in still-water conditions. Linking to the soil

erosion problem we therefore here treat a major off-site cost of erosion. The dam is perhaps best seen as a simple hinge point in a disrupted erosion/deposition system (Figure 6.6). For this and other reasons dam siting should be based on far more considerations than the conventional ones of foundation geology and water balance (see Newson, 1994a). Impoundments create discontinuities in the river channel ecosystem; adopting the River Continuum Concept of Vannote *et al.* (1980), Ward and Stanford (1983) refer to the Serial Discontinuity Concept by which the position of an impoundment within the continuum (of biotic and nutritional change) resets the channel downstream to an analogue of the river headwaters. After a certain recovery distance the continuum is reset until the next impoundment downstream.

Because sediments as well as water are stored by reservoirs there are considerable changes in erosion and deposition downstream of dams; channel capacity is adjusted to the transport of both water and sediments (Chapter 2) and both are altered. Alterations also occur to water quality as a result of storage or of the uses to which the stored water is put. Impacts tend to decay downstream as natural tributaries restore the original flow regime and through time as a new equilibrium is reached, but not before primary effects of flow and sediment regulation have produced secondary and tertiary effects on the river environment (see Figure 6.9).

6.2.1   Soil erosion and reservoir sedimentation

In Chapter 2 we saw how difficult it is to predict fluvial sediment transport; however, in cases of severe practical need, such as soil erosion and reservoir

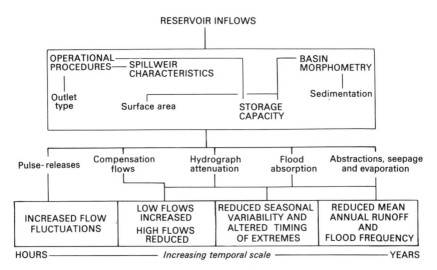

Figure 6.6 Influences of river regulation (by dam construction) on downstream flows (Petts, 1984)

sedimentation, simple empirical devices are better than ignorance. For the Kesem dam feasibility study (described in Chapter 5) the author compiled basin sediment yield estimates using all the available empirical techniques – the bewildering array of outcomes is shown in Table 6.8 (the precautionary principle demands that the most extreme estimate be accepted if valid). In Chapter 2 we also stressed the importance of natural processes of sediment storage in the basin sediment system; these tend to mitigate the impact of erosion (natural or accelerated) in source areas. The process of sediment delivery is that which converts erosion products into a sediment yield and the sediment delivery ratio links the two (Walling 1983; see also Walling 1994). Delivery ratios have been researched since the early 1950s with a view to allowing downstream impacts, notably reservoir sedimentation, to be calculated from erosion measurements (or predictions – using the USLE for example). The source area erosion rate is considered as unity (or 100 per cent) and the intervention of storage downstream reduces this value as area increases; area is the most common variable used in delivery ratio predictions (see Figure 6.7a) but multiple regression equations also exist and there have been many attempts to explain the variability shown in Figure 6.7a.

Walling (1983) attempts to shed some process 'light' on the black box model of catchment behaviour inherent in the delivery ratio. Ratios are subject to variability because of the sorting of different sediment sizes, inter-storm

Table 6.8 Sediment yields to the proposed Kesem Dam, Shoa Province, Ethiopia

Method	Suspended sediment yield tonnes per year	Bed material yield tonnes per year	Total sediment yield tonnes per year	Time to half fill reservoir capacity (years)
1	5,370,250	–	5,370,250	135
2	5,946,426	2,378,570	8,324,996	87
3			9,448,890	53
4	4,007,026	–	4,007,026	181
5	1,352,226	–	1,352,226	536
6	5,741,039	–	5,741,039	126
7a[a]	2,569,924			
7b	4,127,298	–		
7c	–	3,406,566	7,533,864	96
7d[b]	17,736,263	–	17,736,263	41
7e	603,033	–	603,033	1,202
7f	2,201,743	–	2,201,743	227
7g	661,485	–	661,485	1,096

Source: Newson (1986)
Notes Groups 1–6 are calculations based upon internationally recognised formulae.
[a] group 7 are based upon empirical fieldwork data from the Kesem catchment.
[b] 7d–g are based upon soil eroison data and various values of delivery ratios.

variability of erosive sources and re-mobilisation of previously stored sediment from sources downstream (see Figure 6.7b–d and Chapter 2).

A further property of the erosion-deposition system which is of critical importance to predicting reservoir sedimentation rates is the *trapping efficiency* of the impounded waters (Brune, 1958). Controlling variables include:

(a) Ratio of reservoir capacity to catchment size.
(b) Detention time of water in reservoir.
(c) Mixed variables such as
   —inflow volume/reservoir volume;
   —detention time/velocity of flow through system.

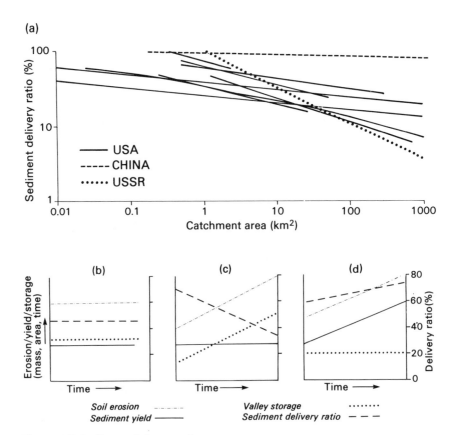

Figure 6.7 Sediment delivery ratios:
      (a)   Variability of empirically derived curves (Walling, 1983)
      (b–d) Schematic illustration of the role of remobilised sediments in affecting delivery ratios. Periods of erosion lead to storage (c) and subsequently release (d) of eroded sediments (Boardman *et al.*, 1990)

It is also important to calculate the location and thickness of deposited sediments within reservoirs. The successful operation of off-takes to supply, valves and power turbines, is fundamentally upset by sediment accumulation either because of the simple mechanical obstruction or because of abrasion damage. Upstream impacts are also produced by reservoir sedimentation – the backwater effects of deltaic deposition encourage deposition within the inflow channel and flooding may then occur through loss of channel capacity.

Increasingly, dams regulate sediment accumulation in their reservoirs by flushing and sluicing (Bruk, 1985). For example, on the Blue Nile the Roseires Dam forsakes the storage of water during the early part of the flood season to escape the first 'flush' of sediments; later the clearer inflows are stored as the valves close.

6.2.2   Regulated rivers below dams

We have already encountered many of the technical problems of regulating rivers (Chapters 4 and 5). It is difficult to ignore the recent prominence given to the construction and operation of large dams in relation to irrigation problems in the developing world; nevertheless river regulation has many purposes. There are many means of achieving it and it has a very long antecedence. Broadly speaking such a situation was inevitable for Man's habitation of earth, considering the very small proportion of usable water on the planet which flows in river systems (3 mm depth equivalent) and the very short residence time it has in those systems (two weeks). If one further adds the repeated tendency for development processes to curtail major elements of natural storage such as wetlands and floodplains then the need to interrupt the natural flow of rivers and its pattern in space becomes inevitable. From the unsuccessful Sadd el-Kafara dam onwards (i.e. for nearly 5,000 years) the major form of river regulation – the dam – has spread to the point at which Lvovitch (1973) estimates that 15 per cent of the stable discharge of world rivers is provided by the 4,000 km^3 of water now impounded behind them.

If one adds the extent of river margins protected by flood banks or in which groundwater or wetlands are manipulated (or indeed where profound changes to land use have modified flows) it would be difficult to find a 'natural' river. By the year 2000 it has been estimated that two-thirds of the world's runoff will be controlled even under the narrow definition of reservoir regulation. By that time the current potential availability to each of us of eight times more water than we need will have been halved by population growth, so measures to ensure its delivery will be crucial.

Beaumont (1978) catalogues the growth in the number of dams built across the world between 1840 and the 1970s; this assessment puts the peak of dam building at 1968 but, since development is accelerating, as is the problem of food supply and aridity, the phenomenon is possibly one brought about by the increasing size of dam schemes. Changes to design have allowed earth-fill, a relatively cheap technique, to replace masonry or concrete structures. Unit costs are lower and maintenance less problematic with big dams. Finally, problems of gaining acceptance and land for dams are sometimes easier when a single dam is proposed and the prestige of size cannot be discounted in this respect. Nevertheless, some notable large dam proposals have, according to the International Rivers Network, recently been cancelled or delayed by the action of popular protest movements: for example, Silent Valley, Kerala, India; Franklin, Tasmania; Hainburg/Nagy Maros, Danube; Kings River, California; Serre de la Fare, Loire.

Dams are built for several purposes; most are multi-purpose or are adapted to be multi-purpose. They supply water from storage to domestic, industrial and agricultural users, either by feeding into *direct supply* pipe-lines or by *regulating* the flow of the river below the dam site (i.e. by using the river as a natural pipeline). Dams also house turbines for generating electric power from the controlled gravitational flow of water down a steep hydraulic gradient. They may also produce advantageous flow regimes for users of the river (navigation) or neighbouring land (flood control). Finally, they may have subsidiary aims of environmental enhancement, for example, by aiding fisheries or maintaining wetlands but as yet only small dams are built to improve habitat conditions. Some large dams capable of regulating rivers have also produced episodic benefits for recreation (e.g. canoeing) and for pollution control (flushing accidental spills).

While the regulation river flow regime has major benefits, especially for water suppliers, regulation brings environmental impacts whose full duration and scope are now being recognised and quantified. Use of the river as a 'cheap pipeline' permits a higher yield from a given storage volume because regulation patterns can be varied with natural conditions. Regulation allows flood control to 'ride on the back' of other aims because both storage volume and downstream flows can be manipulated. Regulation encourages multiple use and re-use of water within the same river basin.

The environmental costs of regulation tend, however, to ruin the impression held formerly that rivers can be 'cheap pipelines'. Even without river regulation the obstruction represented by dam construction to the downstream system of water, sediment and solute transport and to the both-way migration of living matter can have serious consequences depending largely on the siting and operation of the dam in question; each is different and generalities are only now appearing as the literature grows and syntheses are made (e.g. Petts, 1984).

6.2.3   Impacts of regulation

The impact of a dam upon the physical, chemical and biological environ-ment downstream varies mainly with the way in which the dam is operated

Plate 6.2 The control of river flows by valves: inside Clywedog Dam, mid-Wales, regulating the flow of the River Severn (photo M. D. Newson)

with regard to the *timing* and *magnitude* of water releases (and partly with the depth from which releases are made). Since river channels tend to adjust to flood flows of a magnitude which occurs approximately annually it is important to note the general reduction of floods downstream of any storage (including natural lakes). Reviewing a large number of UK reservoir release strategies, Gustard *et al.* (1987) calculate a 26 per cent reduction of mean annual flood downstream.

However, where can this or any other flow distortion be measured on the downstream river reaches? We clearly need to consider an impact decay downstream as 'natural' tributaries gradually restore the original regime to the river. Petts (1979) provides considerable help in his depiction of the role of unregulated tributaries (in terms of both flow and sediment contributions) and the way in which a morphological response (channel capacity) responds through time (Figure 6.8). The time dimension is vitally important as the river system re-establishes a new steady state to reflect the balance of regulated and natural inputs at all points downstream (Petts suggests that morphological responses may be minimal by the stage at which the reservoir catchment area is less than 40 per cent of the total catchment area to that point). Figure 6.9 combines space and time in a conceptual framework for the range of physical, chemical, biological and off-stream impacts of regulation.

Returning to the temporal pattern of releases, Gustard *et al.* (1987) document the predominant types found in the UK (Figure 6.10a) and tabulate the frequency of use by reservoirs. They also graph the frequency of reservoirs by purpose and Figure 6.10b shows the domination of direct

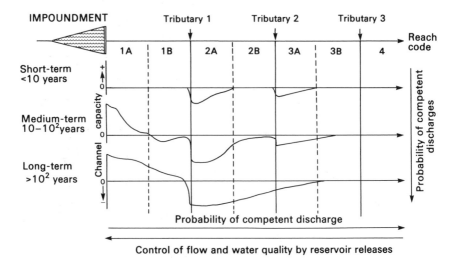

Figure 6.8 The role of tributary flow and sediment inputs in the adjustment of fluvial morphology to regulated flows below reservoirs (Petts, 1979)

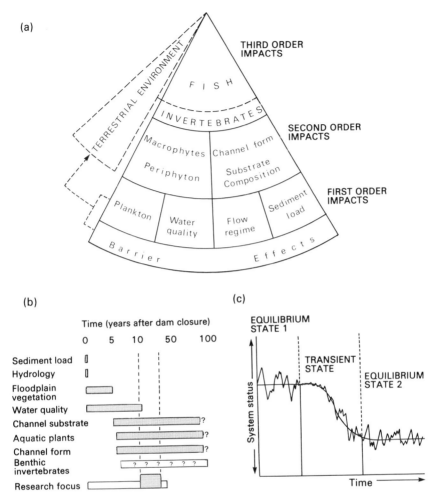

Figure 6.9 River regulation and equilibrium:
    (a) Chained impacts: first, second and third orders (Petts, 1979)
    (b) Reaction times recorded in the literature
    (c) A hypothetical general model of adjustment through time (Petts, 1987)

supply reservoirs, a legacy of the tendency to 'dam each valley' in parts of industrial Britain to obtain sufficient supply. Very little attention was given in the early approaches to setting compensation water rules (see Table 6.9 and Chapter 1) for the protection of fisheries or wildlife below such reservoirs. More recent, larger schemes have been dominated by a regulation operation (some of the older dams have also been modified in terms of operation) and some schemes have called in surveyors of impact down-stream. For example, an extremely comprehensive survey has been

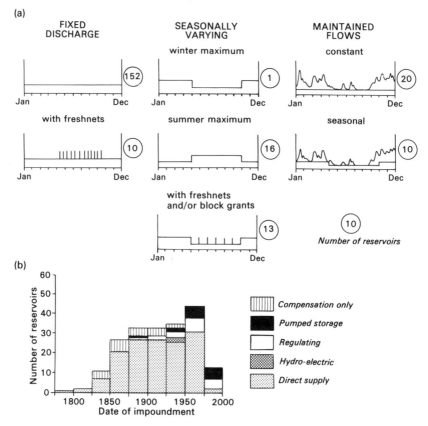

Figure 6.10 Patterns of UK river regulation:
   (a) The river flow regime below typical schemes
   (b) Changes in the aim of schemes built since 1750 (after Gustard
       et al., 1987)

conducted on the River Wye, already the site of impounding reservoirs, to establish the baseline of biological habitats prior to a new dam which has not yet been built (Edwards and Brooker, 1982).

However, such foresight is rare and is defined largely by the delay in dam construction; elsewhere in Britain little monitoring was commissioned before dam construction although major releases have produced *post hoc* evaluations of the key processes at high flows after dam construction (e.g. Petts *et al.*, 1985; Leeks and Newson, 1989). Another notable exception to the lack of impact predictions in the UK is the case of Kielder Reservoir, Northumberland. Prior to its construction, surveys were conducted of the stability of the bed sediments under regulating flow conditions and of the invertebrates living above and below the dam site (Boon, 1987) mainly because of the importance of salmonid fisheries on the river.

Table 6.9 Summary of river uses which are considered when assessing compensation flow requirements

 1 Existing licence holders to abstract water for agriculture, industry and water supply.
 2 Dilution of point source effluents; river quality objectives; public health.
 3 Power generation – with regard to daily and seasonal demands.
 4 Navigation – maintenance of adequate minimum depth and lockage.
 5 Riparian rights, e.g. stock watering and household purposes.
 6 Migratory and coarse fish.
 7 Angling.
 8 Plant and invertebrate ecology – nature conservation.
 9 Amenity – canoeing, swimming, bank-side recreation.
10 Maintaining natural beauty.

A more recent *post hoc* evaluation of Kielder's impacts was commissioned as the result of a change in the operation of the dam; because water *supply* demand has largely proved insufficient for the deployment of regulation releases for downstream transfers and off-takes, hydro-electric turbines were installed and a diurnal pattern of release waves is now the norm (Johnson, 1988; see Section 6.2.5 below).

The Kielder study is a vignette compared with the potential duration and extent of the eventual adjustments to regulation; full biological adjustments may take decades and be symptomatic of a 'complex response'; particularly if the sediment system and channel morphology are involved (Petts, 1979). As Petts (1984) stresses: 'Completely adjusted systems have been observed only rarely and most studies relate to the period characterised by transient system states' (p. 258).

This conclusion has very serious implications because the decadal timescale is now seen as one over which both economic and climatic systems change profoundly; it is unlikely that Kielder will be the only dam to change its operating pattern in response to outside pressure, thereby initiating another transient phase. Another question raised by climatic change is the safety and permanence of dams; will future dam decommissioning require impact assessment? Impact studies of dam disasters are, regrettably, fairly common in the geomorphological literature (Costa, 1988).

6.2.4   Regulation in tropical catchments

Turning to river regulation in the developing world, Kariba Dam on the Zambezi River was the first large developing world scheme to involve river regulation (in 1958). Although Leopold and Maddock had written the first fundamental assessment of the physical impacts of regulation in 1954, Kariba has only been evaluated from an environmental viewpoint in retrospect, along with Volta (Ghana), Kainji (Nigeria) and the Aswan High Dam (see Chapter 5) by Obeng (1978). Obeng's review of these impoundments deals with the more spectacular effects on communities, health, seismic activity, growth of macrophytes, and fisheries with only a passing mention of effects on channels. There is a perceptual point of difference here between developed and densely settled countries and the developing world: in the latter the types of impact described by Obeng dominate those of regulation, partly because they are major impacts but partly because the riparian zone downstream is not the densely owned, exploited, enjoyed zone it is in the developed world. The impacts of regulated flows downstream are more often quoted in terms of coastal sediments and coastal morphological change. For example, the Akosombo Dam on the Volta River in Ghana, 110 kilometres from the coast, has reduced sediment inputs to the Guinea Current to the extent that in places 100 metres of coastal retreat has occurred in five years.

Of the small collection of developing world studies of river regulation effects two are particularly valuable because they stress the importance of the environmental conditions of the tropics. Dealing with the savanna conditions of northern Nigeria (wet season June–September) Olofin (1984) describes the geomorphological response of gullies to regulation below the Tiga Dam (completed 1974). Because the Tiga reduces annual flows to a quarter of their natural value, gully tributaries have incised their beds to reflect the lowering of the effective base-level to their development; incision of almost a metre was achieved in three wet seasons with obvious effects on the sediment load of the river in the regulated reach and on the morphology of the floodplain, its habitat and communications.

Hughes (1990) reports that large areas of evergreen forest, a relatively common vegetation type in arid and semi-arid Africa, are vulnerable to the altered regime of regulated rivers. The Tana River of Kenya (see Section 5.5.2) has a 6 kilometre-wide floodplain in its lower reaches upon which the forests are of conservation interest and of resource value to the tribal inhabitants. Both maximum and minimum flood levels and frequencies are found to control sensitively the occurrence of the forests; 'natural' rates of meander migration also help to maintain the floodplain vegetation. Consequently, the impact of the water resource developments built and proposed on the Tana is likely to be adverse.

6.2.5   Improving the efficiency and accountability of river regulation

A major technical problem encountered with river regulation is that of 'regulation losses'. Reservoir loss occurs because a large tract of open water is exposed to evaporation or because macrophyte plant communities colonise the reservoir and transpire without limitation. Regulation losses comprise 'operational losses', inherent in the time delay between the requirement for water use at a downstream site and the opening of a valve in the dam control room, and 'natural losses' which occur by seepage into the floodplains alongside the regulated river. Research in Britain (Central Water Planning Unit, 1979) has established from field experiments under drought conditions that natural losses are less than 20 per cent even in conditions and locations favourable to seepage through river banks. While a release down the River Severn in 1975 was 'wasted' by approximately 20 per cent, another trial on the Tees in 1976 (a hard-rock and boulder clay basin) recorded only 2.5 per cent losses into the floodplain. As well as geological conditions differing between the two sites so did the height of the release wave (60 cm in the upper Severn, 20 cm in the upper Tees).

By varying the pattern of release waves it is also possible to predict the timing of downstream arrival and therefore reduce 'operational losses' (Figure 6.11). The CWPU research presents four methods of prediction, of which the simplest is illustrative of the importance of the discharged flow from the dam, namely:

$$\text{Wave speed} = 1.4 + 6Q_{pr}/Q_{av} \ (\text{km hr}^{-1})$$

where Q_{pr} is the peak discharge of release
Q_{av} is the average river discharge.

Clearly, therefore, 'tweaking' reservoir release patterns is in the interests of more efficient use of the river channel and 'playing tunes' on a suitably flexible arrangement of valves at a dam allows more equitable use of regulation flows. By 'equitable' one hopes to reconcile the different requirements of water management and the physical, chemical and biological elements of river habitat. These ambitions are incorporated in the modern concept of 'ecologically acceptable flows' (Petts et al., 1995 and see below).

For example, recent research on the River Tyne, UK, has demonstrated that use of a scour valve (set low in the dam and therefore apt to discharge water which is silty and out of temperature regime with natural flows) to generate hydropower can significantly influence habitat downstream:

(a) Fine deposits blanket the bed material, reducing the number and diversity of invertebrates.
(b) Warmer-than-ambient water emerges from the dam in spring, disrupting the normal seasonal pattern of fish emergence from eggs.

Releases from a valve nearer the reservoir surface might not produce these effects but would be much less efficient for power generation. The Tyne study also concludes that the diurnal pattern of releases, which rises in 2 hours and falls in 1.5 hours over a tenfold flow range, produces fluctuating velocity conditions for micro-organisms and an oscillation of bed material significantly strengthening the fabric of gravel riffles; salmonid fish find

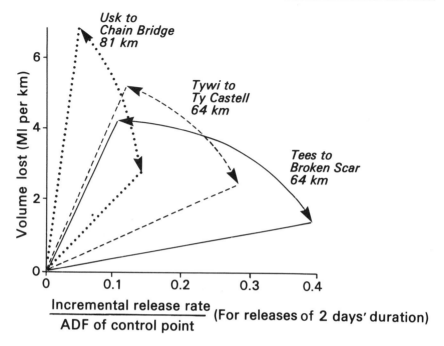

Figure 6.11 The relationship between release volume and timing and the operational loss of reservoir release water to floodplains and evaporation (ADF = average dry-weather flow) (Central Water Planning Unit, 1979)

greater difficulty in excavating egg-laying sites. Alongside these, albeit currently restricted effects, the North Tyne also shows degradation and armouring for 7 kilometres downstream because of the obstruction of bedload supply and considerable growth of tributary bars where 'natural' inputs of gravels still occur into a flow of reduced competence (Petts and Thoms, 1987; Sear, 1992).

There is rapid technological innovation in the selection of 'environmentally acceptable flows' in rivers affected by regulation schemes. Geomorphologists are conducting impact assessments on the sediment transport impacts of high regulated flows and they are combining with freshwater biologists to identify those habitats which typify rivers of different morphological characteristics. Petts *et al.* (1995) describe an approach used for a regulation scheme in lowland Britain based on a hierarchical survey of the impacted channel. In addition they use the hydraulic habitat model PHABSIM (*p*hysical *hab*itat *sim*ulation) to predict ecological conditions at varying flows. For PHABSIM to be successful it is necessary to know the requirements of each species of instream biota which is considered sensitive (i.e. rare or economically important). Gore *et al.*

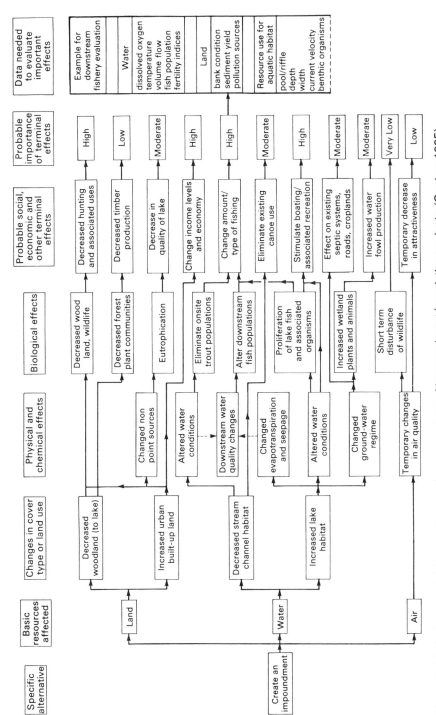

Figure 6.12 Environmental impact diagram for assessment of impoundment/regulation projects (Canter, 1985)

(1992) use PHABSIM to simulate the conditions for fishes and hippopotamuses in the Kruger National Park of South Africa.

6.2.6 Future of regulation

It is unlikely that we can proceed to manage rivers without dams and, since we constantly explore the capacity of the environment to provide resources, we need, along with asking the questions, 'Is the scheme necessary?' and 'Is a dam the only answer?', to carry out very far-reaching proactive *impact assessments*. Engineers must be aware that virtually every site and system is different when local sensitivity to change and local requirements for operational patterns are superimposed. A modular approach may be possible but even this will work only if our knowledge is resourced by comprehensive and long-lasting *post hoc* audits on regulated rivers, stretching even to the 'finer points' of river conditions under which certain operational patterns are permitted and to the siting of new dams in the optimum position with regard to 'natural' tributary inflows. Canter (1985) offers the impact matrix for impoundments modified here as Figure 6.12.

Finally, inevitably, *people* affected by dam schemes need an active and continuing role in management – beyond the flooded area and the irrigated area and including, especially, riparian interests. Lambert (1988) describes a prototype institutional framework; the Dee Consultative Committee, set up by Act of Parliament to influence the operation of a major regulating scheme in the UK, is dominated by technocrats but nevertheless admits that the operators of the scheme need inputs of advice (and even censure) from outside their ranks.

6.3 IRRIGATION: LAND, WATER AND *PEOPLE*

Agriculture still dominates mankind's use of freshwater (Table 6.10). The lion's share of the water use is by irrigated agriculture (Table 6.10b), (Figure 6.13). No other economic activity uses as much water per unit area as agricultural applications; only evaporation from the surface of large reservoirs such as the Aswan Dam 'wastes' so much water. Yet the 15 per cent of world agriculture which benefits from irrigation contributes 40 per cent of the world's food supplies (Earthscan, 1984).

Irrigation can be defined as a human intervention to modify the spatial and temporal distribution of water occurring in natural channels, depressions, drainageways or aquifers and to manipulate all or part of this water to improve production of agricultural crops or enhance growth of other desirable crops. This simple, technical statement, when superimposed upon the drastic needs of the world's drylands and all developing countries for food to feed growing populations and the human tradition of 'hydraulic civilisations' helps explain why governments so avidly pursue the development of 'irrigation potential'.

Table 6.10(a) The use of the world's freshwater in the 1980s: broad geographical regions

Geographical region	Annual withdrawals		Percentage distribution of water used		
	Cubic metres per head	As % of 'available' water[a]	Domestic	Industry	Agriculture
World	660	8	8	23	69
Africa	244	3	7	5	88
North and Central America	1,692	10	9	42	49
South America	476	1	18	23	59
Asia	526	15	6	8	86
Europe	726	15	13	54	33
USSR	1,330	8	6	29	65
Oceania	907	1	64	2	34

Note: Estimates refer to various dates in the 1980s.
[a] Available water includes internal renewable water and river flows from other countries.
Source: Water Resources Institute and United Nations Development Programme, 1992.

Table 6.10(b) The use of water in different countries by individual domestic users compared with that on 1 hectare of irrigated land in a semi-arid country and comparing earning capacity – 1991 estimates (per head)

	Developing countries	Egypt	UK	USA	One hectare in irrigated agriculture
Water use/ year (m^3)	<100	100	500	2,00	15,000
GDP/year (US$)	1,500	700	16,000	23,000	1,500–5,000

Often this 'irrigationism' (Adams, 1992) takes a political hold with no regard for the river basin context of large-scale irrigation or for the real needs of those who will work the crops to be fed. Adams is vitriolic, from personal experience, about the impact of large-scale irrigation schemes in Africa:

It has been good for international consultancies and engineering contractors, and for the experts who jet out from northern universities to hold forth on the future of Africa. It has suited young states eager to establish control over remoter regions, and to harness and direct energies within the country to predictable and controlled ends.

(p. 156)

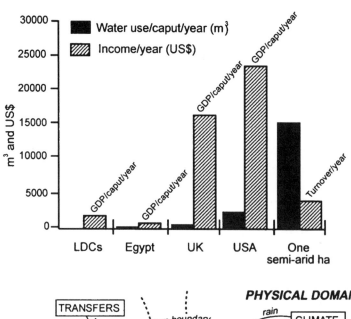

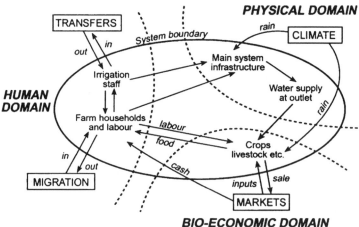

Figure 6.13 Irrigation and its water use:
      (a) Water use and returns in agriculture – developing and
          developed world (after Allan, 1992)
      (b) The domains of irrigation and their linkages (after Chambers,
          1988)

Reasons for the failure of large-scale schemes are given by Carter (1992) –
see Table 6.11.

    To be more positive (because the crisis of irrigated agriculture is perhaps
the most immediate of all global environmental crises), Carter (1992) lists
the 'true' parameters of irrigation potential (Table 6.12) and these form the
urgent research agenda for hydrologists, agricultural scientists but, crucially,

Table 6.11 Reasons identified for poor performance of large-scale irrigation schemes

Problems at conception and planning stage
- The inadequacy of conventional criteria for project acceptance
- Lower yields than anticipated in project plans
- Generally over-optimistic projections of benefits at conception/feasibility
- stages
- Inadequate time allowed for project preparation
- Insufficient account taken of land tenure and water rights
- Irrigation activities not integrated with rainfed cropping and other farming
- and income-generating activities

Inherent difficulties
- The complexity and problematic nature of irrigation as a form of agricultural intensification
- Problems associated with the farmers' reduced independence on becoming irrigators
- Problems of soil suitability and soil variability

Design and construction stage flaws
- Inadequate design, requiring subsequent major changes
- Poor on-farm development
- Inappropriateness of imposed cropping calendars

Operation and management difficulties
- Poor management of irrigation agencies (government departments)
- Problems involved with the management and maintenance of irrigation technology
- Poor distribution of water within the scheme
- Poor on-farm water development
- Inadequacy of extension services
- Public health problems
- Problems of salinisation and water-table rise, necessitating costly drainage and reclamation strategies
- Poor level of involvement by farmers

Financial/ economic problems
- Very high development costs per hectare
- Inadequate importance attached to funding of post-construction activities
- Heavy costs and low level of performance of irrigation agencies
- Difficulties of collecting water charges

Source: Carter (1992)

social scientists. As we saw in the case of soil erosion, the identification of environmental resources and hazards is part of a vernacular knowledge. Add understandable self-interest and one has the perfect reason for full consultation, rather than control, for the implementation of irrigation schemes: 'The degree of people's interest in, commitment to and willingness to invest in irrigation will depend on how it is perceived to enhance or diminish their

Table 6.12 Two paradigms of the concept of 'irrigation potential'

The simplistic myth
● Suitable land + Proximity to water = Irrigation potential

A closer approximation   If the following apply:
● water supply is a major constraint to production *and*
● soils are suitable *and*
● water and affordable, maintainable water delivery technology is available *and*
● farmers wish to irrigate *and*
● appropriate farmer and/or government organisations for water management exist or can be developed *and*
● inputs – fertiliser, seed, pesticides – are available *and*
● credit for input supply is available *and*
● crop prices are attractive *and*
● markets are accessible *and*
● issues of land tenure and water rights have been satisfactorily resolved *and*
● proposed irrigation activities are environmentally sustainable

then irrigation development may be a viable option.

Source: Carter (1992)

lives.' Table 6.13 draws up the social balance sheet of large-scale irrigation for the poor farmers of the developing world. 'Irrigation is not necessarily beneficial' (Guijt and Thompson, 1994), yet is seen as a development panacea (Turner, 1994, introducing a theme issue of 'Land Use Policy' on small-scale alternatives).

Planning a large-scale irrigation project requires inputs from a large number of disciplines. Basic physical inputs include:

● Hydrometric data for the catchment and 'command area' served for irrigation.
● Design of dams and diversion structures.
● Design of canals and distribution systems.
● Erosion protection of canals and distribution systems.
● Soil moisture and crop yield studies.
● Infiltration and evapotranspiration data.
● Drainage/wastewater investigations.

This agenda is clearly daunting for a developing-world nation without the use of overseas consultants; it is also daunting for the world of applied research – the knowledge-base for irrigation is constructed by an estimated world research and development budget of 0.0005 per cent of agricultural turnover (compared with 1 per cent of turnover in the oil industry). We are also emphasising in this treatment the wider linkages to development, thus nesting the irrigation scheme in the river basin far more comprehensively than merely via the physical considerations listed above. Chambers (1988) writes of three domains of irrigation systems: physical, human and bio-economic. Figure 6.13

Table 6.13 Gains and losses by the poor under large-scale irrigation

(a) GAINS

Type of gain	Who gains	Under what conditions
Employment in construction	Male and female agricultural labourers	Labour-intensive schemes
Levelling off employment in agriculture	Male and female agricultural labourers	Irrigation-induced intensification
Increased wage rates in agriculture	Male and female agricultural labourers	Irrigation-induced intensification
Growth in non-farm employment	Male and female agricultural labourers	Irrigation-induced intensification
Return migration	Male and female agricultural labourers	Irrigation-induced intensification
Lower food prices	All sections of rural society	Cash payments
Non-agricultural uses of water	Those close to canals etc.	Year-round access to water

(b) LOSSES

Type of loss	Who loses	Under what conditions
Increase in land prices	Marginal famrers bought out. Landless tenants displaced	Irrigation-induced agricultural intensification
Competition for rainfed farmers	Marginal rainfed farmers	Irrigation-induced agricultural intensification
Increased unpaid workloads for women	Women	
Increased water-borne diseases	Particularly agricultural workers	Lack of preventive health measures
Labour displacement	Mechanical and agrochemical substitutes	Outweighs benefits of productivity
Waterlogging and salinity	Small farmers	Faulty irrigation design/operation

Source: Chambers (1988)

Table 6.14 Comparison of major methods of irrigation

Bases of comparison	Canal irrigation			Aerial irrigation
	Flood irrigation	Furrow irrigation	Sub-irrigation	
1 Total capital costs 2 Total annual costs 3 Crops	Low Low Pastures, grain	Low Medium Row crops e.g. corn, cotton, potatoes, vegetables, orchards	High Very low Annual root crops, vegetables, orchards	Medium High Pastures, vegetables, orchards, nurseries
4 Soil type	Adaptable to most soils but fields may be small if infiltration rate high	Adaptable to most soils having good lateral moisture movement characteristics	Must be capable of capillary rise into the root zone. Must allow good lateral movements of moisture	Adaptable to most soils
5 Topography	Slopes capable of grading to 1% max	Slopes varying from 0.5% to 12%	Fairly uniform slopes preferred	Adaptable to most slopes which could be farmed
6 Water required	Large streams	Fairly large streams are sometimes necessary depending on the number and size of furrows	Small flows may be utilised. Groundwater or farm dams	Small flows may be utilised. Groundwater (pumped)
7 Soil or crop damage	Over-watering may cause salting, puddling or surface crusting	Furrow erosion is a big problem	Extremely unlikely	Both are possible if water drops are large

Source: adapted from Withers and Vipond (1988), table 7.8; based on Weisner, C.T. (1970) Climate, Irrigation and Agriculture, London, Angus and Robertson.

shows the interactions which should be given equal weight in the investigation of the feasibility of a scheme and in its detailed planning.

Chambers' study is of canal irrigation – the most common form of gravitational supply in constructed large-scale schemes. Table 6.14 shows, however, that there are three other main forms of irrigation scheme based upon how the water is eventually applied to the crop. Flood irrigation mimics the natural 'bulk' application of water from river to land but at the field scale it is obviously the least controlled. At the sophisticated end of the spectrum are sub-irrigation, with water being applied directly to the root zone of the crop, and sprinkler irrigation, both of which require a considerable infrastructure of pipes. The costs of infrastructure figure highly in the assessment of benefits of irrigation and energy costs are especially punishing (e.g. in pumping groundwater from depth or forcing surface water through sprinklers).

The distribution of irrigation water, as Chambers explains in close detail, is where technology, economics and politics confront each other in space and time. The volume and timing of the supply of water in a canal scheme (see Figure 6.14) can make or break the efforts of the farmers. Too much irrigation water can be as damaging as too little, and water applied too early or too late means that optimum yields are missed.

As well as irrigation schemes having the potential for social 'inefficiency', their efficacy depends on the degree to which the water supplied from river, dam or pump translates into improved agricultural production; at every stage in the water supply process to the fields there are opportunities for malfunctions and transmission losses. Chambers' tabulation of data for seven schemes in India shows that, while transmission losses were anticipated (2–10 per cent) actual losses were greater in every case (6–40 per cent). It has been estimated that the gross efficiency of irrigated agriculture worldwide is 37 per cent. A notable exception to these dismal figures is that of irrigated agriculture in Israel which, as well as making more and more use of recycled wastewater, is using minimum-waste delivery systems, such as trickle irrigation, to deliver the water needed (and the fertiliser) according to biological and meteorological needs (Plate 6.3 and Figure 6.15).

Adams (1992) extends the typology of irrigation schemes to 'indigenous irrigation' in Africa. There is a continuum of traditional techniques from adaptation to natural flood patterns in wetland areas to complete water control. His main division is between flood cropping, stream diversion and lift irrigation which in total comprise 47 per cent of Africa's irrigated area. He warns, however, that Africa has a relatively small total area of irrigated land compared with South Asia (India, Pakistan, Bangladesh and Sri Lanka).

Returning to the social problems of large-scale irrigation development, Davies (1986) infers that one of the modern world's first large-scale

Plate 6.3 Trickle irrigation with fertilisers automatically mixed, Israel. Inset: The proliferation of farm dams in the Western Cape, South Africa, has depleted river flows and severely damaged ecosystems (photos M. D. Newson)

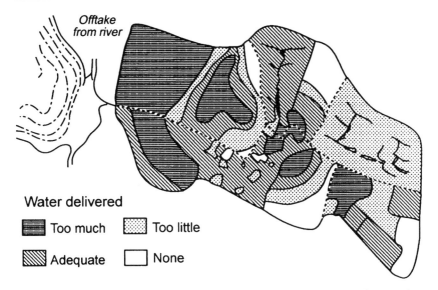

Figure 6.14 Canal irrigation: distribution problems with increasing distance from the river offtake (after Chambers, 1988)

schemes was in fact a social success; the Gezira (Sudan) scheme (which is often quoted as a model in a similar way to the TVA for basin authorities) was developed after completion of the Sennar Dam on the Blue Nile in 1925. The crops were cotton (for export and cash income) but also sorghum (Sudan's national food staple) and hyacinth bean (for fodder); rotations involving fallow mirrored local traditions. Davies says that the scheme has performed well because it provided food and security for humans and animals, it kept cultivation methods as simple as possible and incorporated a profit-sharing arrangement which gave incentives to the people. It also eliminated the possibility of large landowners taking over. Land reforms are often a prerequisite of successful irrigation resettlement, as Cummings *et al.* (1989) make clear for Mexico. Rich farmers in irrigated areas own more than four times the land of poor farmers; rich farmers are allowed to select the best land, leading to a 'crazy quilt' pattern of holdings; overpopulation has been encouraged by settling immigrants on small plots; and rich farmers sink groundwater boreholes to increase their irrigation power, putting at risk regional development of aquifers. The last point is a telling one in the context of river basin management and planning; if private water developments are sanctioned as part of making irrigation attractive there are immense problems of coordination when drought threatens or resources near their limits. Lemon *et al.* (1994) use social surveys on the Argolid Plain of Greece to uncover the decisions made about irrigation by local farmers, their crop water requirements and other significant data which should be

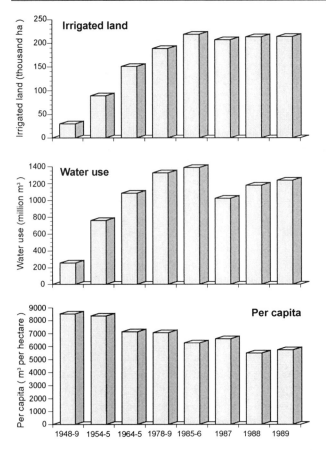

Figure 6.15 Water use in Israel, 1948–89 (from data in Kliot, 1994)

part of the planning framework for irrigation schemes within water-based development as a whole.

Just as it is misguided to think we can do without large dams in future, so we have not seen the end of large-scale irrigation, and one alternative, that is, small-scale irrigation, should not be seen as a universal solution. The essential elements of small-scale irrigation are that plots are small and, more important, that small farmers control the works and that those works are at a technological level which they can operate and maintain (Carter, 1992; Carter and Howsam, 1994). The differing perceptions of irrigation schemes between men and women can be revealing (Guijt and Thompson, 1994). These authors also emphasise that the common features of participatory approaches include group dynamics and flexibility. 'Appropriate technology' for small-scale irrigation clearly needs formalising and handbooks are beginning to appear (e.g. Stern, 1979). A danger of perception of 'progress' and 'efficiency' by state water managers when dealing with traditional or modern small-scale irrigation is brought out by Haagsma (1995). While the crop choice, water use, operation and maintenance of small schemes on Santo Antao (Cape Verde) may appear to be 'primitive' the overall efficiency is high when considered in relation to river basin resources and resilience in the face of water shortages. The state should, says Haagsma, 'play a facilitating role, enabling farmers to re-assume greater responsibility for local water management'. It is, however, critical to gain a detailed understanding of indigenous irrigation practices before endorsing them – an academic fascination with them is not sufficient in itself (Adams *et al.*, 1994).

Drought is a hazard to dryland communities, especially those in the southern hemisphere where rainfall variability is at its global peak (see the discussion of Australia, Chapter 4.4). All natural hazards may be thought of as comprising a risk term and a vulnerability term; the risk is probabilistic (for example, we speak of the 100-year event) but vulnerability is increased through poor hazard perception (for example, hazardous locations, activities) and inadequate warning systems. Drought hazard protection is a lot less well developed than flood protection, possibly because large-scale development in drylands is a new phenomenon and is occurring principally in those countries which do not have the capacity built for systems such as long-range weather

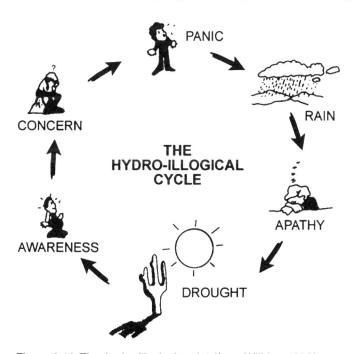

Figure 6.16 The 'hydro-illogical cycle' (from Wilhite, 1993)

forecasting. However, there are also technical difficulties not associated with flood protection – the complex causation for rainfall deficit, the slow build-up of damage and the lack of communications systems.

Drought has no single technical definition; we may envisage a time sequence from meteorological drought (a rainfall deficit for a certain period), through soil moisture deficit (involving evaporation loss – sometimes called 'agricultural drought') to various measures of extremely low groundwater levels or river flows. If poorly designed irrigation schemes increase vulnerability to drought their shortcomings are conveniently confused by politicians with the 'technical' drought problem. As Bandyopadhyay (1987) puts it: 'The ambiguous term "drought" has been rather freely used in our country (India) to explain away the process of ever increasing water scarcity' (p. 12). National governments are frequently caught in the 'hydro-illogical cycle' as regards drought hazard (Wilhite, 1993, and Figure 6.16) and the international community may complicate matters by giving aid during the 'panic' stage which induces, once rain falls, 'apathy'. This is totally unlike the adaptive strategies of diverse uses implemented by indigenous peoples in drylands. Mortimore (1989) made an intensive study of the adaptation of the Hausa, Ful'be and Manga communities in Nigeria to the droughts of the 1970s and 1980s; he concludes that:

> Arid and semi-arid environments call for special adaptive skills on the part of those who live there. The twentieth century has unwittingly transformed the socio-economic milieu of the inhabitants of these environments .... Therefore, ways must be found of supporting productive communities in such high-risk environments.
>
> (p. 230)

For both developed and developing countries the World Meteorological Organisation called for the development of drought plans in 1986; progress has been slow because governments are not inherently interdisciplinary in knowledge or intersectoral in action – a similar problem afflicts river basin management and planning. Yet drought relief programmes are infinitely more expensive (Wilhite, 1993). Australia has, since 1990, operated a policy which reduces relief in favour of advance preparedness, education, decision support and weather forecasting. In the USA many states have drought contingency plans, but in most countries reviewed as case studies in Wilhite's 1993 volume there has been little or no progress. In Chapter 5 we saw that India's drought problem has brought little by way of sustainable government response.

Researchers in the UK have recently recommended the following areas for technological improvement in coping with drought:

- Climatology – monitoring, forecasting.
- Hydrology – efficient water use.
- Ecology – primary productivity in drought, resilience.

- Pastoral farming – land tenure, markets.
- Arable agriculture – separation of drought, desiccation and degradation phenomena.
- Social factors – food models, institutional structures.
- Risk assessment – vulnerability mapping.

(Drought Mitigation Working Group, 1993)

Barrow (1993) sums up the global problem as follows:

> by the start of next century much of the world's easily developed irrigated land will have been opened up. The problem will be how to intensify, diversify, extend and sustain the production of numerous, remote, small cultivators who will probably have little or no money, no access to groundwater or supplies from canals and streams. The answer must largely be the development and promotion of improved rainfed cultivation, runoff collection and concentration and the use of ephemeral flows and floods, together with the cultivation of new moisture-efficient crops and improved varieties of existing rainfed crops.

(p. 197)

Quite possibly, genetic engineering, together with indigenous soil and water management, holds the key to world food supplies in the next decade.

6.4   CONSERVATION AND RESTORATION OF RIVER CHANNELS AND WETLANDS

'River restoration has a short history but is experiencing a steep growth curve' (Newson, 1992c). In the first edition of this book it was possible to relegate restoration to a small mention near the end (p. 317). This 'steep growth curve' has continued, forcing the topic into the mainstream of technical issues in river management. In addition, wetland scientists have also begun to carry out restoration schemes. Both movements suggest that managing river basin systems along ecosystem lines can now incorporate a specific technical approach to the spontaneous regulation functions of the Marchand and Toornstra model. The danger might be that developers can risk damage or destruction of these functions in the knowledge that, after a 'quick profit' they can be restored, a deviousness which has afflicted certain terrestrial habitats. A realistic balance may be that if damage or destruction is an inevitable corollary of development, an equivalent, compensatory, capacity is created in the same system by the developer. All too often, however, this 'restoration' is merely rehabilitation or enhancement and the costs of so doing are out of proportion to the achievement. Nevertheless, losses of the spontaneous regulators in the developed world have been massive and in some senses 'every little helps'.

Gore's (1985) volume on river restoration used case studies from the USA where the preoccupation was then with mitigating the impacts of surface mining. He emphasised that stream ecosystem structure and function were controlled by energy source and water quality as well as flow regime and habitat structure. He equated river restoration with 'recovery enhancement', reflecting the prevailing situation of mining impacts. Hasfurther's chapter in Gore's book represented the sole geomorphological contribution but it is the geomorphological contribution to understanding instream physical habitat and the relationship between channel, riparian, floodplain and valley-floor processes (see Chapter 2) which has grown most in the last ten years.

After an initial overemphasis on meander planforms and the riffle-pool sequence as vehicles of channel restoration (see Hasfurther, 1985; Keller, 1978), geomorphological knowledge is now seen as an important input to channel management for flood defence, water resources, pollution control, conservation, recreation and amenity functions (Brookes, 1995a). Predictive capability has extended to many more morphological and sedimentological aspects of the 'natural' channel, classification schemes allow a context to be set for engineering in different parts of the basin and the ecological significance of fluvial features/dynamics is beginning to be understood (see Newson, 1995). Perhaps the most significant demonstration of confidence has been the publication of *The Rivers Handbook* (Calow and Petts, 1992; 1994), the *Rivers and Wildlife Handbooks* (Lewis and Williams, 1984; Ward *et al.*, 1994) and the titles selected for Hey's recent contributions on 'Environmentally sensitive river engineering' (Hey, 1994) and 'River processes and management' (Hey, 1995).

To obviate the need for river channel restoration it is essential to be proactive in river basin development where, in most cases, some form of

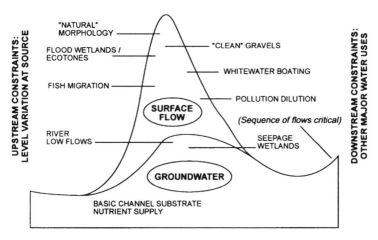

Figure 6.17 Competing uses for segments of the flow hydrograph – developed world. Note that there are also upstream and downstream user constraints

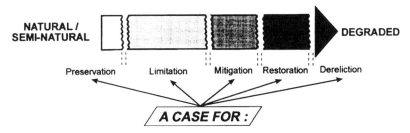

Figure 6.18 The case for conservation: matching effort to system conditions
(after Boon, 1992)

engineering operation will change the natural dimensions and features of the
channel, however large (consider the flood action plan proposals for
Bangladesh, see Chapter 5). 'Environmentally sensitive river engineering'
therefore seeks a compromise between retaining those dimensions and
features, the need for flow conveyance (principally floods) and the
requirements of instream users (see Figure 6.17). Once made confident
about the knowledge-base in fluvial geomorphology, civil engineers are
willing to incorporate predictive techniques to mitigate damage, though in
some cases there will be controversy about the balance between conserving
naturalness and facilitating development in less hazardous conditions, that
is, in some cases settlement, industry or agriculture must move to less
vulnerable locations in the interests of conserving rivers. With this in mind,
conservationists are evaluating and registering their sites of special interest
(Boon, 1992). Boon lists the motives for river conservation as maintenance
of earth's life support systems, practical value (e.g. erosion control),
economic importance (minerals, tourism), scientific research, education,
aesthetic and recreational value, and ethical considerations. He establishes
(Figure 6.18) a range of management options which depend on the
evaluation of the existing condition of the river system in relation to
the 'natural' (effectively the semi-natural because of mankind's all-
pervading influence on rivers). Schemes to assist this evaluation are now
appearing, such as Rosgen's classification of channels in the USA (Rosgen,
1994) and *River Habitats in England and Wales* (National Rivers Authority,
1996).

   As yet, documentation on restored river channels is sparse, most schemes
being as Brookes (1995a) puts it an amalgam of the scientific and prag-
matic. This is Brookes' experience of participation on the European
Communities project in Denmark and England (Brookes, 1992; Brookes and
Shields, 1996); the River Restoration Project has taken two English lowland
rivers, the Cole and the Skerne, and 'restored' 2 kilometres of each, partly
by excavating new channel and partly by changing channel profiles and
vegetation. The rural Cole has been the simpler case because no housing was
threatened in the floodplain; the Skerne flows through urban Darlington

Plate 6.4 The restored River Skerne, Darlington, UK, showing the introduced meander and rehabilitated floodplain (photo copyright AirFotos, Newcastle upon Tyne)

(Plate 6.4) and it was necessary to both maintain flood conveyance and protect urban infrastructure. Practical experience (added to that of Brookes in his work for the National Rivers Authority and in Denmark) suggests that monitoring is essential if the apparent success of the geomorphological design procedures is to be translated into ecosystem recovery – at the moment this is an act of faith. Table 6.15 lists the guidance from geomorphology which Brookes considers important to restoration. The UK River Restoration Project utilises, for example, multistage channels, meanders, pools, riffles, shoals, islands, backwaters, bays and substrate improvements as part of its site works.

It is more difficult to convince river basin managers that channel changes result not just from engineering intervention but from flow changes brought about by river regulation (see above) and by land use in

Table 6.15 Geomorphological guidance for river restoration

Type	Content	Limitations
Channel profiles (long-section and cross-section)	Knowledge of variation of channel morphology along a reach, particularly in relation to pools, riffles and planform	Sparse data on typical width – depth ratios for a range of channel types
Low-flow width in channel design	Best obtained from neighbouring natural section of same slope and geology	Site-specific measurement required. Natural widths for a range of channel types related to catchment area largely unavailable
Design and location of pools and riffles	Information on topographical, sedimentary and flow characteristics, size, location, spacing and slope values at which they occur	Knowledge base particularly for gravel-bed rivers. Limited knowledge of adjustments during and after flood flows
Substrate reinstatement	Reinstatement of gravels for different channel types (either to remain static – armoured/segregated bed – or to be mobile)	Most knowledge is for mobile gravel-bed rivers
Prediction of channel changes	Use of historical records, maps or surveys to predict nature and locations of future channel change (including lateral and vertical change)	Imperfect knowledge for a wide range of channel types (e.g. sand-bed rivers). Assumes change at a site will be an ongoing process
Bank erosion/protection locations	Good knowledge of location of natural bank erosion mechanism, especially for gravel-bed rivers. Some understanding of how artificial influences (e.g. boat-wash erosion) modify or initiate patterns of erosion	Bank erosion mechanism not fully understood for all key channel types

Source: after Brookes (1995b)

Table 6.16 Human-induced river channel changes

Cause of change	Channel character	Average change ratio		Minimum change ratio		Maximum change ratio		Total no. of studies
		UK	Other	UK	Other	UK	Other	
Urbanisation	Width	1.16	–	1.16	–	1.16	–	
	Depth	1.06	–	1.06	–	1.06	–	2
	Capacity	2.50	–	2.50	–	2.50	–	
Reservoir construction	Width	0.85	1.14	0.29	0.01	1.49	1.56	
	Depth	0.92	3.42	0.34	0.01	1.62	8.10	489
	Capacity	1.25	–	0.29	–	2.29	–	
Channelisation	Width	1.33	–	1.00	–	2.02	–	
	Depth	1.06	–	0.59	–	1.67	–	57
	Capacity	1.39	2.70	1.00	2.70	2.53	2.70	
Land-use change	Width	1.41	0.94	0.96	0.80	1.88	1.05	
	Depth	–	2.31	–	1.58	–	3.97	221
	Capacity	–	2.15	–	1.53	–	4.11	
Water transfers	Width	0.57	–	0.57	–	0.57	–	
	Depth	0.84	–	0.84	–	0.84	–	
	Capacity	–	–	–	–	–	–	

Source: Gregory (1995), 'Human activity and palaeohydrology', in Gregory, Starkel and Baker (eds) *Global Continental Palaeohydrology*.

the contributing catchment. Brookes and Gregory (1988) graph a 'channel change ratio' in relation to indirect impacts such as afforestation and urbanisation; Table 6.16, taken from Gregory (1995), lists the causes of change. It is clear that true, sustainable river restoration can only proceed if the catchment context (particularly of the sediment system – Sear, 1994) is appreciated. The expenditure implied by this 'purity' of approach may mean that we are likely to be most frequently offered enhancement or rehabilitation as options to assist the basin's spontaneous regulation functions.

Losses of wetland area have been a spectacular accompaniment to the settlement of the humid-temperate developed world. The huge losses which have occurred in the USA are now linked to the severity of the 1993 Mississippi floods (see Chapter 4). Losses in Europe have also been spectacular (Baldock, 1984; for a world review see Maltby, 1986, Duggan, 1990). It is now appreciated that wetlands offer many more spontaneous regulation functions in pristine river basins than the storage of flood waters. Table 6.17 shows that they play a major part in a range of management functions in the undeveloped basin (see also Table 3.11).

The restoration of wetlands varies according to the basic controls on their origin and dynamics; simple flooding is often not their *raison d'être*: raised and blanket bogs – a feature of cool humid upland areas – require inputs from precipitation. They are ombrotrophic and only rainfall water quality is

Table 6.17 Functions, related effects of functions, corresponding societal values, and relevant indicators of functions for wetlands

Function	Effects	Societal value	Indicator
Hydrologic			
Short-term surface water storage	Reduced downstream flood peaks	Reduced damage from floodwaters	Presence of floodplain along river corridor
Long-term surface water storage	Maintenance of base flows, seasonal flow distribution	Maintenance of fish habitat during dry periods	Topographical relief on floodplain
Maintenance of high water table	Maintenance of hydrophytic community	Maintenance of biodiversity	Presence of hydrophytes
Biogeochemical			
Transformation, cycling of elements	Maintenance of nutrient stocks within wetland	Wood production	Tree growth
Retention, removal of dissolved substances	Reduced transport of nutrients downstream	Maintenance of water quality	Nutrient outflow lower than inflow
Accumulation of peat	Retention of nutrients, metals, other substances	Maintenance of water quality	Increase in depth of peat
Accumulation of inorganic sediments	Retention of sediments, some nutrients	Maintenance of water quality	Increase in depth of sediment
Habitat and food web support			
Maintenance of characteristic plant communities	Food, nesting, cover for animals	Support for furbearers, waterfowl	Mature wetland vegetation
Maintenance of characteristic energy flow	Support for populations of vertebrates	Maintenance of biodiversity	High diversity of vertebrates

suitable. By contrast fenland ecosystems require a water quality (high in bases) characteristic of groundwater. Both these (extreme) cases prove that merely to obstruct drainage already installed to drain wetlands does not restore them. Even where damming, flooding or 'irrigation' is used to restore wetlands a complex seasonal regime may be necessary to suit the needs of the biota comprising the original ecosystem. The recent volume by Wheeler *et al.* (1995) explores the approaches and relative success rates of restoration in the wide variety of wetland sites in the temperate zone; there is also, however, a critical need to conserve wetlands in dryland river systems (see Hollis, 1990) and in humid tropical basins. The USA has now initiated a major programme of wetland research to create a functional assessment of wetlands, linking their dynamics and impact to their landscape position in river basins in an attempt to economically value and ecologically protect these vital natural regulators (National Research Council, 1995). In Israel, the Huleh Valley wetlands of the Jordan Valley have been rehabilitated after it was realised that peat drainage was unsustainable; many Israelis are now, however, worried that the wetlands are threatened by recreational developments.

It is important that river restoration and wetland restoration sites should be linked if possible – an easier task for floodplain wetlands than for upland ombrotrophic bogs which act as the source areas for small headwaters.

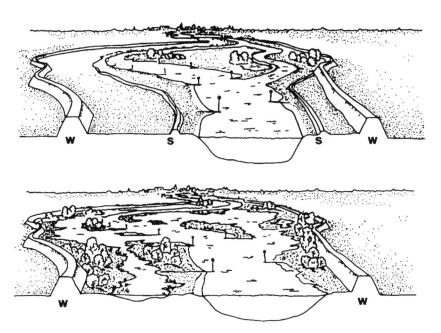

Figure 6.19 Restoration of Netherlands tributaries of the Rhine (S = summer, W = winter dykes; the summer ones are removed to restore the floodplain)

Championing the integration of rivers and wetlands, Petersen *et al.* (1992) lay out an agenda in which the valley floor is returned to its pristine condition via the following steps:

(a) Buffer strips.
(b) Revegetation of riparian zone.
(c) Horseshoe wetlands to intercept existing drains.
(d) Reduction of channel bank slopes to reduce erosion.
(e) Restoration of meanders.
(f) Restoration of riffle/pool sequences.
(g) Restoration of wetland valley floors and swamp forests.

Many practical restoration schemes are meeting many of these criteria, as evidenced by the Rhine scheme shown in Figure 6.19.

The joint restoration of rivers and wetlands as adjacent ecosystems has a further justification in that it creates 'ecotones', defined by Decamps and Naiman (1990) as consisting of 'active interactions between two or more ecosystems (or patches), with the appearance of mechanisms that do not exist in either of the adjacent ecosystems' (p. 2). Pinay *et al.* (1990) offer

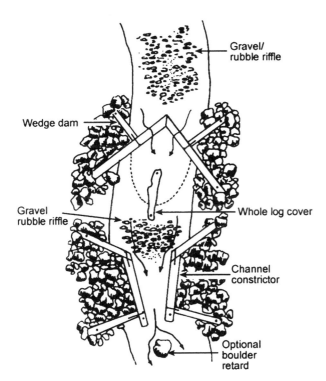

Figure 6.20 River rehabilitation via instream structures – frequently used by fisheries interests

practical proof of the effectiveness of ecotone processes, for example, in reducing nitrate concentrations in the Garonne system; an intact channel-riparian wetland interaction is also important to the nutrient spiralling which helps in the chemical equilibrium of channel flows. Riparian ecotones have achieved most attention in attempts to extend the principles of in-stream ecosystem protection to other parts of the valley floor landscape (Malanson, 1993).

Perhaps the best (most realistic) appreciation of the potential of river and wetland restoration can be gained from serious attempts at definitions. The US National Research Council (1992) offers: 'return of an ecosystem to a close approximation of its condition prior to disturbance. Both the structure and functions of the ecosystem are recreated' (p. 18). It also stresses that what cannot be achieved is 'setting the clock back' to the conditions and timescales through which the ecosystem evolved – imperfections are therefore inevitable. Brookes (1995b) tabulates other, similar, definitions but helpfully offers two alternatives to restoration: rehabilitation (return to pre-disturbance for a limited number of attributes) and enhancement (improvement of one or more attributes). Kern (1992) chooses 'rehabilitation' because the German *Leitbild* vision of channels, following improvements in water quality, is to improve the structural and aesthetic qualities. Fisheries managers have been particularly active in 'enhancement' (e.g. Hunt, 1993), installing channel structures to improve flow character-istics for the sometimes confused purposes of improving the fishery and improving fishing (Figure 6.20).

The role of wetlands in improving food production in the developing world is becoming increasingly realised as a major policy plank in sustain-able river development. Acreman (1994) shows how, when wetland products are properly costed, their performance is superior to that of crops from large-scale irrigation schemes; once again, community management and an enhanced version of indigenous, traditional skills make for the best outcomes. Rydzewski (1990) aligns the benefit/cost ratios of various irrigation strategies in the developing world and finds that flood recession and flood spreading represent sustainable use of floodplains, with ratios of between three and five times those of modern, large-scale irrigation.

6.5   CLIMATIC CHANGE AND RIVER BASIN MANAGEMENT

Man-induced climatic change is a major issue for environmental science in the 1990s; identified as a by-product of the burning of fossil fuels as early as the 1930s, the crisis of confidence (in science) now facing the world's political leaders is one of trusting the evidence of progressive climatological trends and, furthermore, of taking actions which curtail economic growth by accepting the causal interpretations of change. At the World Climate Conference, in Geneva in 1990, there was a good measure of agreement between climatologists in 'scenario setting'; this is achieved by examining recorded trends (in CO_2 concentrations and global mean temperatures) and using global climate models (GCMs) to extrapolate into the next century. The conclusions presented to the 1990 World Climate Conference and their implications for rivers and the coast are as shown in Table 6.18. The International Panel on Climate Change updated its findings in 1992 (Houghton *et al.*, 1992) but this report mainly substantiated, rather than refined, the earlier predictions. Difficulties continue to beset the modelling of climatic variables 'downstream' of global warming (which seems increasingly obvious from collected data). Prediction of the components of the water balance and of their extreme values is especially difficult.

While hydrological predictions have been made from GCMs and some trends in river behaviour already appear conclusive to journalists (e.g. droughts in the USA and floods in Bangladesh), hydrologists find the incorporation of global water balancing in models very difficult indeed. Furthermore, the response by water managers to global change will produce feedback effects.

6.5.1   Asking questions of data and models

The most recent international meeting on the prediction of hydrological changes (Solomon *et al.*, 1987) raised the following apparently insoluble questions:

(a) Is climate variability distinct from climatic change?
(b) What are the causes of long- and short-term climatic change?
(c) To what extent are people inadvertently causing additional variability or triggering change?
(d) What are the best ways to treat data to detect and quantify variability and change?
(e) What are the quantitative relationships between climatic and hydrological variability/change?
(f) How can Mankind cope with change?

In dealing with the prime question of the inherent variability in elements of the hydrological cycle, McMahon *et al.* (1987) stress the sensitivity of Australia and Southern Africa to the variability of precipitation simply because of the high evaporative demand in those regions (Figure 6.21). These authors have used 30,800 station years of river gauge data (average thirty-three years from 938 stations worldwide) and the records of 424 raingauges. A much fuller analysis is in progress and is expected to reveal the importance of the large semi-arid interiors of Australia and Southern Africa in promoting efficient evaporation and of the indigenous plant and animal species in utilising episodic and extreme supplies of water.

Table 6.18 Range of climate changes

Phenomenon	Projection of probable global annual average change	Distribution of change			Significant transient	Confidence of projection		Estimated years for research leading to consensus
		Regional average	Change in seasonality	Interannual variability		Global average	Regional average	
Temperature	+20 to + 5°C	−3 to + 10°C	Yes	Down?	Yes	High	Medium	0–5
Sea level	+10 to + 100 cm		No	?	Unlikely	High	Medium	5–20
Precipitation	+7 to + 15%	−20 to + 20%	Yes	Up	Yes	High	Low	10–50
Direct solar radiation	−10 to + 10%	−30 to + 30%	Yes	?	Possible	Low	Low	10–50
Evapotranspiration	+5 to + 10%	−10 to + 10%	Yes	?	Possible	High	Low	10–50
Soil moisture	?	−50 to + 50%	Yes	?	Yes	?	Medium	10–50
Runoff	Increase	−50 to + 50%	Yes	?	Yes	Medium	Low	10–50
Severe storms	?	?	?	?	Yes	?	?	10–50

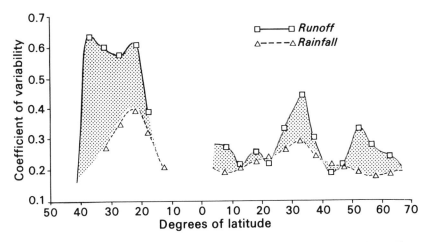

Figure 6.21 Rainfall and runoff variability by latitude (McMahon *et al.*, 1987)

The Vancouver conference at which McMahon *et al.* reported their study used six classifications of contribution:

(a) Identifiable hydrological patterns (such as those of McMahon *et al.*).
(b) Proxy (historical) data.
(c) Stochastic models of change.
(d) Sensitivity of river basins to change.
(e) Sensitivity of water resource systems to change.
(f) Man's influence on change.

Each contribution appeared prepared to admit to the likelihood of a change in the hydrological cycle towards higher activity rates as a result of warming. Evaporation is likely to increase and therefore precipitation will also increase. There the consensus ends because there is no firm guidance on the regional variability of the changes, the timing of change and the importance of extreme conditions, the effects of the response to change in the atmosphere or by natural plant covers, crops or water management systems.

6.5.2   Reconstructing past changes

The basic river basin system of water and sediment/solute transport has proved sensitive in the past to climatic change (in so far as we can identify this from proxy data). From records preserved in sediments, morphology and human cultures it is possible to invoke analogues in the past for those changes in major variables such as temperature now proposed for the immediate future. For example, Newson and Lewin (1991) use analogies between current trends and the warming in Britain following the Little Ice

Table 6.19 Evaluation of scientific approach methods for studying impacts of climatic change on abiotic processes

ANALOGUE STUDIES

Advantages
1 The (bio)geological record provides a solid base for establishing trends between the 'end-members' of process-response systems.
2 Palaeogeomorphological data allow one to study the response time and recurrence time of a specific process and thus could indicate threshold values.
3 They are useful for validation and improvement of existing climatic models.

Disadvantages
1 Time resolution in the (bio)geological record is yet too small for impact assessment on a decenium scale.
2 Records made within contemporary time are too short to indicate the pattern of landscape evolution.
3 Short-term event sequences may provide a misleading impression of the long-term variability of a process.
4 Contemporary data do not cover the whole process-response period and thus do not offer information on threshold values.

PRESENT-DAY ON-SITE MONITORING STUDIES

Advantages
1 Possibilities of quantification and statistical manipulation.
2 They offer detailed knowledge of local significance on impact–process–response.
3 They provide knowledge about the physical basis of processes.

Disadvantages
1 Limited representation for larger regions and limited extrapolation in time.
2 Records are often too short with the possibility that extreme events with longer recurrence intervals are not covered.
3 Methods are time and money consuming.
4 Low representation.
5 Extrapolation in time and space is hard.

MODELLING AND SIMULATION

Advantages
1 It allows quantification and statistical manipulation of past, present and future climate patterns and their impact.
2 Relatively low costs are involved.
3 General insight can be obtained of process–responses to changes in climatic parameters.
4 Input of data and usage of parameters can easily be simulated if real figures are not available.
5 Quick results.

Disadvantages
1 Oversimplification, due to the limited number of parameters which can be simulated.
2 Output generally depends too strongly on how the input variables are chosen.

Source: Eybergen and Imeson (1989)

Age; this occurred during the eighteenth and nineteenth centuries and was accompanied by increased storminess and flooding. These authors refer to the 'forcing function' of climate in controlling river basin evolution; land use tends to bear the imprint of climatic variability in which the record of extreme conditions is most marked.

Eybergen and Imeson (1989) are more cautious than Newson and Lewin, stressing the difficulty of using analogues but also listing the problems associated with monitoring and modelling the processes of river basin change (Table 6.19). Eybergen and Imeson set up a research agenda which urges attention to:

(a) A synthesis of the available knowledge of processes sensitive to climatic change.
(b) Quantification of the key interrelationships.
(c) Identification of the key meteorological variables.
(d) Identification of the role of extreme meteorological events.
(e) Specification of threshold effects.
(f) Reassessment of monitoring programmes.
(g) Quantification of response times and timescales of impacts.
(h) Identification of the main climate-sensitive river basins.
(i) Comparison of climate and anthropogenic signals.

6.5.3   Change, management and resource stress

The Vancouver symposium pointed up that while changes in the geomorphological aspects of river systems are likely, bringing adaptive problems for flood control, reservoir design and irrigation, this may be a rather narrow perspective from a humid temperate standpoint. There will clearly be profound changes to glacierised river systems if melting increases, considerable movement in the boundaries of arid lands, extensive alterations in the volume of lakes and reservoirs and changes in water quality. Peterson *et al.* (1987) argue convincingly that the position of regional atmospheric systems off the west coast of North America exercises considerable control on runoff patterns; sequences of dry and wet seasons are especially important in explaining water quality variability.

In terms of the water resource management problems posed by climatic change, attention has inevitably focused on those regions which are already marginal in terms of coping with drought or flood; it is no accident that a comparison has been made between the Colorado basin, USA, with available water supplies halved over fifty years, and the Ganges/Brahmaputra in Bangladesh with a markedly higher flood risk over the same period. Westcoat (1991) reports on an assessment of management problems in the Indus basin; in the developing world there are huge technical, financial and institutional constraints on adaptive response in sensitive resource systems. Westcoat also

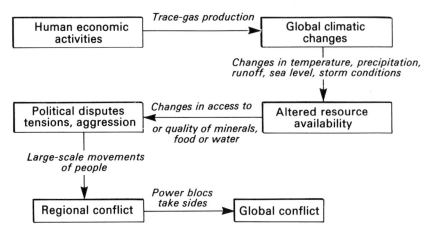

Figure 6.22 The critical path to global conflict over water and other major
resources following climate change (after Gleick, 1987)

treats the international political dimension (the upper Indus is disputed) and
the role of Muslim attitudes to the scientific evidence for climate change.
Riebsame (1992) stresses the need for a planning system which offers a
'response repertoire' for major world basins. Both authors agree on the need
for incrementally increasing public focus on practical implications of change
and maximum information provision. Gleick (1987) considers that international conflict may well be the major result of CO_2-induced warming

Table 6.20(a) Rivers with five or more nations
forming part of the basin

River	No. of nations	Watershed area (km^2)
Danube	12	817,000
Niger	10	2,200,000
Nile	9	3,030,700
Congo	9	3,720,000
Rhine	8	168,757
Zambezi	8	1,419,960
Amazon	7	5,870,000
Mekong	6	786,000
Lake Chad	6	1,910,000
Volta	6	379,000
Ganges–Brahmaputra	5	1,600,400
Elbe	5	144,500
La Plata	5	3,200,000

Source: Gleick (1987)

Table 6.20(b) Per capita water availability in selected countries dominated by
international river basins

Country	$10^3 m^3$ per year per person[a]	Land area in international river basins (%)[b]
High water availability		
Bangladesh	12.1	86
Brazil	35.2	61
Burma	27.0	73
Cameroon	18.8	65
Colombia	34.3	64
Ecuador	29.9	51
Guatemala	13.0	54
Kampuchea	10.9	87
Lao People's Dem.	59.9	94
Nepal	9.4	100
Venezuela	44.5	80
Low water availability		
Afghanistan	2.5	91
Belgium	<1	96
Botswana	0.8	68
Bulgaria	2.0	79
Czechoslovakia	1.8	100
Egypt[b]	<1	30
Ethiopia	2.3	80
Germany (Dem. Rep.)	1.0	93
Germany (FRG)	1.3	88
Ghana	3.4	75
Hungary	0.6	100
India[b]	2.3	<30
Iraq	1.9	83
Israel[b]	0.4	6
Jordan[b]	0.2	6
Kenya	0.6	64
Luxembourg	2.8	100
Pakistan	2.7	>75
Peru	1.8	78
Poland	1.3	95
Portugal	3.3	56
Romania	1.6	98
Spain	2.8	57
Sudan	1.2	81
South Africa	1.4	66
Syria	0.6	72
Togo	3.4	77

Source: Gleick (1989)

[a] Data are for internally available renewable water resources. Some of the largest
developed countries, such as the Soviet Union, Japan, Australia, New Zealand, Canada,
and the United States have few or no internationally shared rivers.

[b] Note that some regions with tensions over water may have only small areas that are in
international basins, for example, Israel, Jordan (the Jordan River). India and Egypt are
other examples. Some developed countries such as the Federal Republic of Germany,
Israel, and Belgium have very low per capita water availability, while some developing
countries, such as Nicaragua, Ecuador, and Indonesia have high per capita water
availability. A realistic assessment of water availability must consider total water supplies,
timing of water availability, quality, location and political allocations.

(Figure 6.22) with many international river basins the scene of potential warfare over scarce resources (see also Chapter 5).

Tables 6.20a and b illustrate the inevitability of a water management crisis in many parts of the world: nearly fifty countries have more than three-quarters of their land area in international basins. Over 200 river basins are multi-national including fifty-seven in Africa and forty-eight in Europe. In an adjacent article to Gleick's, Milliman et al. (1987) quantify the additional pressure on land and population likely to be suffered by Egypt and Bangladesh as the result of losses of delta land to sea-level rise. Between 15 per cent and 30 per cent of habitable land could be lost, especially where the rise of coastal waters is accelerated by depriving the deltas of sediments as the result of dam building on the contributing rivers.

The Bangladesh floods of 1987/8 are briefly described in Chapter 5, together with the proposed engineering and environmental response. At the other extreme the drought which affected the USA in the same years is documented by Ross (1989); the extreme low flows reached across 60 per cent of the USA and into southern Canada with major river systems flowing 45 per cent below median discharge – exceeding the drought of 1934 during the 'dustbowl years'. The American Association for the Advancement of Science has concluded that it is the impact on water resources which is the most critical aspect of global climate change for the USA (Waggoner, 1990). Recommendations from the Waggoner volume include:

- All government agencies should consider climate change in setting scenarios for water issues.
- Governments should encourage flexible institutions, including markets for water allocation.
- The appropriate units for management may be 'problem sheds', i.e. water source/use units.
- Climate change will favour those who invest in variability and then in change; research is vital.
- Water conserving technology should be a focus for research and development.
- Those researching climate change bear a special responsibility to society.

The importance of the USA for grain production has a further influence on world flood resources (and hence water management) in the event of more frequent droughts of this severity, although some agronomists point to the improved conservation of water practised by plants in an atmosphere which contains double the present amount of carbon dioxide (Allen et al., 1991). Individual crop responses are important; Allen et al. show how corn and winter wheat require less irrigation water in a 'greenhouse' world. Mearns (1993) cautions that it will be the prevalence of extremes which will condition world food production after global warming has run its course. Temperature changes may be relatively well simulated by present models

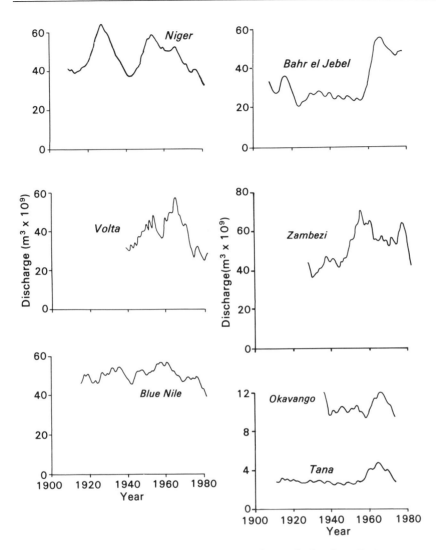

Figure 6.23 The variable evidence of climate change in the river discharge
records of Africa (from Sutcliffe and Knott, 1987)

but precipitation extremes are the least well understood. Improved use of anomalous conditions (e.g. sea surface temperatures and the El Niño oscillation) may be the best way to forecast such extremes and plan cropping accordingly. Even so, crop failure is likely to be more common.

Perhaps the longest available records specific to an important, managed river system are those from the Nile. The Roda 'nilometer' (Figure 1.2) record dates back to AD 641; it has been shown to exhibit periodicities of small amplitude. However, a wider variability of flood levels dates from

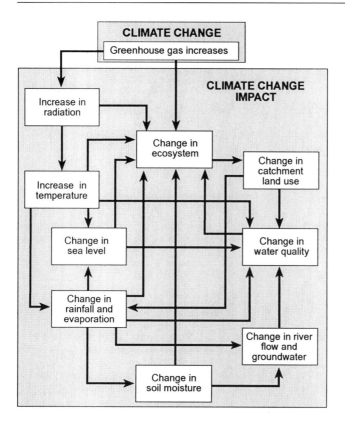

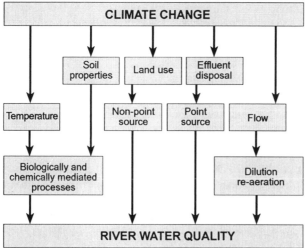

Figure 6.24

	Water Resources	Water Quality	Flood Defence	Fisheries	Conservation	Recreation	Navigation
Change in temperature	■	■	■	■	■	■	
Change in rainfall	■	■	■		■		
Change in evaporation	■				■		
Direct effects of CO₂	■	■			■		
Change in river flows	■	■	■	■	■	■	■
Change in groundwater recharge	■	■			■		
Change in water chemistry and biology	■	■		■	■	■	
Change in storminess			■		■		
Sea level rise	■	■	■		■		
Response of NRA and other agencies to climate change	■	■	■	■	■	■	■

Figure 6.24 Climate change and river management in England and Wales:
    (a) General context and linkages
    (b) Impacts on water quality
    (c) Impacts on management functions for each element of change
        (National Rivers Authority, 1994)

1741 (T. E. Evans, 1990) and the period 1822–1921 shows the largest fluctuations of any of the 100-year periods. The fall flow records at Aswan date from 1871 and exhibit a falling mean discharge:

1871–98	102 Mm^3
1899–1971	88 Mm^3
1972–86	77 Mm^3

Hulme (1990) uses general circulation models to construct future climatic scenarios for the Nile Basin; increased evaporation is likely to be a feature and more important than changes in precipitation, although these may well produce relative changes in the flows from the Blue and White Nile, requiring flexible management. Egypt's existing and proposed methods of coping with periodic droughts in the Nile basin are considered by Fahmy (1992). Sutcliffe and Knott (1987) clearly indicate a considerable variability in the pace and direction of contemporary change in African hydrology (Figure 6.23).

    In the UK the use of analogues of past cool and warm periods within the climatic database allows Palutikov (1987) to predict a general increase in wetness for the north of England. However, reduced river flows in southern England may afflict those regions where, currently, there is a steep increase in demand for water resources – summer droughts may become the norm, perhaps forcing the UK government into the demand management strategies which are a corollary of sustainable management.

A more recent analysis of the implications for river basin management in Britain has been carried out by the National Rivers Authority (Arnell *et al.*, 1994). It uses a matrix approach to assessing the sensitivity of all river basin (and coastal) management functions and also to the reliability of the predictions so far available. The report concludes that changes in the Authority's coastal flood protection responsibilities will accompany sea-level rise; fluvial changes are much less certain but will involve the need to cope with more variability, that is, a change in the regime of rivers and groundwater levels. All aspects of the hydrological and marine system are influenced (Figure 6.24). There are also possible impacts on river water quality and therefore on stream ecosystems and fisheries. An emphasis is put on preparedness in all the operational functions of the Authority as predictions become more certain through time. As an indication of the difficulty of making these predictions for a small landmass such as Britain, three papers deal with the rainfall, evapotranspiration and flood runoff modelling techniques available (respectively, Rowntree *et al.*, 1993; Lockwood, 1994; Beven, 1993). Rainfall patterns are confirmed as those originally suggested by Palutikov – general increase but highly seasonally and regionally distributed. Predictions for evapotranspiration are complicated by cloud cover, vegetation change (affecting albedo, stomatal resistance, the balance between transpiration and interception). Floods are complicated by antecedent conditions, catchment scale and land use but the biggest deficit is in lack of predictability of extreme rainfalls which overcome all these factors.

6.5.4   Global approaches

Clearly, a global approach to predicting the effects of climate change on the hydrological cycle is desirable and it has been a tradition of Soviet, and now Russian, hydrologists to provide worldwide assessments; their territory spans almost every world climate. In this tradition Shiklomanov (1989) develops the picture of future world water use first put forward in the *Global 2000 Report to the President* of the USA (Barney, 1982) which predicted 'serious water shortages in many nations or regions' (including Africa, North America, the Middle East, Latin America and South Asia).

Shiklomanov, besides providing estimates of the demand for water in the year 2000 for twenty-six regions of the world, deals separately with the extra evaporation brought about not only by global warming but also because of the increased 'irretrievable losses' of water through consumptive use in agriculture, especially in irrigated agriculture. The extra evaporation effectively redistributes the extra use: the global cycle is conservative (Table 6.21). However, regions gaining from extra precipitation and runoff are not necessarily those using most. Australia and South America do not

Table 6.21 Changes in precipitation and runoff from continents affected by human activity

Continent	Mean total annual runoff (km³)	Irretrievable additional water losses (km³ year⁻¹)		Total additional precipitation (km³ year⁻¹)		Volume of additional runoff (km³ year⁻¹)	
		1980	2000	1980	2000	1980	2000
Europe	3,210	127	222	60	173	19	55
Asia	14,410	1,380	2,020	790	1,320	306	512
Africa	4,750	127	211	185	245	21	36
North America	8,200	224	302	110	338	34	104
South America	11,760	71	116	0	0	0	0
Australia and Oceania	2,390	14.6	22	0	0	0	0

Source: Shiklomanov (1989)

recoup from the increased circulatory volumes which amount to 17–34 per cent. As the author observes:

> Thus a change in the evaporation regime resulting from human activity may lead in future to some transformation of the ratios between water balance components in various continents and large regions. The quantitative evaluation of these events relative to vast regions is of great scientific and practical importance for the future planning of large-scale projects for the national water resources development of a global nature [sic].
>
> (Shiklomanov, 1989, p. 515)

Clearly, since one of the themes of this book is the influence of land use, we cannot ignore the fact that changes in natural land covers brought about by climatic changes and the cultural response by agrarian and forest productive systems may be crucial to good predictions of river response. Canada has already drawn up comparative maps of ecoclimatic zones before and after a doubling of CO_2 (Rizzo, 1988) but it is as yet impossible to model river flows to incorporate such wholesale shifts in some of the controls.

While the major control of river basin dynamics is climate the cultural influence on the hydrological cycle can be profound; in some scenarios painted of climatic change it is the 'knock-on' effects of changing settlement and food production patterns on river basins which are proving difficult to determine. For example, the change in patterns of global food production which would be brought about by the 'retirement' of the arid Mid-West of the USA (Dallas, 1990) could have a degrading effect on river basins thousands of kilometres distant from those re-created 'buffalo commons'.

6.6   CONCLUSIONS

This chapter has made clear that an era dominated by simple engineering 'solutions' to river basin management has ended, in favour of technical solutions which have multidisciplinary or interdisciplinary authorship. A partial selection of 'new ways' has been inevitable. However, each demonstrates another important modern component of technical management: it is inefficient, ineffective and often invalid on its own, requiring (as shown classically by soil erosion control and efficient irrigation) large inputs of social science and vernacular science too.

As the damage done by unsustainable development to river basin ecosystems becomes clearer (in terms which are costly to those who have benefited from that development) restoration, rehabilitation and enhancement are bound to figure highly in future. The problem underlying future schemes, in terms of the long timescales demanded by thinking in terms of sustainability, is the uncertainty about climate change. This uncertainty, too, will impose a social component and make the science/politics interface more active than ever before; the next chapter turns to aspects of this interface.

Chapter 7

Institutional issues in river basin management

River basin management will inevitably be delivered through institutional structures. In this chapter we refer mainly to the agencies which are set up specifically to manage river basins, though the management of the resource 'package' inherent in a basin will clearly involve the whole panoply of society's structures. Agenda 21 states (Chapter 18.21) that,

> Although water is managed at various levels in the socio-political system, demand-driven management requires the development of water-related institutions at appropriate levels, taking into account the need for integration with land-use management.

The Agenda also provides good leadership in respect of the stakeholders involved; Figure 7.1 illustrates the range of society's groupings deemed to have key roles in sustainable development, and underlying their involvement are the basic human institutions of finance, education, law, etc. As Priscoli (1989) puts it:

> Institutions are the embodiment of values in regularised patterns of behaviour.
> The institutions and organisations that supply and distribute water resources reflect society's values towards equity, freedom and justice.
>
> (Priscoli, 1989, pp. 33, 34)

While many water professionals may prefer the stability of their occupational remoteness from everyday issues, the importance of their field makes this an impossible luxury:

> The significance of modern science and technology is that we now know well the potential for degrading the water and for safeguarding it, and this sharpens the *social and political challenge of water policies.*
>
> (Kinnersley, 1988, pp. 6–7; emphasis added)

As was noted earlier (Chapter 1), a feature of the hydraulic civilisations of prehistory, due mainly to their complete dependence upon irrigation, was a very strong social structure in which the most basic, vital commodity was a

ELEMENTS OF AGENDA 21

Figure 7.1 The community scale of Agenda 21 implementation

focus for human organisation. In *Oriental Despotism* Karl Wittfogel (1957) explains the origins of autocratic power through the development of water supply systems. One might speculate that such a strong structure produced:

(a) The technology to exploit the vital resource.
(b) A means of regulating rights to, and use of, the resource.
(c) Protection of the resource against those who had no rights.
(d) Protection of the resource against misuse and deterioration.

At the end of the twentieth century the role and relevance of social and institutional structures is under scrutiny in connection with the whole field of contemporary environmental management. Environmental politics and disputes over basic resources and their despoliation have ensured that societal structures are being re-examined in relation to environmental management of land, air and water. The essential thing about water management with an environmental objective is that the catchment or river basin is an easily appreciated plan projection of the ecosystem requiring management, an area within which the population have one form of common identity, whether they exploit it or not. Consequently, it is suggested that river basins are an ideal unit for many forms of environmental management and for popular participation in that management. In introducing the National Rivers Authority to England and Wales in 1989 its Chairman described it as the strongest *environmental* organisation in Europe; it has now become (1996) the main component of the Environment Agency, all of whose activities (including air pollution and waste controls) are planned around river basin boundaries.

The transition from engineering-dominated *distribution* philosophies to hydrology- (and environmental science-) dominated *collection* philosophies has accelerated in the developed world in the last decade; the American Society of Civil Engineers now speaks of the 'Life' agenda, that is, the

Legal Institutional Financial Environmental

aspects of each water scheme. Of these, the institutional framework is most important since it determines and channels the effectiveness of legal structures and financial processes. Institutions are also important because of the increasing realisation of the necessity to consult widely with the population before environmental policies are implemented. While, for example, purely economic analysis can produce an optimum solution to the allocation of resources in a river basin and purely hydraulic analysis can produce implementation technology, an 'appropriate' design has a host of qualitative aspects which, except in totalitarian hydraulic civilisations, can best be analysed in public.

In this chapter we explore the pattern of institutions relevant to river basin management in the restricted sense of the organisations responsible for that activity. Such organisations will obviously rely heavily on both regulation and economic instruments to carry out their work and in a sense these have become rival guides to institutional behaviour in the developed world with the economic argument being that markets involve people more directly and as freer agents than the framework of law. In order to judge the success or failure of specific river basin institutions it is appropriate to begin with some

Table 7.1(a) The European Water Charter

---

1 There is no life without water. It is a treasure indispensable to all human activity.
2 Freshwater resources are not inexhaustible. It is essential to conserve, control and, wherever possible, to increase them.
3 To pollute water is to harm Man and other living creatures which are dependent on water.
4 The quality of water must be maintained at levels suitable for the use to be made of it and, in particular, must meet appropriate public health standards.
5 When water is returned to a common source it must not impair further uses, both public and private, to which the common source will be put.
6 The maintenance of an adequate vegetation cover, preferably forest land, is imperative for the conservation of water resources.
7 Water resources must be assessed.
8 The wise husbandry of water resources must be planned by the appropriate authorities.
9 Conservation of water calls for intensified scientific research, training of specialists and public information services.
10 Water is a common heritage, the values of which must be recognised by all. Everyone has the duty to use water carefully and economically.
11 The management of water resources should be based on their natural basins rather than on political and administrative boundaries.
12 Water knows no frontiers; as a common resource it demands international co-operation.

Table 7.1(b) The San Francisco Declaration

This declaration calls for a moratorium on all new large dam projects, either planned or under construction, which fail to satisfy all of the conditions listed below. A moratorium should be implemented by all those countries, agencies, and banking institutions involved in financing and building large dams either through loans, the sale of equipment, or the provision of services.

A large dam can only be built if:

1 The *people* affected are included in the planning process and have the power of veto over the project.
2 The people who finance the project (such as taxpayers) and those affected by the dams have total access to *information* on the project.
3 It does not *threaten* national parks, heritage sites, areas of scientific and educational importance, tropical rainforests, or areas inhabited by threatened or endangered species.
4 It improves public *health* and does not threaten to spread waterborne diseases.
5 It poses no threat to downstream *fisheries*.
6 It poses *no threat* to the water quality and water supplies of those living downstream.
7 It poses no threat to *downstream agriculture*, either through increasing salinity or through the deprivation of nutrients.
8 Its associated irrigation works can be guaranteed not to lead to *salinisation* or *waterlogging*.
9 It provides irrigation for local food production and not solely for *export* crops.
10 It benefits large sectors of the population rather than just the *urban élite* and export industries.
11 Its operation and maintenance is under *community control*.
12 Its energy produced will not be used to fuel *environmentally* damaging activities.
13 It poses no *threat to public safety*, such as inducing earthquakes or the collapse of a dam.
14 A full *assessment* of the short- and long-term environmental, social, and economic effects has been submitted for independent review by a body of experts approved by the communities involved.
15 Available energy-efficient improvements and *water conservation* measures, using the latest technology, have already been implemented.
16 Those people who have suffered the loss of homes and livelihood from completed projects are fully *compensated* with land and other means by the governments and banks that financed those dams.
17 The same governments and funding agencies implement an immediate program to *reforest* those watersheds that have been adversely affected by past water development projects.

very general principles about the management of water as a resource, non-technical but derived from knowledgeable interest groups and activists. We present here both the European Water Charter (Table 7.1a) and the San Francisco Declaration (and the Watershed Management Declaration which forms an addendum to it) (Table 7.1b–c) as statements of principle. Both

Table 7.1(c) The Watershed Management Declaration

1 International efforts must be increased to bring back the *vegetation* that once acted as a groundcover for the river catchment areas. The loss of this groundcover in the last century is a major reason for the depletion of ground-water, soil erosion, droughts and floods in many countries.

2 *Groundwater* must be considered a renewable resource and its use should not exceed its natural recharge.

3 The need for water must first be identified at the *community level*, and any solution devised to meet those needs must include the explicit identification of users and beneficiaries. Solutions must be appropriate to indigenous resource-use patterns.

4 *Local production systems* should be strengthened by phasing out use of capital-intensive, agricultural chemicals, fossil fuel derivatives, and excessive water in favor of low-cost, ecologically safe alternatives.

5 The timetable of a water project should be determined by donor-driven funding cycles. Appropriate development is an economic solution for the long term. Therefore, its planning and implementation must be determined by the *cultural and economic aspects of the community in question.*

6 Reinstitute *traditional methods of water preservation and use.* Rather than building reservoirs bring back methods such as those used in India where forested buffer zones around catchment systems, ponds, water tanks, and wells helped to protect water supplies.

7 *Rainforest preservation* of the earth's great watersheds, such as those of the Amazon and Congo regions, requires our most urgent attention. Rainforests play a crucial role maintaining the health of the biosphere.

8 Legal and political rights to protect the environment are simply not recognised in many countries.
 Therefore we request that all countries
 (a) Create and strengthen *environmental regulations* for water management.
 (b) Democratise and decentralise decision-making for environmental protection and natural resource management. This includes a *public-hearings* process for all project proposals.
 (c) Uphold the *human rights of environmentalists* and water project critics.

9 Create an *International Code* of Waters Resource Management that would provide the legal guidelines for water development and for public interest groups to challenge violations of the law.

10 A compilation of successful sustainable water programs should be prepared and published by the member organisations of IRN.* This can help to encourage the academic community and development experts to re-examine the traditional systems and help to rebuild the self-respect and self-reliance of indigenous peoples.

* International Rivers Network

include the requirement to set the river basin unit as the unit of management. The points given extra emphasis in Table 7.1b and c draw attention to the basic dimensions of people, environment and the long-term, far-flung assessment, monitoring and mitigation of the impacts of water develop-ments. Table 5.1 (p. 152) represents a 'stronger' scientific statement of the ecological principles of river basin development.

7.1   BASIN AUTHORITIES: THE INFLUENCE OF THE TVA

It is appropriate to begin with the Tennessee Valley Authority (TVA) because:

(a) It was formed to ameliorate existing environmental problems.
(b) It has been influential across the world in promoting successor organisations.
(c) It has been the subject of studies which reveal that, by some definitions, it was and is largely unsuccessful.

A British observer of the Tennessee Valley after the American Civil War commented: 'The Tennessee Valley consists for the most part of plantations in a state of semi-decay and plantations of which the ruin is total and complete.'

The extent of the valley (it covers an area equivalent to 80 per cent of England and Wales) means that its environmental problems range from eroded hill farms of the East to malarial swamps in the West. Flat-bottomed boats could navigate much of the 652-mile-long channel except at Muscle Shoals where canal works were begun late last century. During the First World War Muscle Shoals was selected as a site for a hydropower dam and fertiliser plant; it is important to note that this proposal did not survive economic evaluation prior to the 'New Deal' conditions introduced in the 1930s by Franklin Roosevelt. Introducing the Tennessee Valley Authority in 1933, Roosevelt described it as: 'charged with the broadest duty of planning for the proper use, conservation and development of the natural resources of the Tennessee River drainage basin'. (It should be noted that in the USA in the 1930s the word 'conservation' has the meaning of managed use of resources rather than its more recent 'green' definition.)

Much of the criticism of the TVA and its achievements stems from a cultural distrust of large public authorities (as redolent of socialism), and a suggestion of duplicitous power- and finance-brokering by the leading lights (Arthur Morgan, Engineer, Harcourt Morgan, President, and David Lilienthal, Lawyer and Administrator). However, despite recent criticism of the heavy dependence upon dam-building, it is to the integration of land and water management that supporters of the TVA most look for praiseworthy achievements. The 'ruin' observed after the Civil War was largely repaired by improvements to land management and erosion control, with dams and navigation improvements bringing power and salience to a neglected peripheral region.

The TVA supporters would therefore claim that the Authority had trail-blazed the field of *integrated basin management,* led by economic restoration. Recently Downs *et al.* (1991) have referred to the sometimes platitudinous or rhetorical use of the term 'integrated river basin management'. Of twenty-one different approaches analysed, five basic components of integrated schemes are derived: water, channel, land, ecology and human activity. Downs *et al.* prefer to separate the term *comprehensive* river basin management, where several components are involved, retaining *integrated* basin management for schemes where the components interact (though one may lead); they then interject *holistic* river basin management to cover both divisions but emphasising system energetics, change and human inter-actions. Thus, the TVA was a comprehensive organisation but had too little hydrological basis to be integrated and since the working definition of 'conservation' was wholly anthropocentric, it had little concept of sustain-ability. It must also be added that, like many such bodies since the inception of the TVA, part of its attraction was that it gave national identity to a peripheral region.

7.2   DOES AN IDEAL RIVER BASIN MANAGEMENT INSTITUTION EXIST?

Much of the comment by analysts on the TVA and its many imitator river basin authorities has been negative in some respect; often criticism locates near the interface between the technological side of the institutions and the need to be publicly accountable. Clearly the institutions cannot take all the blame since they seldom operate outside the basic policy framework of the host nation, its laws and its financial plans. National traditions of centralist or devolved power broking and decision making also profoundly affect the success of river basin organisations. However, there are systematic problems involved with operating *knowledge-based systems* and, because all future environmental management will fall into this category, it is worth pausing to investigate the power of knowledge about the system to be managed before, in the next section, moving on to how the system is opened to public scrutiny.

If river basins are to be managed successfully there has to be knowledge: this must be both basic geographical knowledge and the technical knowl-edge needed to engineer, control, purify and provide basic resources. In Chapters 4 to 6 we gained the impression that the basic categories of knowledge required are:

(a) A complete description of, and database for, the catchment.
(b) An understanding of the physical processes operating under the boundary conditions specific to the basin.
(c) A breakdown of the resource needs and problems (including conserva-tion) in the basin.
(d) Technological knowledge to manipulate the resources and hazards of the basin.

While *information* may be regarded as inferior in status to *knowledge* there is clearly a justifiable need for information on the current state of any system by those who manage it. The kinds of information available, and particularly the level of sophistication in its gathering, have a considerable bearing upon the flexibility of the institutions which manage river basins. One of the traditional problems of river basin management has been the *interdisciplinary* nature of the knowledge base; information on the basin has, therefore, always been tagged or coloured by a particular application rather than used to create a synoptic view.

There are also new demands upon the kind of information available to be presented to the public. To consult the public on what remain highly technical issues requires innovation in modelling, computing and interactive display. Flug and Ahmed (1990) describe a modelling scheme for reservoir-regulated flows which has a decision-support role in circumstances involving special interest groups. The model functions as a screening tool to identify good and bad flow alternatives. We return to the special role of knowledge in guiding management practice in Chapter 8.

In judging the success or failure of a river basin management institution it is clearly too late to leave analysis until its fiftieth anniversary (as in the case of the TVA); what is required is a checklist characterising the ideal or proven strengths of such agencies. Such a framework is provided by Mitchell (1990), summarised here as Table 7.2.

Table 7.2 Integrated basin management: an analytical framework

Aspects	Components
1 Context	State of the environment (river problems)
	Prevailing ideologies
	Economic conditions
	Legal, administrative and financial arrangements
2 Legitimation	Objectives of basin organisation
	Responsibility, power, authority
	Rules for intervention, conflict resolution
3 Functions	Generic functions, e.g. data collection
	Substantive functions, e.g. resources, pollution
4 Structures	Central versus dispersed (functional/geographical)
	Accountability
	Flexibility
5 Processes/mechanisms	Councils, communities, task forces
	Professional linkages – interdisciplinary action
	Plans and planning processes
	Benefit-cost analysis
	Environmental assessment
	Public participation
6 Cultures/attitudes	Service to public
	Bargaining/partnerships

Source: after Mitchell (1990)

In an empirical survey of integrated river basin management agencies conducted by the Organisation for Economic Co-operation and Development (OECD, 1989) the problem of legitimisation is given special attention. The report includes around 100 case studies and over fifty country reports, allowing OECD to conclude on the relative importance of thirteen administrative/institutional characteristics leading to effective integration of management. The Organisation remarks that this stress on institutions marks a major shift from its traditional emphasis upon economic approaches to achieving efficiency and legislative approaches to resolving environmental problems. In the case of water resource management, however, integration between such diverse strands as pollution control and transportation or hydropower and forestry must be achieved. OECD suggests that the perpetual danger is that of independent, fragmented groups with narrow mandates having a vested interest in a closed decision process.

The thirteen OECD guidelines were derived from four basic dimensions of the 'good' water resource organisation:

(a) Political credibility and legitimisation.
(b) Organisation structures relevant to a spatial hierarchy of issues and functions.
(c) Processes and mechanisms for bargaining, negotiating and planning.
(d) Organisational culture and participant attitudes which will communicate, educate, etc.

Nevertheless, even with the perfect set of attributes, an organisation designed to bring about successful integration and put proactive, anticipatory action foremost will need to improve legislation and also employ 'the economics of integration'. There is considerable current rivalry (arranged largely along left–right political lines) between regulatory and economic approaches to environmental management. Authors such as Winpenny (1994) appear to advocate comprehensive adoption of fiscal measures in river basin management and he roundly criticises the systematic approach to regulatory basin frameworks as an intellectual luxury. Synott (1991), in defining the real openings for economic instruments, exposes the relatively narrow range of their true benefits. Postle (1993) has examined the potential for environmental economics in the costing of environmental benefits and costs of development for the UK National Rivers Authority; she finds that applications are limited and techniques approximate. Resource pricing (including the cost of pollution control and water purification) can, however, be seen as a key element for driving integrated management. The transition to realistic pricing is a special challenge for developing economies which have favoured heavily subsidised water services; Arntzen (1995) notes that in Botswana the impediments to the use of economic instruments include a significant non-market economy, limited expertise and immediate welfare problems for the communities served. Nevertheless, Botswana's

National Water Master Plan is reducing wastage, reducing the use of irrigation water, boosting the development of water-saving appliances and improving the economic viability of recycling. Clearly, benefit/cost analysis at an early stage of water-based developments can avoid unsustainable decisions.

The OECD study promotes a strong movement away, however, from purely economic approaches to river *basin* management. These were formerly related to the development of resource systems rather than their sustainable management but reached a high degree of economic/mathematical sophistication prior to their demise. Krutilla and Eckstein's (1958) study applies economic efficiency as a criterion in three US basins developing integrated hydropower schemes. The generation of power is an ideal vehicle for conventional economic analysis and in such studies 'integration' often refers to the efficient conjunctive use of a number of reservoirs for generation. Even these early economic analyses admitted, however, that 'higher' criteria often applied, for example: 'complementary institutions for group decisions and collective action are required to meet adequately the needs of the members of a free society' (Krutilla and Eckstein, 1958, p. 267).

In the following pages we shall begin to advocate a view that, in order to implement sustainable development of a comprehensive range of river resources and to influence those bodies concerned with land use, land management and other livelihoods a strong agency will be prepared and able to carry out *Catchment Management Planning*. Furthermore, environmental management functions outside the water sector will accrue to such an agency if the planning process becomes established as politically valid.

7.3 CASE STUDY: SUSTAINABLE BASIN MANAGEMENT AND UK WATER INSTITUTIONS

There are many reasons for a special study of water institutions in the UK besides this author's nationality; however, UK water institutions were not featured in Chapter 4 as part of the developed world pattern. As a result of being among the earliest nations to industrialise and simultaneously to urbanise, the UK has a century-and-a-half of institutional developments from which to select trends. It may well be that, despite the OECD's recommendations, institutional controls are always in flux and rapidly evolving; if so, a historical survey of one nation may repay our detailed attention (as did broader historical trends in Chapter 1). This history, recently much reviewed by protagonists and opponents alike of the 'privatisation' of the water authorities in England and Wales, illuminates a number of issues of general relevance to river basin management (Rees, 1989). They are:

(a) Policy responses to the growing list of activities attending the provision of public benefits from water.

(b) Attitudes to private, municipal or basin authorities (with attendant issues of commodification of values and public consultation).

(c) Scales of organisation.

The importance of cultural and traditional factors in river management systems may also be gathered from the UK example: there are entirely separate patterns of organisation in Scotland, Northern Ireland, and England and Wales. Here we are concentrating on England and Wales because of the much heavier population pressures in those two countries.

In Chapter 1 we briefly considered the history of the major evolutionary changes in river basin management in Britain. Rowland Parker's records for Foxton were quoted as an indication that sophisticated *local,* stream-based by-laws were in operation, each of which (whether for maintenance of the channel or pollution control) gave a boost to the integration of legal interests downstream (and, in navigable reaches, upstream too). Kinnersley (1988) also speaks of the Foxton records as illustrating a golden age of simple riparian principles.

Later, as part of the development of common law, this riparian principle required definition and implementation as pressures far greater than 'dunghills' and 'cess-pits' began to threaten the sharing of a fundamental resource. As Kinnersley puts it:

> the sharing of water has long been and will continue to be, in all communities, a matter requiring public and constitutional governance by lawyers, courts and community leaders at various levels.
>
> (Kinnersley, 1988, p. 34)

At the first stage of the Industrial Revolution in England it was the issue of navigation which forced the local by-laws into a larger scale of relevance. The construction and filling of canals also had a profound influence on rivers (Rolt, 1985). Kinnersley views the canals as 'among the first privately financed large constructions intended for use by all comers' (p. 43). Their individual Acts of Parliament were 'cast in a statutory format setting out obligations which protected public and private interests as well as rights necessary for effective co-operation' (p. 43). The canal companies required water from rivers, interconnection with river users and the agreement of landowners for construction; the latter were well represented in Parliament and so the Canal Acts provided a balance between gaining and losing interests and wider public rights. The extension of this institutional structure to municipal water supply and sewerage was, however, much less straightforward. Throughout the eighteenth century commentators made much of the critical problem of the emerging urban centres: a fair supply system of water for both fire fighting (the major physical urban hazard) and human health.

While many authors document the problems of Victorian urban water supply and sanitation as a series of local cameos, two important books develop the critical themes of *local institutional development* (Rennison, 1979, for Tyneside) and the role of health reformers and engineers (Binnie, 1981). Public health was to become the single issue around which the emerging local democracies for cities – the municipalities – could form. Chadwick's *Report on the Sanitary Condition of the Labouring Population of Great Britain* (1842) was followed by the facilitating legislation for private Bills developing corporate powers for water supply. This was in 1847; by 1878 there were seventy-eight municipal water undertakings.

It is worthwhile to examine briefly the reason for the adoption in Britain of water-borne sewerage systems. While the open channels of early Victorian cities were heavily abused by the dumping of every form of waste, the early preference for disposal of human lavatory waste was to land, often outside the city, as in Edinburgh. Until 1818 the law prevented waste other than from surface and kitchen sources from entering sewers. However, the perfection of the water closet (WC) allowed legislation making discharge of cesspool waste into urban drains obligatory (1847). There was clearly none of the anticipatory or precautionary thinking we now find essential when drainage of surface waters was added to this sewerage function.

The rapid expansion of city buildings and roads led to a big surface-water drainage problem, brought home frequently even today when street drains are blocked or surcharged in a downpour. It was logical to channel excess surface flow to the nearest natural river and a major saving in investment could be made if foul sewers were combined with the surface drains such that street drainage could flush the solids along the network of pipes (steep gradients were not available in most cities).

Rivers became the sink for the sewerage systems of the neighbouring cities, once again without considering the immediate or eventual capacity of the river ecosystem to cope. Even if the cholera epidemics became curtailed by the rise of municipal institutions for supply of water, what became of waste was to herald the next scandal – that of stinking rivers. Parliament itself, next to the Thames, had to meet behind sheets soaked in disinfectant in the 1870s and a series of Royal Commissions eventually led to the Public Health Act 1875 and the Rivers Pollution Prevention Act 1876.

These two years mark a very important stage in the development of British institutions for river management:

(a) We see the beginnings of a move from municipal units to river basin units (a law for water supply in rural areas was passed in 1878 and, as Kinnersley records, a geologist named Toplis suggested twelve new river basin authorities in 1879).

(b) The Rivers Pollution Prevention Act 1876 laid down principles of control which reflect a British cultural attitude to statutory law in environmental matters. While it was now to be an offence to discharge waste into rivers (not only sewers but mines, gasworks and other features of industrial growth were also a threat), it was left to the municipal authorities to administer the Act and it was to be a defence that polluters had used the 'best practicable means' of preventing pollution. Since most municipalities were keen to promote further growth at the lowest cost to developers these three words became a polluters' charter. Demand for river capacity was clearly not to be managed during these expansionist days, a situation frequently repeated in the developing world today.

At this stage, therefore, Britain reached back an era to reassess the common law principle of riparian rights; these could not empower the new institutions since they were property rights but, redefined by a series of key judgements, they were to provide some check to river-based developments. It should be noted, however, that riparian rights are not identical the world over; for example, in South Africa there are now attempts to disestablish property-related water rights which are in the hands of a minority of riparian owners in favour of the basic human right to water supply in townships away from rivers – clearly national priorities need centralised apportionment of resources.

There is argument over the classic definition of riparian rights. Kinnersley (1988) chooses Lord Macnaughton's judgement in 1893, while Wisdom (1979) chooses Lord Wensleydale's 1859 judgement in the case of Chasemore v. Richards:

> the right to the enjoyment of a natural stream of water on the surface ... belongs to the proprietor of the adjoining lands as a natural incident to the right to the soil itself ... He has the right to have it come to him in its natural state, in flow, quantity and quality. [For the full definition see p. 16, this volume.]
>
> (Wisdom, 1979, p. 83)

Importantly, groundwater rights were appropriative and judgements made it clear that landowners could not win cases brought against the use of wells or boreholes on a neighbour's land. The law on pollution of groundwater was also unclear. The separation of legal attitudes to surface waters and groundwater in the UK, set down at this early stage, has had unwelcome effects on all aspects of comprehensive water management until recently (see Chapter 3).

Summarising the first phase of management institutions in England and Wales we find that:

(a) They were local in scale (from village to municipality) but this merely gave access to existing power groups and any restrictions on demand for river uses were met with hostility.

(b) Lack of basin scale units hindered any wider application of the 'upstream/downstream' philosophies of riparian rights; essentially the local reach was seen as a self-purifying sink for waste to the cost of downstream communities – only direct damage or 'nuisance' was treated legally.

It became almost inevitable that basin scale authorities would be set up and, as is perhaps the norm in public affairs, it was the crisis in flooding and agricultural land drainage which promoted the next significant institutional changes.

The 1930s and 1940s saw, therefore:

(a) The beginnings of a move from municipal to river basin authorities.
(b) The incorporation of public representation to encompass all relevant related interests.
(c) An increasing role for central planning (and therefore a separation in arrangements for Scotland and Northern Ireland from those for England and Wales).

7.4   RIVER BASIN UNITS: LAND DRAINAGE LEADS THE WAY

An international audience might find it difficult to appreciate the importance of land drainage in the UK without consulting both rainfall and geology maps. The island of Great Britain can be divided into a very wet west and an eastern half which, while markedly drier, has a predominance of low-lying clay land. England and Wales are especially prone to water-logged agricultural soils and to flooding (Figure 7.2) from both rivers and the sea; the first 'Commissions of Sewers' to control local drainage date back to the thirteenth century.

Drainage has, therefore, been of paramount importance in the settlement of and optimum agricultural use of the land in the UK; only recently have conservation pressures forced new attitudes in favour of the retention of remaining wetlands and the new official title for the activity originally (and proudly!) called *land drainage* is now *flood protection* (or flood *prevention* as Hall (1989) and some journalists wrongly record it).

The scientific logic of drainage basin systems makes, as we saw in Chapter 2, the act of draining land in one part of a basin antagonistic in principle to that of protecting downstream areas against flooding. Land drainage interests are powerful enough in England and Wales to have led to the creation of special local institutions. For the 214 Internal Drainage Boards (IDBs) in England and Wales the security of life and economy still centres around efficient land drainage. Their power to levy rates in order to create and maintain arterial drains dates back to medieval times but their formal inception came with the 1930 Land Drainage Act. They have been criticised for being slow to adopt more conservation-orientated practices (Purseglove, 1988) and for being 'exclusive clubs' (Hall, 1989).

Land areas dependent upon complete systems for flood defence and land drainage

Figure 7.2 The drainage problem in England and Wales: flood-prone rivers and wetlands (from Newbold *et al.*, 1989)

Among several Acts of Parliament between the World Wars which attempted to reduce the Victorian legacy in UK water policy, the most far-reaching was the Land Drainage Act (1930) which established forty-seven Catchment Boards. Here at last was, for each major river catchment, an institutional context free of the compartmental strictures of local government yet very widely representative and able to devote investment according to specialist engineering designs across whole catchment areas. The Ministry of Agriculture, which has continued to control land drainage (flood protection) in England and Wales, also ensured a central, specialist coordinating interest.

At the same time the Ministry was also organising catchment-wide Fisheries Boards (from 1923). Fishery interests are a continuing thread in UK river basin management, mainly because they combine the interests of the unpolluted environment (fish act as 'miners' canaries' for pollution) and property rights. They are a valued (in the economic sense) aspect of riparian ownership and rights to fisheries may be sold or leased. Consequently, fisheries, after milling and minor navigations became less important, came to be a major platform from which landed interests intervened in river management.

Kinnersley (1988) records the importance of these developments in membership and financing:

> These new (or in some cases reshaped) boards were bringing together various local interests or lobbies such as elected councillors, landowners, anglers and others directly engaged in river basin activities, in a close relationship with central government also, but in each case for a river-related territory. The significance of the systematic revision of legislation and administration in this way was that, while local influence was being kept very strong, the central government (in the shape of the Ministry of Agriculture and Fisheries) was becoming committed in the 1920's and 1930's to a coherent pattern covering these specialist water functions across England and Wales.
>
> (Kinnersley, 1988, pp. 71–2)

The successes of this phase include the strong local interest created in river management but the failures are obvious – there was little room for conservation and the land-drainage function never really appreciated the relationship between upstream developments and downstream impacts. There was a high level of public investment without a proper analysis of risks and benefits; the owners of fisheries had the most powerful access to the use of riparian rights.

7.5   ISSUES OF RESOURCES AND POLLUTION

The basin-wide needs of fisheries and land drainage for coordinated action to manage resources and hazards led in the UK to strong and purposeful specialist institutions working clear legal and financial systems, but they had far from comprehensive responsibilities even within the water sector. It is interesting therefore to see how the much more generally important themes of water resources and pollution control became administered at the river basin scale.

The 1948 River Boards Act set up thirty-two Boards (plus the Thames and Lea Conservancies). These Boards were to become the administrators of the system of licences introduced first for discharging pollutants to rivers (1951 Rivers, Prevention of Pollution, Act) and, after a further reorganisation into River Authorities, licences to abstract water resources (Water Resource Act 1963). While the concept of *comprehensive* basin management therefore began to grow, the extensions beyond land drainage (flood protection) and fisheries powers for the basin authorities were to be weak, especially in the field of pollution control. Existing dischargers were exempt from licensing, the 'consents' procedures were kept secret (even though the public had a fundamental right to know the sources of river drainage) and the municipal council representation on the Boards often led to a muted approach to improving sewage discharges (mainly from municipal sources!).

For water resources there was a much more positive grouping of interests; few politicians, whether local or national, doubted the wisdom of economic growth in the 1960s. Growth required water resources and the rapidly developing applied science of *hydrology* was now prepared to assess the UK's needs. The 1945 Water Act had regrouped the supply industry and the supply sector was ready and able to distribute large volumes to domestic users (automation and cleanliness in the home was the advertisers' dream), to industry (new large plants placed a huge demand for process and cooling waters) and to agriculture (the drought of 1959 spectacularly checked the post-war productivity surge).

Until the 1960s there were no specialist agencies dealing with water resources. From the Victorian period of health reforms, during which seeking the purity and clarity of upland waters had taken on an almost religious fervour for municipal suppliers, each city wishing to dam an upland valley and pipe its supplies merely took a private Act of Parliament through Westminster.

Not only did the Water Resources Act (1963) perpetuate river basin institutions which were broadly representative in decision-making (thirty-four Boards became twenty-nine Authorities) but water became a resource or raw material like any other: abstractions required consents for which a charge was levied. The scene became set for the commodification of water which was to end in privatisation in 1989. Licensing of abstractions became a public process with advertisement of proposals in newspapers.

The rise of centralism and the continued rise of technocrats in the water industry both received a boost in the Act. It set up the Water Resources Board, a national agency but with non-executive functions, to coordinate water resource policy by advising the River Authorities and collating material from them. The Board began a systematic archive of hydrometric information (the 1963 Act had begun the process of instrumentation in river basins, especially the measurement of river flows).

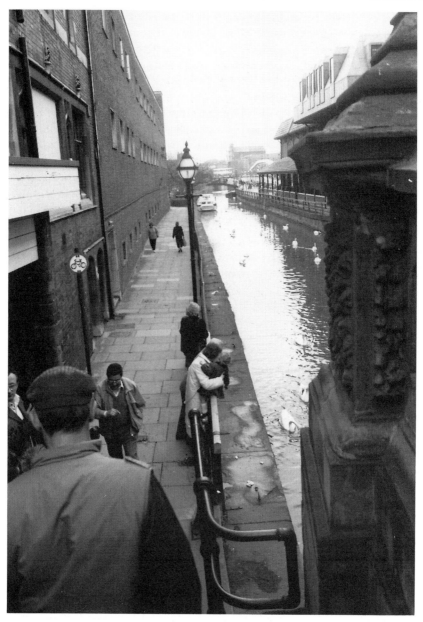

Plate 7.1(a) Flood risk in the centre of Lincoln is high. It has been reduced by creative use of farmland upstream; (b) At high flows water which would normally flood Lincoln is now released onto farmland (photo M.D. Newson)

Plate 7.1(b)

Reviewing the procedures by which water resource planning was achieved during this centralist (and highly technocentric) phase, Rees (1969) pointed to the remoteness and technical domination of the schemes proposed. She revealed the lack of published data on demand and supply, without which, 'no definitive statements can be made on ... future demands for water, or the characteristics of water users, on the possible effects of resource planning and controls, or on the influence of water as a location factor' (p. 2). Her own survey revealed the irony of central planning for a burgeoning demand but the failure of water supply or effluent disposal to influence industrial activity in any major way. Effectively, water supply provision had become an industry in itself. Politically, the centralised era was doomed by demands for more strenuous, more localised, provision of and accounting for water use and a multitude of other river basin activities.

Between the years 1974 and 1989 England and Wales had ten multi-functional Regional Water Authorities charged with operating the human use of the hydrological cycle (i.e. a much more comprehensive brief) within river basin boundaries. Once again the decision-making lacked transparency (Kinnersley, 1988 and Hall, 1989). However, during a period of changing political stewardship it is remarkable that a consensus prevailed. Two reforms were seen as necessary in the early 1970s, one to local government boundaries and the other to control of environmental issues. Almost overnight it seemed that the burden of the UK's water problems had shifted from *quantity* to *quality* of supply.

Environmental interests were, apparently, well protected in the new arrangements if the public put their trust in the technological and spending power of the new Authorities. If they did not have such trust they might suspect an organisational framework which put polluters and prosecutors under the same roof; however, a very generous system of *local government representation* on the RWAs was obtained during the reorganisation – in a way the last throes of the municipal legacy. As an additional reassurance the Control of Pollution Act, passed in the same year as the RWAs began work, made public the consent registers and pollution monitoring data; unfortunately this did not become reality until 1985.

In a celebrated piece of ministerial infighting the Ministry of Agriculture, Fisheries and Food (MAFF) retained control of land drainage (flood protection). The battle between MAFF and DOE, described by Richardson *et al.* (1978), is critical for this review of institutional roles since it effectively prevented complete integration of management and denied an overall environmental 'flavour' to the RWAs. Some of the biggest political battles in their fifteen-year lifetime were to be fought over the production-orientation and farmer-control of the land drainage function within RWAs. This is typical of the tensions in resource management between ministries at national level throughout the world.

Ironically, in the light of the strong political and technical movement away from water resources, the first major practical test of the RWAs was the drought of 1976 (Doornkamp *et al.*, 1980). They came through the water supply crisis 'with flying colours', thanks to some over-provision of storage, to a general neglect of river ecology in seeking Drought Orders (the amount of 'compensation water' allowed from reservoirs to maintain downstream life was much diminished under these Orders) and to some old-fashioned engineering ingenuity in temporarily linking neighbouring supply systems. The only centralised function in the new structure – the National Water Council (which had a pensions and pay negotiating role) – was able to proclaim, 'We didn't wait for the rain' (National Water Council, 1976).

When the newly elected Conservative government of 1979 began its long campaign for executive efficiency in public life (which became a battle against many aspects of local government) it abolished the National Water Council, feeling that the RWAs were now sufficiently competent to manage alone. In a further gesture towards technocratic water management the Water Act 1983 also reduced local government representation (replacing it with consumer consultative bodies), closed meetings to press and public and appointed managerial personalities to key vacancies on RWAs. It would be easy to think that the next major change – privatisation – had begun six years before its enactment.

7.6 PRIVATE OR PUBLIC? ECONOMICS AND ENVIRONMENT AS INSTITUTIONAL FORCES

It is no secret that the concept of integrating river management was championed by the professionals in the RWAs; many made explicitly antagonistic statements during the run-up to privatisation, mourning the proposed loss of total functional integration. However, the RWAs also attracted some detailed political science scrutiny which was less favourable. Interestingly, this was not because of their river basin scale or comprehensive duties but because of their regional autonomy and technocratic monopoly status.

Saunders (1985) placed the Regional Water Authorities alongside the Regional Health Authorities in order to examine the divergence between what he calls the 'politics of production' and the 'politics of consumption'. Local government lost control of both water and community health in 1974 yet, claims Saunders, there was no political analysis of the 'regional state' which was being thereby empowered. He goes on to use the dual politics thesis (production, consumption) and to select the water authorities as examples of the former, subject to influence from those enjoying property rights yet with a large degree of managerial autonomy.

Patterson (1987) is yet more critical of the removal of water management from its democratic, municipal (Victorian) roots in favour of regionalisation in technocratic, remote agencies upon which was then forced the commodification of water in an ill-disguised lead-up to privatisation (following a spell of minimal investment and labour-shedding – see Figure 7.3). In terms of our argument here, over basin management, Patterson claims that while the authorities had integrated control of the water cycle they made little or no attempt to link it to local authority planning, thus they were not 'integrated' in the terms used by Downs *et al.* (1991).

By the mid-1980s the regular surveys of river water quality carried out by the RWAs were beginning to cause concern: a deterioration had replaced the steadily improving trend since the first surveys in 1975. New sources of pollution were partly to blame, such as spillage of slurry and silage liquor from livestock farms and nitrate from arable land; groundwater was becoming polluted and the European Community was steadily tightening the standards for river water, supply water and coastal water. However, a major component of the RWAs' problem remained their own decaying infrastructure of sewerage and sewage disposal. The legacy of the Victorian reformers, when not maintained and replaced by modern investment, became a liability and the progress made by Mrs Thatcher's government in reducing public investment had badly hit the RWAs (see Figure 7.3). A key mistake had been made in terms of sustainability: deterioration of assets had set in.

The Thatcher government had another doctrine in pursuit of economic efficiency: privatisation of large public corporations. Kinnersley (1988)

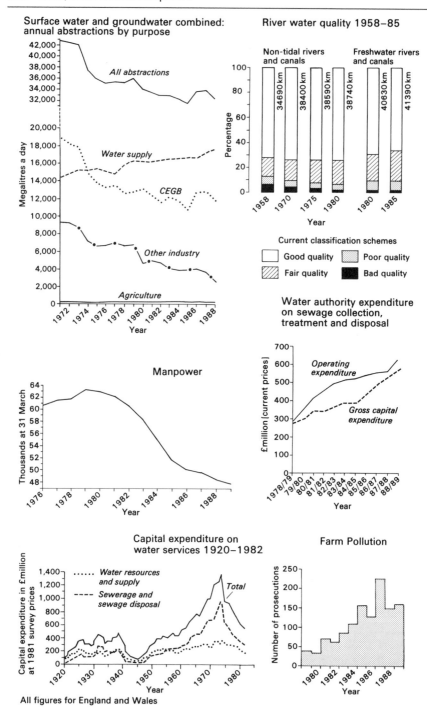

Figure 7.3 Trends in investment, labour and problem abatement in the water
industry of England and Wales at the time of privatisation (1989)
(sources: water industry/DoE)

Table 7.3 Why private ownership?

The Government believes that the privatisation of the water authorities will benefit their customers and employees, and indeed the nation as a whole, in the following ways:

1 The authorities will be free of Government intervention in day-to-day management and protected from fluctuating political pressures.
2 The authorities will be released from the constraints on financing which public ownership imposes.
3 Access to private capital markets will make it easier for the authorities to pursue effective investment strategies for cutting costs and improving standards of service.
4 The financial markets will be able to compare the performance of individual water authorities against each other and against other sectors of the economy. This will provide the financial spur to improved performance.
5 A system of economic regulation will be designed to ensure that the benefits of greater efficiency are systematically passed on to customers in the form of lower prices and better service than would otherwise have occurred.
6 Measures will be introduced to provide a clearer strategic framework for the protection of the water environment.
7 Private water authorities will have greater incentive to ascertain the needs and preferences of customers, and to tailor their services and tariffs accordingly.
8 Private authorities will be better able to compete in the provision of various commercial services, notably in consultancy abroad.
9 Privatised authorities will be better able to attract high quality managers from other parts of the private sector.
10 There will be the opportunity for wide ownership of shares both among employees and among local customers.
11 Most employees will be more closely involved with their business through their ownership of shares, and motivated to ensure its success.

Source: Department of the Environment (1988)

suggests that the government 'stumbled' into privatisation of the RWAs in England and Wales, though Table 7.3 indicates a large number of apparently rational arguments made by government for the policy.

As Kinnersley points out, private companies can produce very efficient water supply and sewerage services but in nations where they do so the infrastructure remains in local government ownership (e.g. France) and there are strong regulatory agencies to maintain standards and to run integrated river basin management. In fact EU rules do not permit private companies to operate the regulatory rules contained in EU Directives.

Rees (1989) develops, at length, the lead-up to privatisation but her main focus is on the way in which true resource economics would impinge on water supply, rather than the impact of regulatory, environmental operations. She was doubtful that the privatised water and sewerage enterprises would achieve improved quality service, renew assets, respond to growth, improve water quality and protect customers against price rises. Price rises sought by the privatised water utility companies have been regulated by the Office of Water Services (OFWAT) since privatisation, in theory offering the public the choice of environmental improvements at a rate their pockets could afford. With so many objectives and so much regulation Rees cannot foresee the pricing of water services as operating any control over demand, even if metering were introduced. In other words 'free market' effects are impotent in a (rightly) regulated social resource.

The government's response to blocking of the move by the European Communities was a dual public and private system, perfectly delimiting the 'rival' approaches to resource management. A National Rivers Authority (NRA) was formed to carry out the regulatory, environmental and flood defence functions for the same river basin areas as the privatised water supply and sewerage authorities. The full list of duties proposed for the NRA is: water resources (planning, licensing), pollution control, flood protection, fisheries, conservation, recreation and navigation. These seven duties are essentially those of any comprehensive river management organisation – the question begged was whether any move towards integrated or holistic management would be written into the foundations of the NRA.

When the National Rivers Authority inherited its new role on 1 September 1989 its staff were described as 'Guardians of the Water Environment'. It had leapt from being a hastily designed alternative to full privatisation to being a vanguard of the rapidly 'greening' Conservative government. It had ten regional units, each with a large measure of public consultations including Regional Rivers Committees, Regional Fisheries Committees and Regional Flood Defence Committees. It also had a firm central administration in London (sixty staff, compared with 6,500 in the regions). Its annual spend was set at £300 million of which £80 million came in from the Department of the Environment direct.

Summarising the build-up of river basin management institutions in England and Wales once more we see that since 1930:

(a) River basin outlines have been retained, with functions progressively added to the point where, in 1974 the Regional Water Authorities were the first comprehensive management bodies. However, the grouping of all functions under one roof led to public fears of monopolistic and technocratic power.

(b) Public representation and central funding were progressively reduced during the period of the Regional Water Authorities, reducing their capacity, for example, to prosecute polluters from outdated sewage treatment plants (often within the same building!); privatisation became an obvious source of investment and of education for the public that water and river quality were not free resources.

(c) National interest, within a regionally devolved system, was retained (albeit progressively reduced) for the purpose of strategic assessments of water resources; it was also critical that national statutory improvements (e.g. Control of Pollution Act 1974) were made to assist the basin managers.

(d) Public interest was retained at the formation of the NRA by keeping regulation functions within the public sector, by restoring higher levels of public representation and by making data services and educational programmes part of the NRA's remit.

(e) The NRA quickly adopted the larger river basin boundaries as the basis for its management structure (within the original regional units). It began to harden the catchment approach in all its duties.

(f) Privatisation is now perceived to have improved the administration of pollution control but financing further environmental improvements is not a matter for public participation – it is rather a matter for the financial regulator and the commercially diversified interests of the water and sewerage companies.

Perhaps the biggest test of the current structure will be when continued drought or drives for 'stronger' sustainability open opportunities for demand management. In recent droughts more than half the population of England and Wales have faced supply shortages (Marsh and Turton, 1996); while the mechanisms of demand management (including conservation and recycling) are well known, their application is rare outside individual housing developments (Guy and Marvin, 1996). During 1996 the privatised Yorkshire Water promoted the development of an environmentally risky water transfer scheme despite its admission of poor planning and infrastructure management. Part of the regulatory procedure of the privatised companies involves judging their performance on the basis of reducing supply shortages and restrictions!

It is interesting to note that the first seeds of truly integrated management (involving land-use planners) and holistic vision for catchments were already planted at the time of the 1989 Water Act. Thames Water Authority (NRA Thames Region, after 1989) began to formalise public consultation in the late 1980s, particularly in connection with flood protection projects. A handbook of their experiences is now available (Gardiner, 1991).

The essence of the Thames approach was and is *appraisal* and *review* which, following the definition of a project, are applied at every stage to the evaluation of options in three fields: economics, engineering and environment. Thames (and now NRA) believe that it is the responsibility of a public authority to:

(a) Demonstrate that all relevant factors have been properly considered at the appropriate stage.

(b) Provide an appreciation of the important impacts of a proposal and its alternatives.

(c) Improve the quality of decision-making.

(d) Ensure efficient implementation of the project.

Project appraisal must be set within the context of integrated river basin management, wherein the changes induced by man tend to change system morphology and the balance of the river environment. Clearly, Thames Water, and now the NRA, have faced costs in this approach, not least in recruiting the relevant disciplines for the promotion of each project. In fact it is, in the view of Gardiner (1988), the interdisciplinary, holistic view of river basins (Figure 7.4) which becomes the major drive for vesting the rejuvenated field of river engineering in the context of project appraisal and management of system equilibria. Woolhouse (1989) confirms the need for both modelling the hydrological effects of new urban developments in Thames Region (where development of Luton, Stevenage and Harlow has had costly effects on flooding) and interaction with planning authorities. The former becomes the vehicle for the latter. Catchment Management Planning had, like the first river basin authorities, been born in the flood protection sector and it is possible that each national situation will spawn such a move in the most hazardous sector of a comprehensive river basin management portfolio.

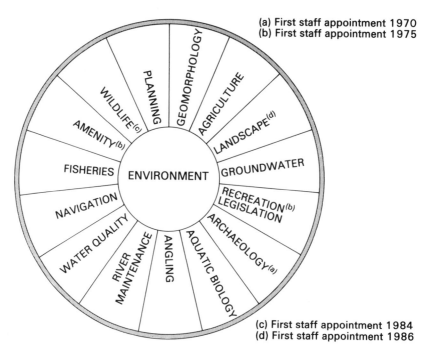

(a) First staff appointment 1970
(b) First staff appointment 1975

(c) First staff appointment 1984
(d) First staff appointment 1986

Figure 7.4 The holistic, interdisciplinary view of river management (modified from Gardiner, 1988)

7.7   RIVER BASIN INSTITUTIONS AND DEVELOPING NATIONS

The historical survey of institutional change in UK river management can be said to have taken the nation's rivers from rural low-density usage to urban, industrial exploitation. There are ways in which this sequence is unlikely to apply to nations experiencing contemporary development of their river basins in the developing world:

(a) The UK is unusual, in that erosion problems are not a contemporary corollary of settlement.
(b) The UK needs little irrigation and its industrial revolution was powered by fossil fuel not by hydro-electric power: the remaining water power interest in fact led to legal protection which is now seen as beneficial.
(c) The earliest movements towards integrated basin-scale management came from land drainage (which is not a problem in most developing countries) and pollution (which becomes a basin-wide problem only after development).

The two most pressing river basin development problems in the developing world are irrigation, to permit rural development, and power generation, largely for urban populations. Both are served best (under existing technologies) by the construction of dams. Because erosion problems are a frequent accompaniment to land pressure problems in the developing world, it is to river basins as sediment transport systems that we should look for guidance on the priorities for institutional action to integrate the interests of land and water users (Newson, 1994a and Chapter 2).

The lesson of these differences should be that UK and other developed-world water professionals enter an entirely new world when they work on development projects. Elliot (1982) points to this confusion of identities but also to the intellectual training of single discipline excellence which is the European tradition in water technology. This, he considers, is a threat to communication between disciplines; therefore we may have *multi*-disciplinary project teams but very little of the *inter*-disciplinary work which is now shown as necessary in both developed world and developing world river management. While the powerful leaders of developing nations often have very little political desire for public participation in river basin development, they should at least expect a good project. Elliot says:

> It is important to emphasise the losses in efficiency that result from this failure of communication. Resources are misallocated and specifications or operating schedules are suboptimal precisely because one discipline cannot fully internalise the complexities and requirements of the other.
>
> (Elliot, 1982, p. 22)

The importance of the signals given by project team operation is that they become embodied in the river basin institutions that are left to manage the systems post-project. Barrow (1987) presents the desiderata for those developing tropical catchments; they are rationalised here in Table 7.4 which demonstrates why the traditional lead discipline – engineering – in water projects needs the support of other skills in education or the interdisciplinary breadth and dialogue requested by Elliot. There is often a need, rarely met by project teams, for a fresh start on issues such as water rights.

Some items on Barrow's list appear to be wishful thinking and clearly represent a view from the cultural background of the developed world. Local consultation may be laudable in the Thames basin but is highly improbable in the Awash. However, properly conducted environmental assessment, matched with economic analyses which include the principles of sustainability (World Commission on Environment and Development, 1987) and intergenerational equity (Dixon et al., 1986), can be further matched with ethnographic observation to moderate rates of change to those appropriate to the cultural environment as a whole. Once established and under local control a tropical river basin development authority can make many mistakes. Despite its mandate to advise the government on 'all matters affecting the development of the Area' and to 'coordinate the various studies of, and schemes within, the Area so that human, water, animal, land and other resources are utilised to the best advantage', the Tana River Development Authority in Kenya has failed to prevent interference between hydropower generation in the upper basin and irrigation in the lower basin and has succeeded in supplying most of the generated power to Nairobi. The most recent Development Plan passes the focus of activity to the Districts as a move from resource conservation to rural development (Rowntree, 1990).

Sadly, the problems are not unique to Kenya. Adams (1992) suggests that river basin planning proved too attractive a model to emerging African governments in the 1960s because of the large-scale, technological and centralised nature of the development it promoted. The use of basin

Table 7.4 Guidelines to tropical river basin development

---

1 Institute comprehensive water resource planning.
2 Institute sound economic analysis of projects.
3 Conduct thorough environmental analysis at the same time as technical design.
4 Open up planning process to the public.
5 Indigenous people affected by the project should be included in the planning process.
6 Provide for independent technical review of projects.
7 Establish links in the international scientific and academic community.
8 Establish links between environmental, human rights and indigenous people's support groups.
9 Establish watershed management and rehabilitation as a priority.
10 Establish enhancement of traditional food crop agriculture as a priority.
11 Educate decision-makers on emerging problems with large dams.

Source: Barrow (1987)

authorities to bypass difficult ministries or traditional laws was common. Adams concludes (p. 127) 'The structures created so far in Africa simply do not warrant [such] hope, demonstrating planning failure on a vast scale.'

Clearly, few developing nations have the capacity to avoid the project cycle and the dependency on external consultants which it normally implies. For this reason Agenda 21 is emphatic that *human resource development* and *capacity building* are central components of the sustainable development paradigm. Only by educating local technicians in, for example, hydrology and by using techniques such as rapid rural appraisal to obtain the vernacular knowledge of the basin concerned can a firm, indigenous institutional framework be established prior to major developments. Institutional capacities should not be created after the project is 'up and running'. River management institutions may have very humble, grassroots beginnings in the developing world. For example, in rural South Africa a 'Conservancy in the Southern Drakensburg' (a coalition of farmers, led by a recreational developer) has as its targets, in order of priority: Cattle theft, trespass, stock numbers, fences, stray dogs, invasive plants, *basin management, erosion control*, sustainable utilisation (e.g. of trout). This is clearly no Tennessee Valley Authority but it may, with government help and advice, become a powerful vehicle for planned utilisation of the land–water interactions of the region.

Adequate consultation with those affected by river developments is an essential process for river basin agencies in the developing world (see Chapter 6), not least because of the shortages of information held centrally; nevertheless, communications and human rights problems frustrate the ideal. At the United Nations Water Conference in Mar del Plata in 1977 the Mexican delegation proposed the following:

> It is recommended that the effective participation of the public is the key for success of programs of water management and that the lack of local participation has frequently resulted in ineffective programs.
>
> (in Elmendorf, 1978)

Mexico has been well served by community-based development since the revolution of 1910; even so, as reported by Elmendorf (1978), 30 per cent of small village water supply systems are inoperative. Maintenance is a key problem and failure to fund maintenance or to equip people in the project community with the necessary expertise once again recalls the need for adherence to sustainable development.

Warford and Saunders summarise the problem in this way:

> No matter how badly (in the opinion of an external appraiser) a village 'needs' a better water supply system, if the population itself does not perceive the value of the system, the usage rate will be low, system maintenance and local administration will be inadequate and vandalism could be a problem.
>
> (in Elmendorf, 1978)

Elmendorf concludes that 'vicariousness' – an ability to imagine how the project will feel to the people of the project area – is a neglected aspect of planning, given that full consultation is seldom possible and that values will need translation into a factual form for the rational choice between alternatives.

7.8   ENVIRONMENTAL ASSESSMENT AND MANAGEMENT OF WATER PROJECTS: WORLDWIDE PANACEA?

Environmental (impact) assessments (EIA or EA) must be more than a cipher in order to free up development funds (it has been the practice since October 1991 of the World Bank and other agencies to insist on EIA before funding projects – see World Bank, 1993), as is shown by Abracosa and Ortolano (1988) for the Bicol River Basin Development Program in the Philippines. The assessment failed to identify problems of flooding in the reservoir basin, fisheries deterioration and the growth of water hyacinth. Failure to include public hearings and to involve the public in assessment are identified by these authors as leading to a very large amount of post-project modification and maintenance.

In May 1996 the US Export-Import Bank removed support from the Chinese plan for the Three Gorges Dam on the Yangtze after an evaluation 'fails to establish the project's consistency with the Bank's environmental guidelines' (*World Rivers Review*, 11/2) Over 1.3 million people were due to be displaced by the scheme. The World Bank has also been prepared to retrospectively investigate human rights records abuses carried out as part of water developments, for example the alleged massacre of 376 Guatemalan Indians in 1980–82 during clearance of the Chixoy Dam basin.

It is very clear that a huge responsibility for effective environmental assessment and the embodiment of precautions revealed as necessary rests with the engineering profession. As Kalbermatten and Gunnerson (1978), introducing the proceedings of a professional workshop on the topic, conclude, it is essential to analyse the pattern of the project 'helix' (Figure 7.5). The engineer needs to 'enter into and support programs for early public disclosure and public participation in the conceptual planning and implementation stages of civil engineering projects' (p. 241) (see also Section 7.10). Post-project appraisal is also essential.

The essential links between land and water developments may be missed by simple environmental assessment; this represents a failure of strategic thinking and a failure to encompass the concepts of environmental capacity. Investment in EA may be made in the political hope that it will bring 'a quiet life' and 'buy off' opponents to projects. From experience with land-use planning in connection with water issues in what was then West Germany, Wertz (1982) concludes that open conflict is the best way to 'settle' rival professional interests and responsibilities. Within such a field of creative tension it would

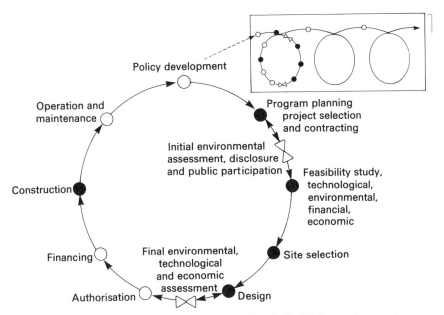

Policy development

Operation and
maintenance

Program planning
project selection
and contracting

Initial environmental
assessment, disclosure
and public participation

Feasibility study,
technological,
environmental,
financial,
economic

Construction

Site selection

Financing

Final environmental,
technological
and economic
assessment

Authorisation

Design

Figure 7.5 The engineering programme planning helix (Kalbermatten and Gunnerson, 1978)

be of value to the water interest to adopt standards to be attained by land-use interests 'even if those standards are not always scientifically dependable' (p. 292). Interestingly for an author representing a successful developed economy, Wertz also concludes that 'it is not advisable to commercialise public interests in water supply and sewage treatment ... public administration and management should, therefore, be given priority over private forms of organisation' (pp. 292–3). Within the appropriate institutional framework, claims Wertz, the one remaining problem is to integrate research knowledge in relation to specific courses of action rather than in abstract:

> Only if exchanges between scientists and practitioners are related to actual projects, while they are being implemented, rather than being treated theoretically, can they be of greatest value. Land-use planners and decision-makers should be included in educational activities and in the exchange of information.
>
> (Wertz, 1982, p. 293)

This voice of experience is echoed by Hellen and Bonn (1981) with their rueful quote that 'the best interdisciplinary activity is often that which goes on in one person's head' (p. 333). In this context, thorough (even bruising!) environmental assessments, which are very broadly scoped, appear to be central to institutional management of developing river basins.

Barbier (1991) suggests that comprehensive identification of components of a watershed system (such as may be identified by EIA) make a very suitable basis for applying the emerging factual basis of environmental economics. In terms of development projects, environmental economics present the opportunity to refine cost-benefit evaluations of project worth. Barbier quotes examples of the cost of soil erosion in Java (both on-site and downstream remedial actions are costed). He further suggests river basin projects as suitable for applying the 'environmentally compensating project to parallel the main project and bring enhancement or rehabilitation to the basin affected'. Tejwani (1993) applies benefit/cost ratios to the watershed management of three reservoirs in India, including in the benefits the improved crops, forest products, employment but none of the 'uncosted' benefits such as conservation and recreation which stimulate proponents of environmental economics in the developed world.

Winpenny (1991) provides numerous examples of how national and project scale accounting, including the costs and benefits of soil conservation, reservoir impoundment, irrigation schemes and water supply schemes permits a more rational approach to water development projects. In a much broader context, Allan (1996) has introduced the concept of 'virtual water' – that water included in patterns of world trade, often at great cost. The concept allows a nation (such as Israel) where water resources are scarce (but which is locked into a world trade pattern of exporting irrigated crops) to elect to import water-demanding produce from rain-fed systems in exchange for more water-efficient exports such as industrial goods. Thus it becomes impossible to separate environmental assessment from world trade patterns; we should therefore seek trade patterns which permit nations to be more sustainable and not to trade in sustainability by importing the products of unsustainable development abroad.

7.9   INTERNATIONAL RIVER BASIN MANAGEMENT

In the context of 'virtual water' all river basins are international! However, we here turn to the specific problems of large basins which cross international boundaries. It is appropriate, once more, to differentiate between problems and policies in the developed world and those in the developing world.

Ten factors which influence the likelihood of achieving workable intergovernmental agreements are:

(a) Severity of the problem.
(b) Degree of technical agreement on the solution.
(c) Geographical balance of the problem.
(d) Law standards and access allowable to foreigners.
(e) Domestic interests and pressures.
(f) Level of preparedness of appropriate institutions.

(g) History of cooperation/conflict.

(h) Relative economic strength/military power.

(i) Third party involvement, e.g. United Nations.

(j) Timing – agreement must precede entrenchment.

Taking two developed world problem basins (the North American Great Lakes and the Rhine) and two developing world basins (the Nile and the Ganges/Brahmaputra), it is easy to see how these factors work slightly differently in the two situations.

In the Great Lakes the situation is covered by an International Joint Commission dating back to 1909. It is a neutral adviser and factfinder staffed by impartial professionals. Since the USA accounts for 87 per cent of the water used from the Great Lakes basin and discharges 80 per cent of the pollution load, it is clear that Canada has a vested interest in the impartiality of the IJC; special boards and advisory groups were grafted on to the Commission to deal with cross-border tensions over pollution. Manifestations of these tensions included environment groups from Canada fighting in the US courts (backed by Canadian government funds) to control toxic dumping and the effects of a stark contrast in pollution control laws (the USA controls at source, Canada uses receiving water standards). However, working in favour of the IJC are the parallel arrangement of the two nations on either side of the water (rather than upstream/downstream) and their similar levels of development.

While the nations of Europe have similar levels of development the River Rhine basin is one in which there are clear source (Switzerland) and receptor (Netherlands) nations and two traditionally antagonistic nations (France and Germany) with intervening riparian interests. Treaties controlling the management of the river date back, through two World Wars fought across it, to 1868. Fisheries and navigation have been regularised for more than a century; however, pollution has proved problematic for several reasons. First, of the four participating nations Switzerland is not a member of the European Union and therefore not bound by EU Directives on the control of pollution. Second, the basin includes extremely important mining and industrial zones relatively high up (e.g. potash mining in Alsace which contributes a third of the Rhine's huge salt load) and very important agriculture low down (in river and altitude terms) in the Netherlands. Nevertheless, the four nations established the International Commission for the Protection of the Rhine against Pollution in 1963 and two important conventions (chemical pollution, chloride pollution) in 1976.

The Commission has shown its appreciation of the basin's ecosystems by advancing the Ecological Master Plan for the River Rhine whose aims are:

- to restore the mainstream as the backbone of the complex Rhine ecosystem and tributaries as habitats for migratory fish;
- to protect, preserve and improve ecologically important reaches.

(Van Dijk *et al.*, 1995)

While the Netherlands is highly dependent on good management of the Rhine upstream it is in no way as vulnerable as Egypt on the Nile. The life of Egypt (and its burgeoning population) is highly dependent on irrigated agriculture (see Chapter 5). A cornerstone to Egypt's survival has been, therefore, friendly relations with Sudan immediately upstream, with whom the first Nile Waters Agreement was signed in 1929. In 1959 a Permanent Joint Commission was set up, gathering data, planning and coordinating. It was this Commission which facilitated the Jonglei Canal project (p. 171). To Sudan the Canal is an equally prestigious contribution to the water resource conservation of the Nile as is Egypt's Aswan High Dam. However, the real problem of Nile management is that while the six headwater nations on the White Nile (Uganda, Kenya, Tanzania, Rwanda, Burundi and Zaire) are apparently amenable to joining a UN compact on the basin, they control only 14 per cent of the flow. Ethiopia, responsible for the 86 per cent from the Blue Nile, is politically, culturally and economically very distinct and refuses to join moves towards integrated basin management.

Cultural and political differences also confuse moves to manage the huge basins of the Ganges and Brahmaputra. In the 1950s India 'stole' water from the Ganges above the Bangladesh (East Pakistan) border to keep open the port of Calcutta by flushing silt into the Bay of Bengal. As Bangladesh became independent in 1971 a Joint Rivers Commission was established but apart from aiding flood control in Bangladesh the Commission has been largely unsuccessful. In 1976 India 'stole' more of the low flow of the Ganges at the Farakka barrage. There is a perennial proposal to divert the Brahmaputra in India across to the Ganges. Bangladesh countered with proposals to dam tributaries on Nepalese territory.

Kliot (1994) has provided the most penetrating analysis of the options and constraints for managing international basins. She offers four principles for interaction between riparian nations:

(a) Absolute territorial sovereignty (each nation develops the river according to need).
(b) Absolute territorial integrity (respect for the needs of the other nations).
(c) Common jurisdiction (jointly agreed management via basin agencies).
(d) Equitable utilisation (sharing the resources).

Clearly (a) and (b) are extreme positions while (c) and (d) are intermediates. Kliot takes the Helsinki Rules (see Chapter 5) and applies them in principle to several of the problem international basins of the Middle East, including the Nile and the Jordan. Table 7.5 here demonstrates the application of the Helsinki Rules to the Jordan. Other authors are less methodical in coming to the conclusion that 'water wars' are the inevitable outcome of failure to reach rational accords of the type advocated by the Helsinki Rules (Bullock and Darwish, 1993; Ohlsson, 1995). Allan and Karshenas (1996) writing about the Jordan basin suggest that the national

Table 7.5 The relative ranking of the Jordan basin co-riparians according to the Helsinki and International Law Commission rules

Features	Lebanon	Syria	Jordan	Israel
Proportion of basin	4	1	2	3
Runoff contribution	4	1	3	2
Climate	4	2	1	3
Use: past	4	2	1	3
present	3	2	1	1
Life expectancy	1	1	2	3
Infant mortality	3	2	1	4
Per capita GNP	n/a	3	2	1
Total debt	n/a	2	1	2
Population (1990)	4	1	3	2
Annual population growth	4	1	2	3
Cereal imports	4	2	3	1

Source: Kliot (1994)

economies of the riparian states have responded to the challenge of the water deficit in 'predictably different ways'. Restating their concept of 'virtual water' they suggest that nations have the option of the 'creation of new livelihoods which use water to greater economic effect'. Clearly, two weaknesses are exposed in the Helsinki Rules by this discussion: they are almost entirely based upon water resource considerations and they neglect that flexibility which exists in the deployment of resources.

7.10   CONCLUSIONS: SUSTAINABILITY AND SUBSIDIARITY – INSTITUTIONS WHICH CAN PLAN BASIN DEVELOPMENT

We have seen that, in addition to decisions over funding and representation, river basin organisations also face profound problems of scale. Paradoxically they can operate best by owning land and/or influencing people to concur with their wishes on land when the scale is very small. That is why many conferences, books and papers are addressed to 'catchment management' or 'catchment control', where the word catchment infers the diminutive of 'basin' (see Chapter 1). Paradoxically, as geographical scale increases to the natural drainage basin outline, including sources and users of water, polluters and conservers of the aquatic environment, the challenge of sustainable management and planning grows, as does the seriousness of the outcome, but there are far fewer means of tackling that task. Schramm (1980) concludes that there is a negative relationship between basin size and the scope of work undertaken by planning institutions. It may, therefore, be an essential feature of large basin management to set goals for small areas, for example the 'Priority Watershed Approach' in Wisconsin, USA (Konrad *et al.*, 1986). The message of the successes won at small scale is that

Table 7.6 Conclusions and recommendations of the UNEP/UNESCO study of 'large water projects' and the environment

1 Water resources projects needed for socio-economic development and the resulting environmental changes are inseparable. Recent procedures for the planning of water resources projects and their assessment have promoted a better understanding of the conflicting nature of this problem and have contributed to improved decision-making as far as the development and management of water projects are concerned.

2 A large number of ecological, financial, social, economic and technical difficulties must be overcome. It is necessary not only to assess but also to manage the environmental impacts of these water projects on both a short and long-term basis. This can only be done if environmental considerations become an integrated part of the decision-making process.

3 It has been recognised that water resources projects have a dual objective nature, namely, to serve both socio-economic development and ecological-environmental development. These dual but sometimes conflicting objectives require trade-offs between them. To do this, alternative options should be considered at various levels of water resources decision-making (policy formulation, planning, design, construction, operation, maintenance, rehabilitation). To obtain the most appropriate option, a compromise between the dual objectives has to be made. This can be achieved by the environmentally sound management of water resources projects which should be oriented towards the establishment of a long-term, dynamic equilibrium between the water project and its environment.

4 The environmentally sound management of water resources projects should be implemented by an active interdisciplinary and intersectorial learning process. Representatives of various interest groups (socio-economic, ecological, technical, legal, local, regional and national) should be involved in the development and management of these projects. A proper institutional and legal framework should be established according to specific conditions. Monetary, measureable and qualitative aspects should be considered on an equal basis.

5 The interaction between water projects and their host river basins should be studied in depth. Both the basin-wide and the regional approach should be basic components of environmentally sound water management.

6 It was agreed that the success of a project is not necessarily related to its size. Both large and small projects have positive and negative elements depending on specific conditions: If the size of the project is one of the reasons for considering an alternative option, a comprehensive analysis of both small and large projects should be undertaken taking into consideration all social, economic and ecological aspects. Both types of project should be developed, planned, operated, maintained and rehabilitated in an environmentally sound manner.

7 To develop alternative options, several specific approaches were suggested. Water demand and consumption control was suggested as an alternative measure for water transfer. The local and regional beneficiaries and the people likely to be harmed by the project should be identified more precisely. Income distribution should also be considered. Land-use planning should be

Table 7.6 (continued)

combined with water resources planning. Incentives to attract people to other areas can also be considered as alternative options. To develop and compare alternative options, the decision support system planned by IIASA [International Institute of Applied Systems Analysis, Vienna] has been supported as a possible tool for the environmentally sound management of water.

8 The costing of environmental impact management as identified by EIA should be integrated into the planning procedure. The application of this approach should be supported by environmental legislation especially in countries where such legislation does not yet exist. In this respect, the overall recommendation of the UN/ECE task force on the application of EIA was supported. It reads: 'EIA should be viewed as an integral part of the project planning process, beginning with an early identification of project alternatives and the potentially significant environmental impacts associated with them and continuing through the planning cycle to include an external review of the assessment document and involvement of the public.'

9 Continuous monitoring of the socio-economic and ecological aspects of water project development and management is strongly recommended with emphasis on pre-project and long-term follow-up monitoring.

10 The UNEP/UNESCO draft methodology on integrated environmental evaluation of water resources development was considered to be appropriate for evaluating the state of environmentally sound management of water projects and river basins. It was recommended that UNESCO and UNEP should finalise this draft methodology and support its further development and application.

11 Better co-operation and understanding between specialists from various disciplines dealing with water projects and problems are needed and strongly recommended.

12 It was recommended that a systematic review of existing methods of evaluating all possible interaction between water management activities and environmental components be undertaken, for example, in the form of a matrix and referring to existing literature.

13 Although in most cases it is not possible or advisable to transfer methodologies for public information and participation from one social environment to another, it is recommended to collect and exchange information on national experiences.

14 As no environmentally sound planning and management is possible without the active positive involvement of professionals including planners, UNEP and UNESCO are recommended to continue and expand their activities related to the incorporation of environmental aspects in the formal education and post-graduate training programmes of engineers and planners. This education should not only relate to the scientific and technical aspects but also to the social and ecological ones.

Source: UNEP/UNESCO (1990)

'ownership rules OK', at least in much of the developed world and it is no accident that the Ontario Conservation Authorities are encouraged at grass roots and can then purchase land over which they want management control. The case of the UK, however, and the recent trajectory of the Catchment Authorities in New Zealand, is that the land planning process is an essential element at the broader scale.

It may not be through the priority interest of the river basin organisation that planning influence is achieved; for example, it was the fisheries and conservation interests in the UK who first persuaded the Forestry Commission to plan and manage plantations with stream sensitivity in mind, yet the water losses by interception are more costly but had no influence. Similarly, the nitrate pollution issue has forced the identification of 'sensitive areas' in the UK agricultural landscape; nitrate is of unproven toxicity yet has achieved a policy change impossible to arrange for the more serious issue of land drainage and flood protection.

No government is going to allow a primacy to its water management organisations; they must achieve influence by various means and it appears that 'sensitive areas' are a good way to publicise the interests of good basin management; they also have a profound educational value.

In time, too, there are problems of scale. Many of the strongest precautionary regulatory activities (e.g. environmental assessment) apply to projects, but not to routines. Recent advice from the United Nations Environment Project (UNEP) and UNESCO (1990) offers an extensive guide to sympathetic management of 'large water projects' – summarised in Table 7.6. Thus, while a large development may be carefully considered by all with a legitimate interest and carefully 'nested' into its environment, a rash of smaller developments over a long time period may escape scrutiny.

In the last five years Catchment Management Planning has become a centrepiece of river management in the UK (Slater *et al.* 1995). Its aims and objectives are listed in Table 7.7; essentially, all scales of development in the basin are considered, thanks to the co-existence of a local government planning system which is open to comment by the public and by 'relevant' organisations such as the National Rivers Authority (now Environment Agency). At the centre of both processes is public consultation, an easier concept to advance than to implement but a considerable evolution from the 'technocratic' tendencies of the engineering project-dominated past.

Further encouragement to the civil engineering profession to formalise public participation in water projects is provided by Priscoli (1989) who sees a continuum between PI (public involvement) and CM (conflict management) – see Figure 7.6. Priscoli reiterates the view that:

Frequently the major problems that engineers and scientists face are not technical. They are problems of reaching agreement on facts, alternatives or solutions.

Table 7.7 Aims of Catchment Management Planning (CMP) as outlined by the National Rivers Authority (now Environment Agency) of England and Wales

Internal aims of CMP	External aims of CMP	Environmental aims of CMP
Translates NRA functional policy into integrated catchment policy	Facilitates the early involvement and participation of community interests	Provides an overview of catchment information
Prioritises needs for the water environment	Creates partnerships with others to achieve effective action	Links land and water management issues
Sponsors integrated action and sustainable development	Guides water companies on future investment	Supports treatment of causes rather than symptoms
Facilitates understanding and ownership of multi-functional working	Guides policies of external organisations, via local authority planning	Sets a consensus-based long-term strategy to guide all actions
Establishes a means for fair and equitable consideration of all water users	Openly identifies and tackles conflicts between competing water users	Ensures timely and appropriate action in the correct location in the basin

Source: after Newson *et al.* (1996)

The engineer, trained and rewarded for technical excellence is frequently frustrated by what are perceived as extra social or environmental design constraints. However, far from constraints, broadening the social objectives of engineering presents new opportunities for engineering service if one makes the effort to look.

(Priscoli, 1989, pp. 31, 36)

Priscoli attempts to guide the engineer through a simple definition of the values which public consulters might bring to discussion of a water scheme (Figure 7.6) and to emphasise that an essential step for the professional is to admit the interactive bias of the situation. Once open for discussion the project loses its Newtonian, mechanistic simplicity and the professional, too, feels lost. 'You need to give the project away,' says one experienced British flood control engineer.

Once the interests of the consultees are established by exploring the salience of values to the particular project, formal methods of interest-based bargaining (e.g. Fisher and Ury, 1981) can be employed to progress to a solution which is both feasible and broadly acceptable.

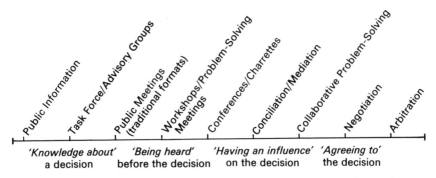

Figure 7.6 Conciliation procedures and public consultation (Priscoli, 1989)

Thus, to follow these principles, it becomes essential for water managers to understand land, its uses, management and planning. To intervene in land issues is a large political step in any system of government; the intervention is highly determined by the context of the information held about land and the policy framework surrounding land. These matters are worth a chapter on their own (Chapter 8).

Sustainable river basin management
Issues of the knowledge base

> Although the river and the hillside ... do not resemble each other at first sight ... one may fairly extend the river all over its basin and up to its very divides.
>
> W. M. Davis (1899)

This definition emphasises for us the huge problems of information, knowledge and influence inherent in our ambition to coordinate the human exploitation of the river basin ecosystem and to make that exploitation sustainable.

The United Nations Conference on Environment and Development (UNCED or 'Earth Summit'), held in 1992, made explicit to all nation states the need to make development sustainable; far from advocating a single operable definition of sustainability there was, within another 'S'-word (subsidiarity), the inherent challenge to define sustainable processes appropriate to particular environments at particular scales. Subsidiarity implies decision making at the appropriate level in a hierarchy of bodies from the international community, through nation states to local government bodies, popular groups and the individual. The word 'sustainable' *is* hard to define; however, many agree that working towards a practical definition is an essential technical and political venture in all fields of environmental management at the close of the twentieth century. It is, however, an appropriate adjective for the type of management which Chapter 7 implicitly prescribes for the river basin system. Pezzey (1989) describes sixty-one versions of a definition and the political, philosophical and scientific doubts which attend the broad notion of sustainable development, first widely publicised by the World Conservation Strategy (International Union for the Conservation of Nature, 1980) but mainly boosted by the Brundtland Report (World Commission on Environment and Development, 1987) which defines it as: 'development that meets the needs of the present without compromising the ability of future generations to meet their own needs' (p. 43).

Pearce (1993) suggests that there are three main forms of sustainability: ultra weak, weak and strong, and that one may judge the stage to which policy has evolved on four dimensions (here modified to policy, economy, society and contact – see Table 9.5). Turner (1993) adds 'very strong sustainability' which has a moral dimension of biotic rights – in other words respect for total ecosystem integrity; this is generally assumed to be politically naive and impracticable in the present era. Strong sustainability puts ecosystem *functional* integrity at the core of its policies, corresponding approximately with the holistic form of management discussed by Downs *et al.* (1991). 'Natural capital' and 'human capital' (what we produce from natural resources) are not seen as interchangeable. Thus in terms of river basin ecosystems the science agenda for strong sustainability would include:

- Ecosystem identification and calibration (to identify capacity).
- Assessment of natural extremes and trends.
- Assessment of system sensitivity to development impacts (natural capital).
- Assessment of risks and precautionary principle for management.

(Newson, 1996a, in press)

The contribution of scientific knowledge is vital in guiding our decisions in all versions of sustainability but it is particularly critical to our choice of how 'strong' we want our policies to be. This chapter is of even greater relevance considering the special position advocated for river basin managers in the vanguard of all environmental management systems by reason of antecedence and opportunity. Paradoxically hydrology is a young science and the need for scientific guidance is urgent; we have an added responsibility therefore to assess the doubts, uncertainties and risks associated with such guidance.

Tollan (1992) has further characterised the ecosystem approach to water management as requiring, before action, synthesis of knowledge from various fields of specialisation; an holistic perspective focusing on links between environmental elements and man; a multi-media approach (land, air, water) which is geographically comprehensive (i.e. whole basins).

His vehicles for delivering the approach are also a useful summary of points raised above:

- Legislation.
- Institutional arrangements.
- Planning.
- Impact assessment.
- Economic measures.
- Ecosystem evaluation/classification.
- Integrated monitoring.
- Public participation.

Can we afford this programme? Perhaps the question should be 'Can we afford to be without it?'

8.1 SCIENCE IN THE 'NEW ENVIRONMENTAL AGE'

While Chapter 7 has extolled the virtues of a rebalancing between technical and popular inputs to river basin management it is clear that basic guidance will continue to come from scientific research. The burden of this chapter is to explore the appropriateness of the knowledge base available for the conjunctive use of the land and water resources of large basins but, before this exploration, we need to address more general problems of the nature of *environmental sciences* and how they interact with the *institutions* whose critical role is now understood (i.e. from Chapter 7).

In one sense sustainable river basin management is a vanguard project in the critical quest for all ecosystem management systems; other than the few world examples of urban 'air basin' management to curtail pollution (recently extended to the European Union in terms of critical air pollution loads for acidification) and ecosystem management in nature reserves, the river basin is uniquely a process-response system with definable, meaningful boundaries often coincident with existing social and administrative limits. The popularity of rivers for recreation in the developed world (see Chapter 9) means that popular environmental campaigns are directed at river managers before the appropriate institutions are aware of problems or at least before they are prepared to take mitigating action.

The high cost of taking mitigating action or of error in proactive action means that from the outset environmental policies in many nations have been cautionary rather than precautionary (see Section 8.3).

For example, in Victorian Britain clear principles were laid down to guide public policy responses to pollution (Department of the Environment/Welsh Office, 1988):

(a) Controls would be applied when the scientific evidence justified it.
(b) Pollution should be prevented at source.
(c) The best commercially viable technology should be used to effect abatement of emissions or discharges.
(d) The polluter should bear the costs of the necessary controls.

Thus scientific evidence is codified as a key factor in prompting response but debate, of course, rages about the stage at which the evidence is 'conclusive'. There are also options to use the same scientific evidence in different forms of policy as evidenced by the contrasts in pollution control philosophies between the European Union and the UK (Newson, 1991). Haigh (1986) describes the pattern of policy responses to the pollution threat from the lead content of petrol; he describes a 'majestic descent' of the lead content in response to a complex interplay of scientific evidence and political and technical activity at a variety of scales. Ashby (1978) has also noted a repeatable pattern of 'ignition from public opinion', 'examination by scientists' and 'formulation of political action' by a combination of evidence and advocacy.

At this stage we may note the ironic situation in which science is credited with a key role but not as an anticipatory and radical force in the same way as technology. For these reasons we need to examine the nature of environmental science.

8.2   THE ENVIRONMENTAL SCIENCES

The popular definition of science plays up its objectivity and exactness; science has become in the twentieth century utterly confused with technology, to a point where 'technocentrist' positions on environmental management (O'Riordan, 1977) are optimistic that the continued success of research and development will ensure technical solutions to resource and hazard problems.

Environmental sciences fall into the popular and political images built for the laboratory sciences and this leads to considerable confusion over the incorporation of research results into policy; by comparison, the incorporation of laboratory science into technology is extremely simple and is mainly concealed from public scrutiny. Conventional research and development processes have tended to support the growth of economies, diversifying, extending and modernising the range and capability of manufactures and services. The short history of the environmental sciences has, however, shown them to be markedly different:

(a) They have tended to be identified with the sounding of alarms about the effects of economic development.
(b) Their stated results, or the conclusions made from them, have been contentious and much of the resulting public debate is between scientists with opposing results or interpretation.
(c) Environmental sciences have lacked the methodological rigour of the established sciences, being forced often into extensive modes of inquiry, and uncontrolled or at best statistically validated frameworks.
(d) Environmental sciences have tended to group together in pursuit of trans-boundary problems, particularly during the late 1980s with the rise of global scale environmental challenges. Headings such as 'Earth and Atmospheric' or 'Terrestrial and Freshwater' sciences have appeared on doors and on letterheads. In addition, the social sciences have been drawn into an increasingly holistic framework widely regarded as appropriate by both philosophers (Bunyard and Goldsmith, 1988) and managers (Gardiner, 1988) of environmental systems.

A further characteristic, particularly of environmental science integrated with development studies, is its willingness to form alliances with 'vernacular science' or 'indigenous knowledge' systems, in other words the cultural, informal information (often value-laden) which is in the possession of those groups impacted by the development. We investigate the characteristics of this significant dimension below.

How well equipped is environmental science to provide knowledge for practical management of natural systems? Because river basins are, and have been, in the vanguard of system management we can partly answer this question by following the development of hydraulics and hydrology (see Chapter 1). However, it is first necessary to consider the relationship between providers and users of knowledge and therefore between science and society.

8.3 'SCIENCE SPEAKS TO POWER'

The field of metascience (the science of science) has been a very popular area of enquiry for philosophers in recent years. Perhaps initially inspired by the environmental damage brought about by some facets of applied science and technology (e.g. nuclear power) but also by the way in which positivist approaches to social systems were applied in an unquestioning way by policy-makers, the metascientists have been able to capitalise on a period of profound introspection on the part of scientists themselves, largely brought on by a rapidly falling resource base for research.

Ziman (1984) attempts a guide to the contemporary status and problems of *applied science*, though without any special attention to environmental science. Ziman's view is that, despite a treasured perspective of science as a distinctive and wholly objective philosophy, independent of its material base, this perspective has never been representative outside higher education. He suggests that epistemological, occupational and societal functions are always combined in a scientist. For long periods (e.g. 1850–1950) societal functions of science are stable; an important aspect of stability is autonomy. However, *external steerage* has now largely taken over (Kogan and Henkel, 1983) and this has implications for what is researched, the manner of research and the interface with policy.

Collingridge and Reeve (1986) are even more dubious of the role of science in relation to policy rather than technology. In a study which gives its title to this section, they unpick what they call the 'myth of the power of science' in relation to policy-making, that is, that 'whatever information is needed to reduce uncertainty in making a particular policy choice, science can meet the challenge' (p. 2).

They claim that this myth is perpetuated as recently as the introduction of environmental assessment in development programmes and conclude that:

> Contrary to the myth of the power of science there is a fundamental and profound mis-match between the needs of policy and the requirements for efficient research within science which forbids science any real influence on decision-making.
>
> (Collingridge and Reeve, 1986, p. 5)

Collingridge and Reeve see research proceeding best when scientists are allowed autonomy in their choice of problem, allowed to work in single disciplines and allowed to reach a consensus with low error costs. However, the more relevant the policy field, the more likely will be criticism of technical arguments. The status of scientific knowledge as simply a stage of negotiation among scientists which has reached consensus is therefore directly threatened. Similarly scientists become frustrated with the fact that policy-making does not involve a synoptic rationality in which exact bricks are built into a carefully planned wall of knowledge. Policy-making therefore becomes incremental rather than fundamental and this aspect of the political filtering of research results is poorly appreciated by scientists, including the present author.

Three years after the acceptance of a paper on the hydrological impacts of afforestation in the UK by Calder and Newson (1979) by the hydrological community, the Secretary of State for the Environment in the UK government made the following statement in Parliament: 'As regards afforestation, its percentage and its effect on catchment areas ... I am advised there is a lack of clear scientific evidence' (*Hansard*, 21 March 1980). In later papers (Newson, 1990; 1991; 1992d), therefore, the author has tried to put the knowledge base of hydrology into a policy context. These papers illustrated a slow but measurable policy readjustment to the original scientific research.

One weakness of Collingridge and Reeve's arguments for the field of research on the natural environment is that they ignore the interdisciplinary nature of a science like hydrology; contributions by generalists in this field have been far more important than those in the medical and psychological research fields they cover. Nevertheless, it is highly appropriate to the subject matter of this chapter to consider the need for knowledge in river basin management to be constrained by the restrictions raised by Collingridge and Reeve. It is especially pertinent to consider the concept of managing the uncertainty which applies to all scientific findings; we may speak, therefore, of the *error costs* of changing policy in response to faulty scientific guidance. This will apply particularly in river basins to problems of scale and of experimental control which arise in hydrology.

To indicate that error costs may be an unneccessarily pessimistic image of scientific guidance we need to briefly review a more optimistic innovation in former West Germany's public policy-making which subsequently caught the attention of the Royal Commission on Environmental Pollution (1988).

The German *Vorsorgeprinzip* (precautionary principle) may be defined as taking integrated steps to protect the environment from processes of degradation which can be identified by research but about whose precise operation and impact there is still scientific uncertainty. If applied to river basin management the precautionary principle could have no other outcome than land-use planning because it would be important to apply anticipatory controls to maintaining both quantity and quality of river flows and to managing the effects of extremes and responses to climatic change. Such planning is already part of public policy in some countries (see Chapter 9).

8.4   ENVIRONMENTALISM, ENVIRONMENTAL SCIENCE AND RIVER BASIN SYSTEMS

Many of the oft-quoted visionaries of the environmental movement have made observations which describe the relationship between the land and water resources, and their human exploitation, within river basins. As explained by Chapter 1, what allowed them to bring out the connectivity of basin systems (before aerial and satellite photography and environmental science 'revealed all') was the depiction of land and channel components as integral on plans and maps – dating back to Leonardo da Vinci and beyond.

The origin of the modern phase of 'basin scale environmentalism' is reputed by many writers to have been 'The Alpine Torrents' controversy (Glacken, 1956) of the nineteenth century in Europe. In 1797 a French engineer (Fabre) linked the sequence of damaging floods from rivers draining the Alps to deforestation of the headwaters. Fabre listed seven kinds of disaster which resulted in:

(a) The ruin of the forests themselves.
(b) The erosion of mountain soils and consequent destruction of mountain pastures.
(c) The ruin of settlements near streams.
(d) Instability of channels.
(e) Litigation over channel migration.
(f) Siltation lower down rivers.
(g) Diminution of runoff from springs and subsoil.

Fabre's work produced a series of studies in France which eventually evoked a policy response from the French government. It introduced a project of *reboisement* (reafforestation) during the nineteenth century (1860 and 1882). The Austrians and Italians also faced the same problem. Alexander von Humboldt pronounced on the causative link between catchment mismanagement and environmental stress, using lake levels in the New World, Asia and South America as evidence:

> by felling the trees which cover the tops and sides of mountains, men in every climate prepare at once two calamities for future generations: want of fuel and scarcity of water.
>
> (von Humboldt, 1852, quoted in Kittredge, 1948, p. 9)

A further profound influence at about this time was the landmark volume, *Man and Nature, or Physical Geography as Modified by Human Action* by George Perkins Marsh (1864). Widely acknowledged as the first major influence on Western environmental concern, especially by geographers, this book records Marsh's extensive travels in the Alps. He was despatched by President Lincoln as an American ambassador to several countries with environmental problems in the Mediterranean and in Alpine Europe. He gave very high prominence to the dangers of deforestation and became instrumental as a diplomat and bureaucrat in drafting laws on irrigation (in France and in California); he also influenced the British Parliament in a policy of reafforestation for India.

Marsh focused much of his factual reporting of deforestation on the Alpine zone of Europe, linking deforestation with both flood and drought; in places his observational data and bibliographic enquiries appear to yield contradictory evidence on the precise hydrological effects of a forest cover but concluded overwhelmingly that they were beneficial. It is interesting to note a recent and contentious entry into the debate over deforestation; Metailie (1987) concludes, from a study of the Pyrenees, that nineteenth-century catastrophes from flooding can be blamed on exceptional climatic sequences and natural geological and geomorphological proclivities rather than 'anthropic' erosion which was limited in extent. The same sort of battle between alternative explanations rages in connection with the Ganges (see Chapter 5).

The Swiss established a paired catchment study in 1900 to investigate the effect of forest cover on runoff. They concluded almost entirely beneficial effects of tree cover – balancing the extremes of flow and attracting precipitation. The results were reported in the USA by Zon (1912), two years after the establishment of the Wagon Wheel Gap paired catchment experiment (including forest felling – whereas the Swiss had compared catchments with 98 per cent and 30 per cent covers). Zon reported that: 'Accurate observations ... established with certainty ... Forests increase both the abundance and frequency of local precipitation over the areas they occupy' (quoted by Kittredge, 1948). After less than fifty years of a plethora of US paired catchment experiments, Hibbert (1967) summarised the results of thirty-nine such studies as concluding that forest reduction increases water yield, and reforestation decreases water yield.

Clearly, these early observers did not have the benefit of quantitative data; these did not become available until the systematic hydrological monitoring of a wide range of land uses in a wide range of climatic and physiographic conditions began during the International Hydrological Decade (1965–74). The Decade established a key distinction for scientists researching options in river basin management: between *representative* and *experimental* basins.

8.5 THE HYDROLOGISTS' STOCK-IN-TRADE: CATCHMENT RESEARCH

Even within the autonomy of science under internal controls there is a profound debate about the appropriateness of one of the basic forms of research used to input to river basin management. We have already referred to the systematic problems of environmental science in framing experiments.

One of these is that it has no unique theories to test but rather explores the boundary conditions which constrain the application of theory within significant combinations of climate and physiography. Enter, therefore, traditional geographical and biological tendencies to induction and classification (Burt and Walling, 1984), which become combined in catchment research frequently being carried out as a series of case studies (Church, 1984). Burt and Walling are very critical of this tendency and, following Kuhn, claim that: 'protracted attachment to the methodological directive "Go ye forth and measure" may well prove an invitation to waste time' (p. 7).

Nevertheless, these authors are hopeful that the phases of classification of catchments and of the identification of processes and their controls are now complete. An experimental approach to catchments can now mean, therefore, the careful testing of models incorporating processes. Church eases this sentiment into two avenues for catchment studies – *exploratory* and *confirmatory*.

It is unlikely that catchment research will avoid a continuing plethora of case studies, simply because land-use and land management effects often require opportunistic research, especially where a manipulation of land is required (part of the original definition of a true experimental catchment). We here begin to move towards the true crux of catchment experimentation – it is almost inevitably an applied science and has, therefore, in addition to the problems of experimental control, problems of relevance (increasingly as funds become scarcer) and the infrequently assessed problem of scale.

During the International Hydrological Decade (1965–74), Ward (1971) produced a concise evaluation of the catchment framework for hydrological work; he wrote from experience, having been among the earliest academic geographers in Britain to set up field hydrological studies. Ward quotes a Texas congressman who proclaimed that an experimental catchment was 'an area drained by a creek of the size that a coon dog could jump across'. More seriously, Ward centres his critical evaluation of the catchment approach on the relationship between the purely empirical findings it provides and two much sterner tests – the ability to extrapolate those findings and an understanding of the processes underlying them. In all, Ward lists five practical problems of catchment experimentation:

(a) Lack of control.
(b) Representativeness.
(c) Accuracy of data.
(d) Data manipulation.
(e) Costs.

In terms of extrapolation, there had been early optimism among field hydrologists that research catchments could be seen as modules, with results being merged and grouped according to relief, soil, climate or land-use protocols, 'grossing up' to the scale of river basin for which management predictions are required. Most of them became less optimistic after their first practical attempts; Amerman (1965) concluded that processes related to the areal pattern of runoff from slopes (such as throughflow in soils and groundwater recharge on slopes of different lengths and geometries) produced an inherent, scale-dependent geographical variability in catchment performance. A further, more obvious source of scale-dependent behaviour is that of the proportion of its travel time spent by an element of runoff on slopes and in channels – the channel proportion increasing with distance from the headwaters. It is no accident that the most successful extrapolations from catchment research have been those determined by atmospheric processes, such as interception loss (see Calder and Newson, 1979). As one 'goes deeper' in the hydrological process cascade, extrapolation becomes much more complex. Turning to problems of providing basic understanding through catchment research, Reynolds and Leyton (1967) were very clear that even smaller experimental areas are needed: 'Watershed experiments cannot provide understanding of the physical processes, for which plot studies must be conducted' (quoted by Ward, 1971, p. 131).

Reynolds and Leyton do not specify the location for plot studies but clearly there is much to be gained from a *nested* approach to scales of study; the geomorphologist's technique of stream ordering can offer a basic principle for this hierarchical approach, one which was to be central to the UK Institute of Hydrology's Plynlimon research catchments (Kirby *et al.*, 1992). As Ward says:

> None of these ... useful ways of supplementing small watershed experiments ... is a replacement for the small experimental watershed, however, since ultimately all other methods must be verified and assessed in relation to watershed data.
>
> (Ward, 1971, p. 131)

This conclusion is valid within the confines of the research approach but still neglects the problem of the next quantum leap – to application. In this respect it is relevant to note that, at the start of the IHD, da Costa and Jacquet (1965) reviewed 252 catchment studies, of which 188 were of areas less than 25 km^2. The size of catchments and the duration of research on them revealed by the Helsinki Symposium in 1980 is shown in Figure 8.1 – they are mainly small and operate for around a decade. Hudson and Gilman (1993) have recently demonstrated the benefits of maintaining the Plynlimon experimental catchments in Wales for a period (twenty years plus) over which climate changes become clear from the assembled data.

At the conclusion of the IHD in 1974, UNESCO implemented an indefinite extension to coordinate further experimental hydrology: the International Hydrological Programme (IHP). Arnell (1989a) reviews the accumulated catchment studies from eighteen nations under the 'FREND' programme (Flow Regimes from Experimental and Network Data). It is important here that 'N' for

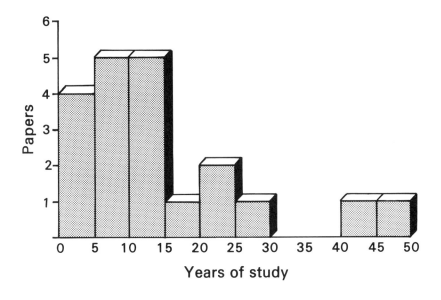

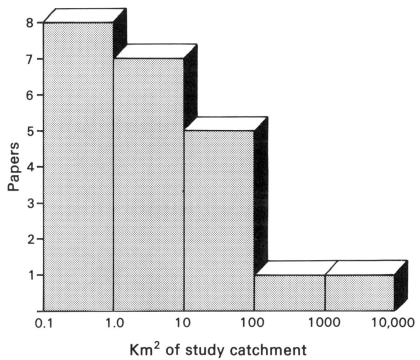

Figure 8.1 The scale and duration of catchment experimentation on land-use effects: data included in papers presented to the 1980 Helsinki Symposium of the International Association of Hydrological Sciences (IAHS)

Network has been included. At the time of the IHD, routine hydrological measurement and data collection were often local and low-tech. Thus catchment experiments were, in addition to their major role, a test-bed for instrumentation techniques and data processing routines. Twenty years later the 'routine' catchment is as well equipped as the early experimental sites, with data processed, checked, used and archived as part of national information systems such as the UK Surface Water Archive Programme (SWAP).

Thus it is now to network data that river managers can look, at least for vindication of an effect first isolated by researchers; the research catchment is still the more likely venue for a true experiment, that is, manipulation of a control such as land use, and has increasing value as a monitoring device as the duration of its records grows and circumstances in policy (and now climate) change.

The relative spatial coverage attained by research catchments and by network measurements is shown in Figure 8.2 for Britain. A further

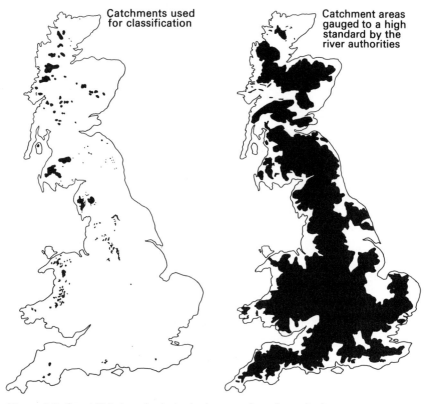

Catchments used for classification

Catchment areas gauged to a high standard by the river authorities

Figure 8.2 Great Britain – hydrological research and monitoring:
(a) The research scale: small, dispersed catchments
(b) The monitoring scale: large, cohesive basins

indication of the growth of network power in the field of land and water management comes from the volume edited by Solbé (1986). Of forty-six pieces of work reported only two make detailed use of research catchment data; seven use plot experiments or lysimeters and the remainder which report data collection programmes utilise network monitoring data, especially in connection with water quality data.

Arnell (1989a) further indicates a shift in scientific attention between the IHD and the IHP. The theme of the Decade, he says in his Preface, was that 'Rational water management ... should be founded upon a thorough understanding of water availability and movement' while in the IHP 'the objectives have shifted slightly towards a multidisciplinary approach to the assessment, planning and rational management of water resources'.

8.6   ALTERNATIVES TO CATCHMENT RESEARCH

Rational management, of course, asks similar questions to those of scientific enquiry but often a preoccupation of the scientist with 'How?' becomes the familiar 'What if?' of the manager.

In an era of fast computation and an explosion in data availability, the conventional answer to such a question is 'simulation'; the mathematical model is in theory capable of replacing the research catchment for a fraction of the hardware investment. Arnell (1989a) includes modelling alongside experimental basins in a list of current hydrological techniques used to predict the effects of land-use change. Also listed are analysis of time series data to detect significant changes of trend (severely constrained by the gradual change characteristic of anthropogenic effects) and regional comparisons, substituting space for time. Both rely on statistical tests for which management tends to have a blind spot; the latter approach is also subject to the problem of transferability faced by catchment results.

Huff and Swank (1985) describe the PROSPER model, incorporating both canopy and (multiple) soil layers which essentially see climate as powering a flux of moisture away from the evaporating surface (the canopy). Interception storage is modelled as an alternative source of the evaporated moisture. The performance of the model on one of the Coweeta forest catchments through a number of years of felling and regrowth demonstrated significant problems of calibrating the leaf area of forest regrowth but as a bulked tool for predicting the increase of water yield on felled catchments in the first year it was more successful. Huff and Swank, however, conclude that the model cannot achieve comparable accuracy to field measurements.

Turning to models of whole catchments, the most usual incorporation of a land-use effect is via a tank concept, one which can be incorporated into a hardware model for educational purposes (Figure 8.3). The parameters of such a model therefore scale the size of the storages in the hydrological

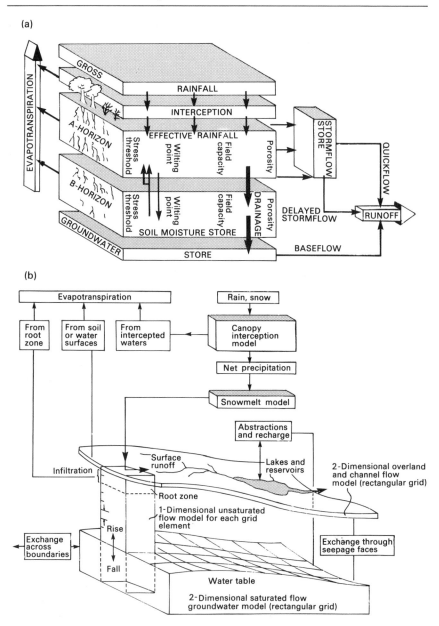

Figure 8.3 Deterministic hydrological models, an alternative to catchment research?

cascade from canopy to channel and the catchment is simply seen as a 'lumped' system of these tanks (Blackie and Eeles, 1985). The interception store is particularly amenable to this treatment but overall there is, of course, the need to fit the model to an actual flow record so as to set the parameters.

By contrast, distributed models are more frequently physically based (Beven, 1985); their parameters are the variables determining the operation of process equations which determine the routeing of rainfall to runoff. Distributed models are therefore useful to land-use hydrology and particularly to catchment experiments because they incorporate a detailed knowledge of actual catchment behaviour and can operate in data-sparse environments and on land-use impacts whose main interest may be in their spatial organisation: seldom will afforestation, or urbanisation, be the 'lumped' phenomenon considered best by 'lumped' models.

8.7   SCIENCE AND POLICY: LAND-USE MANAGEMENT IN RIVER BASINS – FORESTS AND BEYOND

It was important to the progress of integrated land and water management in the USA that Yale established a graduate school of forestry in 1900 and that it promoted the doctrine (attributable to Gifford Pinchot) of 'scientific resource management'. An important graduate of the school in 1909 was Aldo Leopold (Tanner, 1987) who went on to work for the US Forest Service under Pinchot and to develop an holistic approach to land management, a practical evocation of the 'green' attitude of this time. He drew inspiration from the intact landscape and clear, tree-lined rivers of the Sierra Madre and sought to develop a 'biotic view' of land which led him to produce ethical guidance to public policy developments. He was successful as a generalist, claims Callicott (1987), because he provided an impeccable scientific basis for an holistic, systems approach to all landscapes but particularly to those of river basins. As a result land management for the benefit of water resources has taken on an almost evangelical nature in the USA (see Plate 8.1), a point which British foresters, faced with undesirable hydrological implications of their plantations, must find hard to understand and to bear.

The recent compendia of catchment, plot and policy material by Solbé (1986) and Arnell (1989b) allow an insight into the status of hydrological research on issues of land management. The specific problems of environmental science in achieving public credibility, except under a 'precautionary principle', have already been dissected but there are extra difficulties within the phrase 'science speaks to power'. In many ways the case of hydrological research into land use and land management is a case of 'science speaks to the powerless' since it is only rarely that rational management of water has been given the political power to extend to issues of land. In Chapter 7 we suggested that one reason might be the quest for the perfect river basin

Plate 8.1 The American way: forests in support of resource conservation (photo
  M. D. Newson)

institution; in many ways this issue is peripheral, however, to rivalry
between resource managers and property interests.

We have no opportunity to investigate the translation into the nineteenth-
century French policy of *reboisement* (reafforestation) of the observations
by many of changes in the flow of Alpine rivers. Kittredge (1948) lists the
many books and papers published in Europe prior to the policy decision; in
an environmental age we have earlier in this chapter referenced the work of
von Humboldt and Marsh in this connection. However, the telling work,
according to Kittredge was not by an environmentalist but by the engineer
Surell who, in 1841, published a book confirming the observations of his
predecessor Fabre (see Glacken, 1956). We may therefore conclude that
almost half a century of influential, official persuasion was necessary before
policies relating to land were changed. It is also important to note that the
agent of that change was the French government's Forest Department;
Kittredge refers to the Department as being given 'the mission of controlling
the torrents'. It is interesting to note that in neighbouring nations with a
problem of flood-prone Alpine rivers the approach to control was more
direct and structural. Vischer (1989) describes how Switzerland coped with
the apparent increase in river instability in the eighteenth and nineteenth
centuries: here was the birthplace of river 'training' (by analogy one might
call land-use controls 'river education'!).

The gauntlet of responsibility for managing land in relation to the aims of flow regulation passed quickly to the USA where the 1902 Forest Reserve Manual listed the aims of forest reserves as 'to furnish timber' but also 'to regulate the flow of water'. The Wagon Wheel Gap catchment studies in Colorado did not begin until 1909, so this indicates a simple acceptance on faith of the European conclusion. It is perhaps no accident that Kittredge refers to 1877–1912 as the 'period of propaganda'. There were counter-claims against a widespread hydrological influence for forests but in the eastern USA, the region most affected by the activities of the early settlers in clearing trees, the Week's Law of 1911 provided the crucial right to acquire land for afforestation 'for the protection of the watershed of navigable streams' and to 'appoint a commission for the acquisition of lands for the purpose of conserving the navigability of navigable rivers' (Kittredge, 1948, p. 13).

The ability to alter land use and to control land management techniques depends critically on the approach of land-owning democracies to state intervention; the USA became necessarily adept at Federal ownership for the purposes of forest, conservation and erosion management. We must also remember the great importance of river navigation to the spread of the Union and cannot underestimate the undesirable effect of the early felling of forests by European settlers on the erosion of the eastern USA.

From the small amount of historical evidence available on major land policy changes in relation to water management we may conclude that the following issues may be important:

(a) The spread of the observed change in river regime or water quality and the extent to which such changes reflex upon the land interest itself (e.g. through flooding, influence on navigation and trade in land products).
(b) The disciplinary origins of the work establishing the causal links. Government engineers and foresters clearly work from within the policy framework.
(c) The traditions of land ownership and planning in the region/nation affected and the role of federal agencies and corporations as vehicles for change.

These factors may be unique to the problem of forest and water, even to those cases where afforestation has a benign effect on water properties (cf. UK uplands). It is perhaps worthwhile to bring this treatment up to date via the reviews by Solbé (1986) and Arnell (1989b). Solbé's book is divided into four areas of inter-relationship between land and water:

(a) Urbanisation   (6 papers)
(b) Mineral exploitation (4 papers)
(c) Agriculture   (13 papers)
(d) Forestry   (6 papers)

Table 8.1 Land-use effects on hydrology: an international survey

(a) Nations where land-use effect noted

	Canada	China	Finland	France	FR Germany	Hungary	Ireland	Japan	Rep. of Korea	Netherlands	Norway	Poland	Romania	Sweden	Switzerland	UK	USA	Vietnam
Acid precipitation	×																	
Vegetation change		×	×	×	×	×		×	×	×	×	×	×	×		×	×	×
Urbanisation			×	×	×	×		×	×	×	×	×		×		×	×	×
Land restructuring			×	×	×			×			×			×		×	×	
Field drainage						×		×		×	×		×				×	
Mining					×		×	×	×				×	×		×		
Fertiliser application	×		×	×	×		×	×	×	×	×			×		×		×
Waste fill leakage	×				×			×	×				×				×	

(b) Specific impact of land-use effect

	Inputs to catchment	Evaporation/transpiration	Surface/subsurace interaction	Surface processes	In-channel processes	Soil water quality	Groundwater quality	Channel water quality
Acid precipitation	×					×	×	×
Vegetation change	×	×				×	×	×
Urbanisation	×	×	×	×				×
Land restructuring		×	×	×				
Field drainage			×	×		×		×
Mining			×	×	×			×
Fertiliser application	×					×	×	×
Waste fill leakage	×					×	×	×

Source: After Arnell (1989a)

Arnell's table (modified here as Table 8.1) goes into more detail to reflect the considerable geographical spread of his eighteen nations. Since all eighteen nations, bar China, report some impact of these 'in-house' activities by water managers, it is clear that in the development of public policy a particular perception is one of engineers 'in sole charge' of river behaviour. The land users and managers, it would seem, have a ready excuse for inaction through pointing an accusing finger at the water managers; this is particularly true in the case of nations such as the UK in which water quality debates include the added dimension of official responsibility for sewage pollution. The influence of climatic change is also now a potential confusion to policy-makers since it can dominate over both land-use and water-use effects on river regime. For example, Hudson and Gilman (1993) conclude that in recent years rates of evapotranspiration have decreased in the Plynlimon catchments, reducing the water yield differences between forest and grassland; they cannot explain the change but favour an explanation based on increasingly extreme rainfall regimes by season. Climatic change also bespeaks new land uses and management techniques (see Chapter 6).

8.8   POLICY RESPONSES

Arnell's overview of his national reports from catchment research makes interesting reading in relation to public policy reactions to the findings. He concludes that the most widespread human impacts on hydrological characteristics are:

(a) Deforestation.
(b) Irrigation.
(c) Urbanisation.
(d) River regulation.
(e) Use of agricultural chemicals.

He reports:

> The reviews received from the international hydrological community have shown that although there is a general consensus about the types of change resulting from a given activity, the actual degree of change is very variable.
>
> (p. 15)

He also concludes that major river basins perform a kind of smoothing process on the hydrological signals from individual anthropogenic activity:

> Not only do the physical and climatic conditions of basins vary but similarly-titled impacts also take many forms. One activity is rarely performed in isolation, and the hydrological characteristics at a basin

outlet are an integration of the effects of different activities operating at different scales.

(Arnell, 1989a, p. 15)

The policy-maker has, therefore, some excuse for confusion: either land-use effects are confused and overlapping or clear and the source of an insidious chain of 'knock-on' deteriorations in the river environment.

The key to the confusion is twofold: first, land-use effects are regionally adjusted in their impact by major variables such as climate (e.g. forests behave differently in wet, dry and snowy climates) and second, their impact will depend on the sensitivity of the river basin resource system considered (e.g. in relation to the degree of control on basin behaviour already exerted by climate and water management).

Hydrologists have been slow to make their findings clear to policy-makers; it is not just a case of there being few appropriate laws or institutions to bring about land/water control in most developed nations. None of Arnell's authors put their reports of hydrological experiments in the context of policy, though the report from Germany at least begins with a tabulation of national land-use categories and recent rates of change. With the exceptions of Australia and New Zealand, the choice of research catchments as far back as the launch of the IHD in 1965 was without reference to the potential use of results in a policy context.

A typical outcome of the neglect of policy links in the UK is the retrospective multivariate analysis of over 300 research catchments in which acidification has been studied (Bull and Hall, 1989). This study groups the catchments statistically so that six groups are linked for future data collection programmes.

While Arnell's authors report as scientists to a scientific peer group in UNESCO, those contributing to Solbé's (1986) volume, while also scientists, reflect far more the results of the relationship with policy (i.e. power) described by Collingridge and Reeve (Section 8.3). They also contain a high proportion of applied scientists, engineers and managers. Thus, under urbanisation, we read of tests of control structures (Coombes; Hellawell and Green). Hamerton lists the Acts of Parliament relevant to his water quality study, Howells and Marriman list legal difficulties as well as chemical determinands and Worthington seeks regulatory remedies to halt farm pollution.

Interestingly this largely British book contains 'the farmer's view' as a chapter while the contribution by Phillips suggests that better engineering will clean up rivers draining farming regions. A similar theme of 'managing through' is set by British contributions on the (deleterious) effects of conifer afforestation (chapters by Binns and by Mills).

In complete contrast, chapters from the USA on farm pollution control (Konrad, Baumann and Ott) and forest management (Ponce) are much more prescriptive and describe rational policy structures reflecting the impact of those activities on river basin management.

Clearly, Solbé's collection is of more direct use to policy-making and equally clearly it contains controversies and contradictions of exactly the type predicted by Collingridge and Reeve for work produced in a policy context. Possibly the knowledge base for river basin management is destined to remain divided into a set of concentric rings around the core of policy implementation.

The salient features of the present state of the upland management debate indicate that at the stage where there is no policy, for example on rural land-use planning, there will be considerable rejection of research results. O'Riordan (1976) suggests that 'Where problems pose solutions which challenge the dominant values and rules of political consensus, substantial power may be directed simply at keeping this challenge out of the political arena.' Under such circumstances the researcher may well feel jealous of the effect evoked by qualitative rather than quantitative evidence, such as in the case of the high impact of the conservation lobby in the uplands (see publications by Nature Conservancy Council [1986] and Tomkins [1986]). O'Riordan (1976) finds society's fears of rationality predictable; as a direct result, 'policy making is basically a political process'.

At a second stage one can detect that society 'feels a policy coming on'. Hydrologists may well feel resentment that moves towards policy will be led by other issues, principally nature conservation and agricultural production. Thus, once again, numerical inputs to a rational model for land use are swept aside by fiscal, social and even ideological considerations.

In the uplands the activities of single-interest agencies clearly need to become broadened and coordinated while the value of land is manipulated to achieve some form of planning. Roome (1984) concludes that there must be 'fundamental change to the distribution of rights, whether voluntarily accepted or legally enforced'. Clearly, we are some distance now, both conceptually and methodologically, from the interception of rainfall!

Returning to the assertion in Chapter 7 that a valid river basin management organisation should demonstrate its legitimate interests in land use by becoming involved with catchment planning we now need to ask what special demands on science (of all varieties) does such a venture make.

8.9 THE STRUCTURE OF IMPLEMENTATION: LAND-USE CONTROLS IN RIVER BASINS

Chapter 7 shows how the former National Rivers Authority in England and Wales evolved a system of influencing land-use decisions from the perspective of their flood defence function. Catchment Management Planning, a much broader process, has now reached a critical stage of implementation (Figure 8.4), but it is not a device unique to the NRA. For

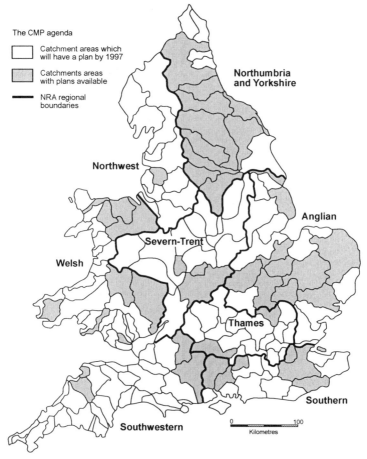

The CMP agenda

☐ Catchment areas which
will have a plan by 1997

▨ Catchments areas
with plans available

━━ NRA regional
boundaries

Northumbria
and Yorkshire

Northwest

Anglian

Severn-Trent

Welsh

Thames

Southern

Southwestern

0 ━━━━ 100
Kilometres

Figure 8.4 The National Rivers Authority programme of Catchment Management
Planning in England and Wales

example the New Zealand Catchment Authorities have been involved with
local government planning since the 1980s (Otago Catchment Board, 1986;
see also Figure vii). Buller (1996) compares the NRA's planning approach
with that adopted in France since the Water Act of 1992. France has six
regional river basin agencies (Loire, Seine, Somme, Rhine, Rhone,
Garonne); within each a SAGE (Schéma d'Aménagement et de Gestion des
Eaux) plan is being prepared for significant individual basins (e.g. the Isère)
under the guidance of local water commissions comprising representatives
of central and local government and water users. The plans are largely
driven by water quality interests rather than ecosystem protection and Buller
also warns that the basis of devolved power in France is political, not
environmental – hence fragmentation may frustrate holism. Buller makes

no comment on the role of scientific knowledge in French catchment planning but it, too, is fragmented thanks to the long history of local government management of water supply, water quality and flood protection. In the UK sixty years of basin authorities, nationally coordinated, and thirty years of harmonised monitoring and catchment research offer a sounder basis for holistic planning.

To develop an influential position with regard to land-use planning those concerned with river basin management have a tough problem. It may be compartmentalised as a series of gaps, between knowledge and applications (Figure 8.5). Since most research is done on small research catchments the prediction gap is that between such small entities and the larger areas for which practical measures are needed. To be prescriptive, however, a wide range of these larger units must be addressed – the extrapolative technique must therefore have comprehensive scope, the models must be robust. Finally, a policy gap may exist between the boundaries of those agencies (e.g. local councils in the UK) with land-use planning powers and the outline of the river basin unit. In the first edition of this book we noted the pathways in which the hydrological findings about the impact of conifer forests on upland water quantities and quality opened a doorway between water managers and land managers. Only an outline is repeated here.

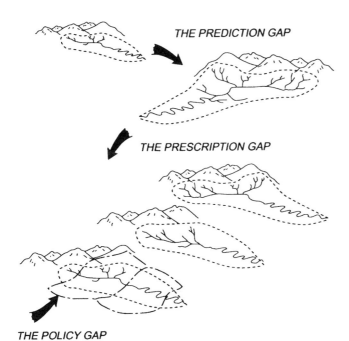

THE PREDICTION GAP

THE PRESCRIPTION GAP

THE POLICY GAP

Figure 8.5 Gaps in the application of hydrological research to Catchment Management Planning

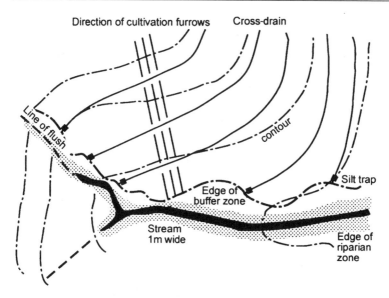

Figure 8.6 Guidance on cultivation furrows, cross-drains and buffer zones (after Forests and Water Guidelines, 1993)

There are two essential prerequisites of policy information before investigating the uptake of hydrological guidance in land use and land management in Britain.

(a) Rural land-use policy is not achieved directly but by market interventions for produce or incentives offered by the Common Agricultural Policy of the European Communities.
(b) Outside towns there is little land-use planning, exceptions being national parks and land for nature conservation.

Thus, in a sense, the UK cannot use hydrological advice except and unless the government forestry and agriculture agencies take action (mainly fiscal action via grant control) or the planning process for towns incorporates the professional guidance of water managers. Movement is now brisk on both fronts and this brief review explores the factors which lie behind recent policy developments (see also Newson, 1988, 1990, 1996b in press).

Two major inconsistencies in the attitude to upland forest land use taken by the water industry itself are important as background. First, British water engineers and scientists are not accustomed to invoking catchment area processes to explain river dynamics. Second, and more specifically, for at least forty years the water industry gave a cautious welcome to conifer plantations on catchment areas. Newson (1986) reviews the history of attitudes within the water industry and how research results, principally those of Law (1956), eventually impinged on decision-making. Prefacing the 'era of trees on catchments', the Gathering Grounds Committee (Ministry of Health, 1948) decided that while trees did not attract rainfall they did protect upland reservoir catchments against erosion. The Committee reached this conclusion on the basis of a trawl of the qualitative opinions of experts, much of the information coming from abroad. Much of it was inappropriate but several well-known reservoirs became surrounded by conifers in an effort to blanket them off from human and livestock influences.

Law's (1956) results, which 'proved' that trees 'use' more water than rough moorland, did not lead to a change of policy but to an intensification of research. Indeed there was no public policy to change, catchment area land use being mainly decided by individual water suppliers, who desired to own whole catchment areas in order to prevent public access and agricultural improvement.

In 1979 the results of the important Plynlimon (mid-Wales) catchment experiments were published, confirming and extending Law's adverse conclusions about the reduction of water yields when the uplands are covered by mature conifer plantations (Calder and Newson, 1979). In the late 1970s the outlook for timber production (Forestry Commission, 1979) suggested that Scotland would bear the brunt of new planting. Already the Scottish local councils and hydro-electric boards were identifying local hydrological problems in connection with afforestation. Since research results were not available in Britain on the effects of a 'natural' vegetation dominated by heather or of frequent snowfall (both conditions likely to make Scottish afforestation unique in its hydrological effect), the Institute of Hydrology set up a paired catchment study on the Plynlimon model at Balquhidder, Perthshire. Work began in 1980. By now the water quality dimension was becoming much more important to water industry perceptions of land use and land management (see Youngman and Lack, 1981). Issues not strictly related to water supply, such as fisheries, also began to surface (Harriman, 1978) and none of these research results was favourable to forestry.

The last fifteen years have brought further research results which sustain a critical attitude by the water industry to proposals for upland afforestation, principally on the grounds of erosion, acidification and discoloration.

Legislative support for land-use control in the UK has grown, principally in other contexts (e.g. to reduce nitrate pollution from agriculture). It is now possible to consider protection zones and environmental quality standards (for certain pollutants). In addition, and in contrast, voluntary guidelines for foresters have been published to reduce conflict. These standards (Forestry Commission, 1988, 1991; Forestry Authority, 1993) are used directly by the Commission but also indirectly in the Commission's judgements on the suitability of applications by private forestry for government grants. They are also likely to be used by forest developers facing the new legal

Table 8.2 Allocation and accommodation options in water-related land-use and land management policies, UK

Option	Policy	Aims
Allocation	Water industry forest policy	To alert forest industry to sensitive upland reservoir catchments
	Critical loads policy	To prevent afforestation of areas where forest cover would exacerbate acidification of freshwaters
	Catchment Management Planning	To influence development controls related to sustainable use of the water environment
	Nitrate Sensitive Areas/Vulnerable Zones	To change farming practices to reduce leaching of nutrients to groundwaters, surface waters
	Groundwater Protection Zones	Protection of important groundwater sources from pollution from any source
	Industrial Protection Zones	Protection of rivers and aquifers from spillages of pollutants
Accommodation	Code of Good Agricultural Practice	Definition of protective measures in the use and storage of agrochemicals and farm products
	Forests and Water Guidelines	Definition of good practice in relation to water at every stage of the commercial plantation cycle
	Use of buffer strips in farming/forestry	Protection of streams using non-agricultural activity in riparian zone

requirement in the UK for an environmental assessment on new plantations larger than 200 hectares.

Newson (1996b) sets out the new policy environment for the commercial forestry cropping cycle in UK upland catchments; it is increasingly influenced by the water sector through Catchment Management Plans (see below), the Environment Agency's Forestry Strategy and the industry's own Forest Redesign Plans which aim to mitigate all adverse environmental impacts. In the UK there are signs of a swing in commercial forestry interest towards the lowlands where a tree cover is normally seen as *comparatively* beneficial to the environment where it replaces intensive agriculture.

It is useful to maintain and develop the dichotomy between issues of land allocation and issues capable of resolution through existing legal or technical fixes. We may set up two scenarios for decision-making:

(a) The *allocation* option, whereby a 'keep-off' attitude to catchment areas may be followed by the water industry, armed with maps of sensitive areas. A refinement of this approach might be to plan catchment land use rationally on the basis of land capability assessments and hydrological predictions.

(b) The *accommodation* option, whereby land is allocated by a combination of 'free market' forces and a technical dialogue between the forest and water industries. Water and timber are harvested from the same land but both industries accept that higher costs may be involved, e.g. for greater care in preparing ground for afforestation, for leaving large strips of land unplanted or for higher levels of water treatment from upland sources.

Table 8.2 shows how both options have come to be used in recent years in the UK. It is an extremely important context for the successful implementation of catchment management planning that scientific evidence of damage has been used in policies, for example, to protect groundwater, to minimise the risk of accidental pollution, and so on. The role of the European Union's Directives on these topics has also been important, although the UK government has always employed scientific surveys to locate the application of such policies to the optimum sites (some would say to minimise expenditure). Furthermore, in the production of guidelines to good practice (see Figure 8.6) the integration of professional data and standpoints has been something of a breakthrough in the style of British administration.

8.10   KNOWLEDGE AND POLICY CONSENSUS: SCIENCE SPEAKS TO *PEOPLE*

Throughout the latter stages of this book frequent, and at times slightly patronising, reference has been made to the involvement of the general population in sustainable management (of all resources – hence the Agenda

21 references – see Figure 7.1). We now need to investigate, briefly, the nature and role of what ordinary citizens have as knowledge about river management and, more importantly, what they *count* as knowledge about the problem. O'Riordan and Rayner (1991) have elevated this knowledge to the status of 'vernacular science' but it is more commonly referred to as 'indigenous knowledge'.

Adams (1992) gives a concise general definition of the value of indigenous knowledge by writing that 'African land users have often had (and still have) a better understanding than anyone else of what can and cannot be done in their environment' (p. 37). He goes on to bracket 'specific techniques of the management of land and water resources' with 'the cultural and socio-economic systems in which they are rooted'. We appear, therefore, to have a ready-made holism in indigenous knowledge, one which, unlike academic knowledge, does not suffer interdisciplinary tensions. Adams further emphasises the ability of indigenous knowledge to be responsive to changing conditions, which is why he prefers the term to 'traditional knowledge'. This dimension makes popular consensus imperative in both the developed and developing worlds. Formal science is too uncertain about the risks involved in environmental management and the trends in the boundary conditions to lay out its traditional prescriptions.

Another feature of indigenous knowledge which has become apparent from more modern forms of extension services in developing-world land management has been its ability to accept and exploit the appropriate in modern technology from outside – IF this is not imposed as a complete solution. Adaptability and resilience are essential in sustainable basin management and it may be that modern systems of information technology are the most likely forms of 'technology transfer' to benefit local communities (rather than the construction 'mega-projects' of the past). The notion is often developed through the medium of decision support systems (DSS).

DSS is defined by O'Callaghan (1995) as 'computer-based information systems that combine models and data in an attempt to solve poorly structured problems with *extensive user involvement*' (p. 15, emphasis added). O'Callaghan and co-authors describe a DSS to assist in debates about catchment land use in the UK. It employs hydrological, ecological and economic models (each appropriate to the scientific methodology of its founding discipline) and a geographic information system (GIS). A graphical user interface represents the true technological achievement of modern DSSs (McClean *et al.*, 1995) because it allows the user to select questions, data and models and explore the outcomes interactively on an attractive screen layout. Davis *et al.* (1991) depict some of the questions and options raised by their DSS for catchment land use in the hinterland of Adelaide, Australia. The user is asked, for example, to specify a view on whether dairy farmers should install soakaways for waste in areas of more than 900 mm per annum precipitation – or build full wastewater treatment lagoons. These authors emphasise the need for users to be fully informed about the sources and reliability of data used in DSS and the impact of compliance with control scenarios.

It is clear from recent volumes on water management that experimental DSS is maturing and taking a place in public debate about planning options. In the USA there are support systems for wetland conservation and regional water use, for relating storage and utilisation of river flow in the Colorado and as part of a popular campaign to improve water quality in the Tennessee Valley Authority area (Fontane and Tuvel, 1994).

What is not clear from the literature is the degree to which DSS has penetrated truly democratic pathways; most people in the developed world encounter IT in their daily lives but it is a critical political decision as to whether the popular culture of the VDU is overtly and honestly deployed in planning resource allocation and use.

8.11 CONCLUSIONS: INTERVENING IN LAND – A POLITICAL TEST OF KNOWLEDGE

The author well remembers the chill realisation that his work was politically unrecognised and, perhaps, unknown to an otherwise successful politician (see Section 8.3). We may conclude that the location of the research effort within water management, the degree to which it is funded and the degree to which it is 'controlled' (towards practical answers) is a critical aspect of the ideals of this book. If we wish for 'hydrologic civilisation', hydrology must be done, well done and exposed to popular and/or (depending on the national context) political scrutiny. As many scientists would conclude, this requires an educational movement among politicians; many of the conclusions from hydrological research are as difficult for politicians to comprehend as is defining sustainable development. In other words, while spectacular results may convince the politician, they are seldom forthcoming; instead hydrology tends to produce 'depth charges' or a 'delayed fuse' when it comes to impact.

Because hydrology is a regional science, results often differ between research sites – this is notably true of forest hydrology (Newson and Calder, 1989). A far larger problem is that land-use influences are apparently subordinate to direct flow management influences and to climatic change in gross effect on rivers. The scientific agenda is moving strongly to changing environments at present and away from man-made manipulations. However, the polluting effects of unwise land use are well known, as is the conservation loss produced by some forms of catchment 'abuse'. In addition, catchment land use or management can be used to mitigate the effects of climate change – but the processes involved must be understood and deployed.

The lessons are surely that hydrologists must start with policy in mind; just as engineering hydrology was adjusted to the service of society, so

Table 8.3 Linking small-area research to large-area application

Methods of extrapolation[a]	Methods of incorporation[a]
1 Replication of research in other environments	1 Education – broad approach
2 Pooling data from individual research efforts; synthesis	2 Technical education of practitioners – 'good practice'
3 Use of geographical predictions	3 Fiscal manipulation of land-use financial support
4 Mathematical modelling of processes	4 Proscription of damaging operations (plus prescription of beneficial ones)
5 Natural 'demonstration' of effect – hazardous event	5 Protection zones→whole-basin planning

Note: [a] While the columns do not cross-correlate, they both represent an ascending sequence of demonstration and action.

environmental hydrology must agree a certain subordination to human needs. It must anticipate needs. For example, at the time of writing it would seem obvious to put more effort into groundwater research since, during periods of environmental change, the groundwater store provides continuity and survival. Fundamentally, there must be an evolution of interdisciplinary research, taking as its example the inherent holism of the 'vernacular science' which yields the indigenous knowledge upon which river basin management ultimately depends.

Finally, as Table 8.3 shows, there must be a continuing research field which attempts to link the outputs of research from small catchments to devices and structures which control the processes of accommodation or allocation in large basins.

Chapter 9

Land and water

Towards systems of management in a period of change

9.1 FUTURE OF THE RIVER BASIN IDEA

This book follows in the footsteps of a geography text, *Water, Earth and Man* (Chorley, 1969), which brought together authors from the physical science, social science and humanities areas of geography to set down what were at that stage separate agenda items, components of an integrated approach to river basins. In the final chapter of *Water, Earth and Man*, O'Riordan and More (1969) set a very perceptive agenda for the future development of water resources:

> Thus, whereas the vehicles of water-resources planning are becoming more massive and complex, the requirement for their manoeuvrability is also increasing, and the aim of all future planning is to produce a large-scale and completely integrated scheme capable of constant re-evaluation.
>
> (O'Riordan and More, 1969, p. 572)

These words were written before the arrival (in policy terms) of the 'New Environmental Age'; without superhuman foresight the authors could not have judged how great the need for manoeuvrability might become; for example, in 1969 climatic change was being written of but was mainly played down by official agencies. *Water, Earth and Man* was published four years before Schumacher (1973) brought out *Small is Beautiful*, a herald call to a generation to begin thinking in terms of the organisation of human society in units which its members understand and can practically manage with low technologies.

At the United Nations Conference on Environment and Development in 1992, Agenda 21 – the blueprint for sustainability – brought together the biophysical and social aspects of river basin development in several of its chapters (Table 9.1), providing a boost for the concept and a realistic assessment of the huge task ahead for both the developed and developing world. Post-Rio, scale considerations still dominate the geographer's approach to the river basin development problem; the notion that the river

Table 9.1 Agenda 21 policies relevant to river basin management

Chapter 10 Integrated approach to the planning and management of land resources:
● Landscape ecological planning units, e.g. watersheds.

Chapter 12 Managing fragile ecosystems: combating desertification and drought:
● Soil conservation and reafforestation.
● Improved land–water–crop management.
● Anti-desertification measures in national development plans.
● Drought preparedness and drought relief.
● Popular education and participation.

Chapter 13 Managing fragile ecosystems: sustainable mountain development:
● Integrated watershed development.

Chapter 18 Protection of the quality and supply of freshwater resources – application of integrated approaches to the development, management and use of water resources:
● Integrated water resource development and management.
● Water resources assessment.
● Protection of water resources, water quality and aquatic ecosystems.
● Drinking water supply and sanitation.
● Water and sustainable urban development.
● Water for sustainable food production and rural development.
● Impact of climate change on water resources.

basin ecosystem constitutes an obvious 'bioregion' or 'ecoregion', one with which human communities can identify, has become a major justification for river basin institutions (see Chapter 7). Aberley (1993) brings together statements which emphasise this connectivity, for example, 'ecoregional boundaries stand forth as convergent thresholds welcoming us "home".' Subsidiarity in decision making, backed by new devices available to the 'information age' (such as decision-support models) make O'Riordan and More's plea for 'manoeuvrability' realistic and there are unlikely to be 'massive and complex' basin plans suited to all applications. Paradigms have changed as fast as technology since 1969. Ten years ago Falkenmark (1986) set the scene with the words, 'It is urgent to compare needed benefits from land and water development with threats to both the resources themselves and to the biological resources such as flora, fauna and a diverse gene pool' (p. 200).

The reasons for this inherent geographical variability of options for river basin management include:

(a) Differences in scale. In successful river basin management the flow of information is critical (see below) and so scale does not merely become a boundary condition to physical and chemical processes in rivers but a central institutional issue. One of the critical contrasts between the Tyne and the Nile is the presence of nine national boundaries in the latter river's basin.

(b) Differences in trajectory. Here we may include changes through time in water needs, the development process, climatic controls and political controls. Successful river basin management will be the result of considering a range of options to suit the often unique combination of these variables for each basin, though admittedly there will be some basins in which the water stress is so great that emergency, monolithic action is considered essential.

(c) Differences in the environmental capacity and resilience of the system to be developed, in turn leading to different relative needs to manage demand for that capacity. Our ignorance of the particular circumstances of dryland environments is a critical barrier to their sustainable development (Magalhaes, 1994).

It is of interest that the European Communities are at present debating a Framework Directive which includes the acknowledgement that subsidiarity (at the national level) is appropriate for river basin management, despite the need to integrate regulatory functions within management frameworks (Table 9.2).

The United Nations Economic Commission for Europe (1995) has also made a plea for European governments to 'develop an integrated approach to demand management, water allocation instruments, including licencing, minimum acceptable water flow and the pricing of water'. This should be achieved by 'strengthened coordination of the water management activities carried out in key water-related sectors within a catchment area'.

Table 9.2 Principles of European Union water policy (European Commission 1996)

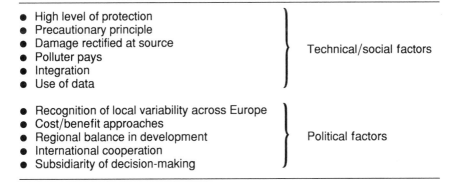

• High level of protection • Precautionary principle • Damage rectified at source • Polluter pays • Integration • Use of data	Technical/social factors
• Recognition of local variability across Europe • Cost/benefit approaches • Regional balance in development • International cooperation • Subsidiarity of decision-making	Political factors

9.2   THE MUTUAL CONSIDERATION OF LAND WITH WATER

It is easy to move from the fact that 98 per cent of rainfall passes over or through land on its way to the river (Chapter 3) to an assumption that land use and management is axiomatic to those who set up water schemes, even where their chosen scale is the river basin.

A major problem here is that of ownership and the history of land use and management; water agencies seldom own land in the river basin and therefore have only indirect controls. We have divided the routes for influence over land into 'catchment control' (allocation) and 'catchment planning' (accommodation) (Newson, 1991). Control implies ownership or legislative circumscription of land use (e.g. South Africa's Mountain Catchment Areas Act, which allows direct intervention in catchments vital to water conservation) while planning describes the consultative indirect manipulation of land management rather than land use. The water interest is sometimes weak, entering the land political ring late, and therefore needs to accommodate to existing patterns in the basin; where it has entered early it has earned a poor reputation for land planning of its own holdings (e.g. careless afforestation, disruption of traditional land rights, salinised irrigation fields, etc.).

Of more direct importance to the theme of this book, however, is that land issues, while all-pervasive and conceptually logical, have many rivals in the perception of water managers interested in the degree of control exercised by all the relevant variables in the river basin environment. Thus land use and management may mean little in some basins compared with tectonic activity (e.g. the Ganges – see Chapter 5); in others climate changes are the main concern. In yet others only urban land use may be important and so the problem of conjunctive management of resources is contained within a small area of point pollution controls or flood runoff detention.

Land use and management therefore has rivals for the attention of river managers, including the strongest of all, namely the influence of river regulation in bringing about fundamental changes to the flow and water quality patterns of up to two-thirds of the flow of world rivers (Chapter 6). This immediately facilitates the traditional engineering approach to river management in which manipulation, 'training' and other interventionist, structural measures come to be preferred because of the degree and certainty of control they offer. The technical fix to such issues has largely arisen from the lack of demand management for such services from the river. In sustainable development programmes demand management is an essential component of respect for ecosystem services provided by critical natural capital.

In terms of catchment management we have, to date, been able to advocate 'suitable' land uses for river basin management and in some cases national planning has concurred, although not at a very large scale. We have been much less successful, however, at removing 'unsuitable' land uses. Nevertheless there are many signs that, for small and *sensitive areas* (both whole catchments and riparian zones of all rivers), there is a tendency for the indirect, non-structural approach to be gaining momentum. This book has been written at a time when public concern for the prevention of pollution is being rapidly translated into schemes of monitoring and control; it is apparent, for example from the Nitrate Sensitive Areas of the UK, that land issues (including rural land uses) will not gain greater prominence in basin management (Newson, 1991). In countries undergoing rapid development and embarking on major water schemes it should not be beyond our wit to warn that a concern for supplies, power, crops and other water benefits bespeaks a relatively rapid (140 years in the UK) transition to the hygienic and then ecological concerns of an urbanised population.

9.3   ECOSYSTEM SERVICES: HUMANS *IN* THE ENVIRONMENT

Naiman *et al.* (1995) tabulate the benefits we derive from freshwater ecosystems – modified here as Table 9.3. Cairns (1995) employs the term 'ecosystem services' in describing the ecological integrity of the aquatic system.

Table 9.3 Benefits humans derive from (intact) freshwater ecosystems

Direct use of surface waters and groundwaters
Preparation of food/drink
Hygiene, waste disposal
Livestock production
Hydropower
Cooling
Manufacturing
Fire fighting
Products harvested from healthy freshwater ecosystems
Fish and wildlife
Riparian products
Wetland products
Streambed minerals and materials
Services provided by healthy freshwater ecosystems
Recreation (fishing, hunting, boating, swimming)
Transportation of goods
Water storage/flood control
Nutrient deposition/waste purification
Habitat for biological diversity
Climatic moderation
Buffering of polluted inputs
Aesthetics and mental health

Source: after Naiman *et al.* (1995)

Use of the term 'benefits' invites the burgeoning field of environmental economics to put financial values on these aspects of intact ecosystems in order to justify their protection (over and above their uncostable value to human society). The value of protecting stream ecosystems by enhancing low flows has been costed by Garrod and Willis (1996), who use 'willingness to pay' criteria based on interviewing riparian communities and visitors to a stream in southern England as a basis for a benefit/cost analysis supporting the costly improvements to flow regime. Pending the 'arrival' of environmental economics those organisations which seek to protect ecosystems (e.g. non-governmental organisations and statutory wildlife agencies) are increasingly campaigning in the water sector. In the UK both the Royal Society for the Protection of Birds (RSPB) and English Nature have recently made policy inputs to water resource development (RSPB, 1995; English Nature, 1996). RSPB advocate site protection, abstraction controls, financial incentives, demand management and planning controls to protect ecosystems.

As an indication of the failure of successive river basin management institutions in the UK to protect river ecosystems over the historical period reviewed here, Table 9.4 lists the losses of the spontaneous regulation functions enshrined in the Marchand and Toornstra model. Even at this late stage in the development process, therefore, 'strong' sustainability is not included in public policy (although attempts to restore ecosystems are becoming more widespread – see Chapter 6).

There is tacit, if not overt, recognition of the ecosystem ideal by many policy-makers. Freshwater biologists are contributing to river management in a much more profound way than they were ten years ago. In many ways they are defining the future shape of the freshwater ecosystem: 'A new discipline of restoration ecology has developed' (Eiseltova and Biggs, 1995). The latter volume describes river and catchment restoration schemes in Germany, Denmark, Sweden, the Czech Republic, the Slovak Republic, Romania, Lithuania and Estonia. They further remark that: 'A healthy river with its adjacent floodplain perform many functions and provide many benefits which, far too late it seems, we suddenly come to miss.'

Table 9.4 Loss of spontaneous regulation functions in river basins: the UK

60.9% of all agricultural land in the UK is drained
35,000 km of river channel in England and Wales are maintained for flood defence
2,361 ha of lowland raised mire lost in England and Scotland, 1948–78
3,370 km² of East Anglian fens lost, 1637–1984
10–20% increase of arable agriculture on lowland floodplains, 1948–78
One-third of all land-use change in England 1985–6 was rural to urban
Between 1945 and 1990 the urban area in England grew by 58%

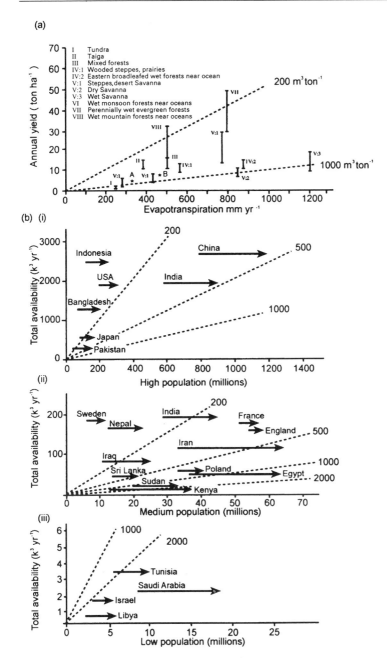

Figure 9.1 Carrying capacity arguments for water use (after Falkenmark, 1989):
(a) Vegetation and water use
(b) Human water use (500 people per unit signifies problems in the next century; 1,000 represents current problems)

Falkenmark (1986) develops a 'carrying capacity' model for human water consumption which links biomass production (in terms of its water needs per tonne of dry matter) with human gross need for water. She uses the basic unit of 1 Mm^3/yr as a guide to the populations which can be supported. In the semi-arid world the total demand is likely to produce a carrying capacity of 2,000 people/Mm^3; interestingly, Israel has already reduced to 500 people per unit through re-use and other efficiency measures. Without such sustainable policies, however, there are sixty-four nations which will not be able to support their population in the year 2000. Falkenmark's analysis is, however, lacking in a quantitative assessment of the biological requirements for water – an essential prerequisite if the model is to be superimposed on the ecosystem model.

9.4   LAND, WATER AND DEVELOPMENT

Development continues. Because of its continuity, albeit at different rates and taking different forms, it is tempting to say that the problems faced by those developing river basin resources in a 'developing' country are no different to those in a 'developed' country. There are good reasons, discussed below, for looking for common institutional problems (see also Chapter 7), but at the outset we may illustrate some typical differences brought about by the physical environment of typical basin situations.

Chapter 5 emphasises that problems of water development in the developing world surround:

(a) The least developed countries.
(b) The drylands and their wetter hinterlands, often mountains.
(c) Major international river basins.
(d) Inter- and intra-national political tension.
(e) Institutional problems of integration and application.

In the developed world, outside drylands (which makes the case of the USA's drylands so intriguing) we have the following problems:

(a) Pollution controls permissive to further development.
(b) Hazard management.
(c) Recreational and conservational priorities.
(d) Problems of inadequate data; decisions must appear rational.
(e) Institutional problems of integration and planning.

Convergence in the last two items of each list is deliberate, despite the dangers to a physical scientist of admitting that the common thread is social! However, since integration, planning and application issues are genuinely interdisciplinary the physical scientist has a right, even a duty, to intervene. This author does so from a background of considerable disappointment at the institutional rejection of certain of his own research contributions, the

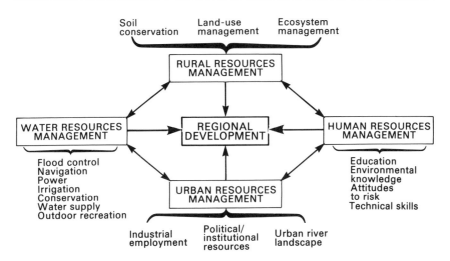

Figure 9.2 The complex of projects involved in river basin management

result of a previously ingenuous attitude to the practicalities of applying knowledge. The real issue, therefore, of sustainable river basin management, wherever it is required, is that of the application of knowledge against an informed perception of the enormity of the integration required (Figure 9.2). Holistic thinking has a special, almost visionary role in helping politicians and people understand that rapid 'fixes' are impossible, but that strategies can, and must, 'keep us on the rails'.

9.5   TECHNOCRACY AND DEMOCRACY: PLANNING AND PEOPLE

The horns of the river basin dilemma are as follows. As soon as one leaves the realms of traditional, local environmental management of the water resource or hazard, that is, when one embarks on development, there is an immediate need for technology and for experts. Society, if it identifies with the project, supports the experts; in the hydraulic civilisations their power was considerable and society was structured around the need to achieve the goals of development. While environmental scientists, including hydrologists, enjoy considerably less certainty in their technology they need the support of society for their agenda by that anticipation or reclamation (see Section 9.6).

Environmental science is highly technical and its proponents find it no easier to communicate simply with those affected by their proposals than their specialist, reductionist colleagues and forebears; there are, however, signs of increasing freedom of environmental information and an increasing effort by specialists and their institutions to provide environmental education. The bridging of the information gulf is essential to the acceptance by those affected by river basin management of the schemes its technologists employ. We cannot dispense with the pure scientist, nor with the engineer. The engineering discipline has dominated water development and basin management throughout history (de Camp, 1990); its support has been tacit because its achievements have been spectacular. However, engineers in particular need to explore a new relationship with people and the education and training of engineers should emphasise this. As Kirpich (1990) remarks: 'the poor performance of engineers in management results to a considerable degree from inadequate education requirements, i.e. insufficient attention to non-engineering but pertinent subjects such as economics, sociology and business management' (p. 846). Most geographers would want another subject added to the list and the present author has often mused on the valency of two disciplines which at first appear to compete on issues of depth and breadth!

It is, of course, also a duty for the rest of us to understand the position of the *engineer in society* (a compulsory paper in the qualification procedure for UK Civil Engineers). As Cosgrove (1990) has stressed, 'the need for visionary engineering is still with us' (p. 11). Society must realise that, outside the prison of its reliable, practical reputation, engineering is speculative and as open to flair and passion as the arts. The world became modernised by engineers and water engineering is often the hallmark of the process. This progress represented an amalgam of engineering skill and the ready response of engineering to the demands placed, without check, by society; there is no reason at all why, under conditions of managed demands, engineering cannot achieve the new agenda items of environmental management.

Consultation over river basin management may be inevitable but it should also be formal; it is a particularly dangerous assumption by environmentalists that consultation does not need rules! Protest movements can often set their own rules, but for an official agency to consult widely needs the formality of the emerging field of *conciliation* (see Chapter 7).

Furthermore, it should not be assumed that consultation is possible under all political conditions. This is a particular difficulty in the politics of development. Nations undergoing rapid development are often one-party states with major regional problems of opposition, even civil war. For this reason the recent tougher stance taken by the major financial institutions to environmental and human rights assessments of proposals is very welcome.

The consultation process also, of course, needs the information to supply from the technocratic to the democratic process or body. As Chapters 6 and 8 show, research agendas need to change to respect the eventual use of the information which is gained. Important here is the need for truly interdisciplinary research activity.

As an indication of the degree to which popular action is now progressing the concepts of integrated river basin management, ecosystem protection and river restoration we need only list the 'grass roots' programmes whose brochures fall on our doormats or whose home pages we access on the Internet: 'Tamar 2000' encouraging better farming practice to improve fisheries, 'Maryland Tributary Strategy Implementation Teams', the Wild Rivers Project, the River Restoration Project, CSIRO's 'Indicators of Catchment Health' programme – and many more. Such popular schemes should not surprise us, neither should their frequent association with fishing – although many 'greens' might consider angling to infringe animal rights. Human use of fisheries was (and should remain) one of the 'extensive exploitation functions' of the ecosystem and should not threaten sustainability, particularly if the restoration programme generating the fishery is rooted in good ecological guidance. Such guidance can often come from indigenous knowledge as evidenced by the many tomes written by those managing fisheries 'from the heart' (e.g. Pease, 1982).

Clearly, projects such as the Maryland Tributary Teams can complement (and even 'show up') agency approaches to the same problem. The Teams are charged with restoration of Chesapeake Bay by cutting nitrogen and phosphate loadings from the river system by 40 per cent by the year 2000, coordinating agency strategies, checking implementation and providing education. This is clearly essential work, given that Falkenmark (1996) denotes 'hydroconservatism' among agencies, resulting from:

- Indifference due to reductionist ideas of water.
- Poor general understanding of the complexity of water issues.
- Fragmented and inflexible administrative structures.
- Inter-cultural difficulties in development work.

Under such circumstances it is useful to read of the partnerships being built to circumvent the blockages in minds, bodies and budgets in, for example, the Mersey Basin Campaign in north-west England, the Tennessee Valley's Clean Water Initiative (Ungate, 1996) and the 'river contract' system in Belgium (Mormont, 1996 and Table 9.5). Mormont quotes the advantages of the Belgian system as empowering local communities, recreating community spirit, being neither 'bottom up' nor 'top down' in direction and seeing no separation between nature and development. In Queensland, Australia, Catchment Care Groups are part of the non-statutory basis of integrated catchment management, even though the Minister for Primary Industries retains the right to appoint the public representatives to the core management boards (Johnson et al., 1996).

The World Resources Institute has made repeated pleas, with audited studies of the outcomes of its recommendations, that the developing world offers a particularly fertile field for the application of participation; the Institute recommends that rehabilitating and improving upon indigenous

Table 9.5 Key features of a river contract in Belgium

Definition	Work programme for the integrated management of a river or part of a river.
	Basis of integration is discussion and negotiation with scientific information in support.
	Local authorities lead but structure involves all interested institutions.
	Diagnosis of problems yields proposals leading to objectives, targets by agreement.
Process	Local authorities lead; regional authority approves/finances.
	First stage is inventory, followed by specialist groups on major problems.
	Proposals need agreement of River Council before incorporation into Charter.
Actors	Procedures must include representatives of all the river users.
	Local agencies are admitted, as are officers of regional administration.
Organisation	The River Committee is central to the organisation and is the decision-maker.
	Committee elects Board to take daily decisions.
	Daily work programme derives from Secretariat.
	Working groups negotiate and generate new proposals.

Source: after Mormont (1996)

environmental management strategies should be a cornerstone of policy for developing-world governments (Mascarenhas and Veit, 1994; Veit *et al.*, 1995; Zazueta, 1995). Bottrall (1992) plays up the cost-cutting element of participation: 'It may be possible to keep the administrative costs of integrated watershed development within acceptable limits by delegating a large part of local management responsibilities to farmers' (p. 89).

9.6   ANTICIPATION AND RESTORATION: A PRACTICAL AGENDA

Environmental Impact Assessment (EIA, or simply EA) has been mentioned frequently in this book, both in a positive and in a negative vein. It is far too early to pronounce EA procedures a success or a failure; they were introduced in 1969 in the USA, from the same stable as an earlier litmus test for development: cost/benefit analysis. Ingram (1990) concludes that the National Environmental Protection Act (NEPA) which established the legal need for EA rang the death knell of the large water project in the USA. The addition of EA to project evaluation by the World Bank (in March 1989) may have a similar effect on developing world schemes; the United Nations

Environment Programme (UNEP) has a programme devoted to the environmentally sound management of inland waters. EA procedures differ wherever they are applied and there is a current clamour for more uniformity, if only among technical practitioners. They are essentially conciliation procedures in which knowledge plays a very large part, if allowed to and if available. Ingram cites the cost of achieving the knowledge for EA as one nail in the coffin of US water development.

There are also humanistic issues in the use of EA; like its simpler predecessor, cost/benefit analysis, it can be used as an alternative to true consultation and may be equally unable to include local factors and values (see Sagoff, 1989). In polarised cases there must be considerable room for the triumph of passion over knowledge as Petts (1990) hints:

> decisions on new water projects will continue to be made without adequate baseline data. Arguably the existence of an undisturbed river valley should be cause enough to guarantee its protection from development, at least until such baseline data are available.
>
> (Petts, 1990, pp. 199–200)

A further, more technical problem with river basin EA is that in river basin management long timescales are essential; Chapter 2 drives home a message from the basic physical science of river systems – geomorphology – that the longer view is essential for basic stability and that stability is a steady state, not equilibrium. Predictions of future changes in the river transport system are extremely difficult to make, especially under conditions of climatic change; the best way in which EA processes can respond is by building in monitoring and evaluation procedures where doubt is honestly expressed in the scientific evidence.

Antle (1983) has suggested that *post hoc* impact analyses will always be necessary in order to refine the process of forward planning in water development, though an intractable element of such analysis is the 'What if not?' question (i.e. What would have happened had not the scheme been built?).

While, in 1978, Kalbermatten and Gunnerson were able to describe EA as exhibiting 'newness and unfamiliarity', they nevertheless prescribe two aspects of the procedure which are still relevant to the success of the principle:

(a) Public involvement and participation is essential to a project's success.
(b) Environmental constraints constitute performance standards.

However, by 1985, Canter's annotated bibliography of published environmental impact analyses (formal and informal) reveals that only 10 per cent were connected with baseline studies or indicators of environmental viability; a similar proportion were relevant to public participation and decision-making, but not at a scale relevant to major international problems of river basin development, nor to the conjunctive consideration of land and water. Only very recently have formal, technical procedures such as decision support modelling been brought to bear on the problem of EA for large international rivers (e.g. Kovacs, 1990; Hartmann, 1990). These authors use techniques such as database management, impact modelling and decision algorithms as a means of producing a flexible-scale and interactive procedure for cross-disciplinary exploration of options.

River restoration capitalises on the innate ability of freshwater biotic systems to recover from damage. The indirect methods of restoration include, obviously, restoration of hydrological stability and the improvement of water quality but direct methods are becoming prominent too: instream habitat structures and management of the riparian zone (see Chapter 6). Restoration is largely successful because it:

(a) Involves a committed professional and public approach.
(b) Uses a systems approach, tackling quite extensive reaches.
(c) Is feasible in terms of land-take and therefore of ownership and control.

The latter point is critical in river basins where ownership confers rights, simply because water agencies, conservation and amenity agencies have very little ownership of land. The valley floor cross-section shown in Figure 2.9 represents a policy agenda for extending restoration campaigns from the open flow channel to the catchment as a whole; this agenda has its spatial implications and controlling functions developed by Newson (1992b).

Widespread public involvement is also critical to restoration; there are now a number of metropolitan restoration schemes in the UK motivated either by those who in increasing numbers live in waterfront locations taken up by developers as part of civic regeneration schemes or by those who are seeking to improve riverside amenity or conservation. The most comprehensive restoration scheme is that for the Mersey Basin; because it takes the basin approach it is multi-agency as well as public (through the voluntary sector). Having spent £0.5 billion on structural improvements during the 1980s the water authorities, local authorities and Shell UK have now come together to make the improvements sustainable; special posts include a 'Waterwatch Officer' and there are joint committees with executive power over interagency action (Mersey Basin Campaign, 1988).

The European Communities' 'Project Fluvius' has brought a programme of river frontage restoration to three European communities: Redbridge in Essex, UK, Romans sur Isère in France and Giannitsa in Greece. It worked by first creating awareness of the river environment in exhibitions and meetings, followed by conferences and comparisons of problems and progress. Once again, in hard financial times, much of the progress was brought about by private initiatives or partnership between the public and private sectors.

Table 9.6 A possible map of sustainable transition

	Policy	Economy	Society	Contact
Stage 1: Ultra weak sustainability	Lip service to integrated policies	*Minor tinkering with economic instruments*	Little awareness/media coverage	Corporatist discussion, consultation exercises
Stage 2: Weak sustainability	*Formal policy integration and deliverable targets*	Substantial micro-economic restructuring	*Wider public education – future visions*	*Round tables, stakeholder groups*
Stage 3: Strong sustainability	Binding policy integration and strong international agreements	Full economic evaluation; green national and business accounts	Curriculum integration; local initiatives	Community involvement in decision-making

Source: modified from Pearce (1993)
Note: Italics indicate the stage reached in the UK with river basin management.

9.7   HOW SUSTAINABLE? A SLOW REVOLUTION

Buller (1996) comments that 'In its current form, catchment planning is a half-way step to sustainable water management' (p. 301). He makes this claim on the grounds that catchment planning:

- gives weight to system parameters;
- negotiates strategies at a local level; and
- integrates the planning and management of land and water.

At this late stage of the book it is appropriate to judge the degree to which river basin management has followed the forms of sustainability laid out by Pearce (1993). Table 9.6 shows how far the National Rivers Authority had reached with the sustainable agenda at the time of its evolution to the Environment Agency in England and Wales.

The caution is clear but the agenda is complex and checks are everywhere in the literature of official agencies. For example, the UK Environment Act 1990 states that the mission of the Environment Agency (which now subsumes the role of the National Rivers Authority) is 'to protect or enhance the environment, taken as a whole, [so] as to make a contribution towards attaining the objective of achieving sustainable development that Ministers consider appropriate' – hardly a revolutionary call!

Nevertheless, post-Rio, the local government sector in the UK has been very active in contemplating and operationalising Local Agenda 21. In this spirit the Thames Region of the National Rivers Authority launched in 1995

Table 9.7 Thames 21 – water quality function

Guidance for development plans
Where development proposals would lead to deterioration in the quality of underground or surface water they should normally be resisted unless or until appropriate infrastructure provisions have been made.

Current sustainability principles for water quality
i   In order to maintain the amenity and nature conservation value of rivers and other water bodies, and the purity of drinking water, there should be no long-term deterioration in the quality of surface and groundwater caused by human activity.
ii Development, including waste disposal, which is likely to place the quality of water at risk should be resisted, at least until any necessary infrastructure to protect water quality is in place.
iii Monitoring the quality of water is essential to determine if waters are of an appropriate quality for their agreed uses, and for identifying where improvement is justified.
iv Initiatives which would lead to an improvement in water quality should be identified and encouraged, and the principles of waste minimisation and good pollution prevention practice should be promoted.
v   Financial penalties should be imposed to deter potential polluters.

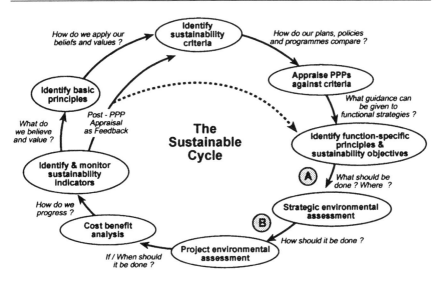

(A) What should go into a Catchment Management Plan? (CMP)

(B) How should the CMP be co-ordinated with local authority development plans (and other plans) to achieve ' Integrated Catchment Management ' ?

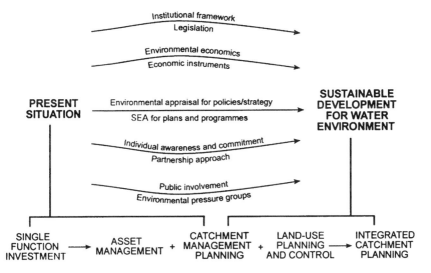

Figure 9.3 Sustainability in practice:
   (a) Sustainable cycles (after Thames Region, National Rivers Authority, 1996)
   (b) Transition to sustainable programmes for UK water (after Gardiner, 1996)

Table 9.8 Water Law Principles (abridged and emphasis added)

A1   It is necessary to recognise the *unity of the water cycle* and the interdependence of its elements.

A2   The variable, uneven and unpredictable distribution of water in the water cycle should be acknowledged.

B1   All water is a *resource common to all*, the use of which should be subject to national control. All water should have consistent status in law, irrespective of where it occurs.

B2   There shall be *no ownership of water* but only a right to its use.

B3   The location of the *water resource in relation to land* should not in itself confer preferential rights to usage.

C1   The objective of managing the nation's water resources is to achieve optimum long term social and economic benefit for our society from their use, recognising that *water allocations may have to change over time*.

C2   The water required to meet peoples' basic domestic needs should be reserved.

C3   The quantity, quality and reliability of water required to maintain the ecological functions on which humans depend should be reserved so that human use of water does not individually or cumulatively compromise the *long term sustainability of aquatic and associated ecosystems*.

C4   The water required to meet peoples' basic domestic needs and the needs of the environment should be identified as the 'reserve' and should enjoy priority of use (the remainder is labelled 'utilisable water' in the publication).

C5   International water resources, specifically shared river systems, should be managed in a manner that will optimise the *benefits for all parties in a spirit of mutual cooperation*. Allocations agreed for downstream countries should be respected.

D1   The *national government has ultimate responsibility* for, and authority over, water resource management, the equitable allocation and usage of water, the transfer of water between catchments and international water matters.

D2   The development, apportionment and management of water resources should be carried out using the criteria of *public interest, sustainability, equity and efficiency* of use in a manner which reflects the value of water to society whilst ensuring that basic domestic needs, the requirements of the environment and international obligations are met.

D3   As far as is physically possible, water resources should be managed in such a manner as to enable all user sectors to gain equitable access to the desired quantity, quality and reliability of water, using conservation and other measures to *manage demand* where this is required.

D4   Water quality and quantity are interdependent and should be managed in an integrated manner, which is *consistent with broader environmental management approaches*.

D5   Water quality management options should include the use of *economic incentives and penalties* to reduce pollution; the possibility of irretrievable environmental degradation as a result of pollution should be prevented.

D6   Water resource development and supply should be managed in a manner consistent with broader environmental management approaches.

D7   The *regulation of land use* should, where appropriate, be used as an instrument to manage water resources.

Table 9.8 (continued)

D8	Rights to use water should be allocated in good time and in a manner which is clear, secure and predictable in respect of the assurance of availability, extent and duration of use.
D9	The conditions under which water rights are allocated should take into consideration the investment made by the user in developing infrastructure to be able to use the water.
D10	The development and management of water resources should limit to an acceptable level the danger to life and property due to natural or man-made disasters.
E1	The institutional framework for water management should be self-driven, minimise the necessity for state intervention and should provide for a right of appeal.
E2	Responsibility should, where possible, be *delegated to a catchment or regional level* in such a manner as to enable interested parties to participate and reach consensus.
E3	*Beneficiaries* of the water management system *should contribute to the cost* of its establishment and maintenance.
F1	Where existing rights are taken away, compensation should be paid.
G1	The *right of all citizens to have access to basic water services* (the provision of potable water supply and the removal and disposal of human excreta and waste water) necessary to afford them a healthy environment on an equitable and economically and environmentally sustainable basis should be supported.
G2	Water services should be provided in a manner consistent with the goals of water resource management.
G3	Where water services are provided in a monopoly situation, the interests of the individual consumer and the wider public must be protected and the broad goals of public policy promoted.

Source: Department of Water Affairs and Forestry, South Africa (1996)

a fully consulted *Thames 21* document in which all the Authority's duties are given listed sustainability principles (see Table 9.7).

Figure 9.3a is taken from *Thames 21* and shows the key questions (and cyclicity) of corporate sustainability. It is crucial that the NRA (now Environment Agency) realises the importance of the links shown to development planning. Gardiner (1996), whose work inspired much of the Thames Region approach, has developed the relationship further in a diagram (Figure 9.3b) which illustrates the general pathways towards sustainable development.

The UK has had a lengthy period of political and economic stagnation; perhaps it is more useful, therefore, to examine the principles and practices being contemplated for water management in one of the newest democracies – South Africa, a nation facing daunting water problems but benefiting from 'fresh start' national attitudes and opportunities. Table 9.8 illustrates the apparently 'stronger' sustainability contemplated for South Africa's new Water Act, an Act which itself is open to public consultation.

Table 9.9 Agenda 21: 'Protection of the quality and supply of freshwater resources' – target costs and sources of finance

Programme area	Total annual budget (US$m)	From external sources (US$m)
Integrated water resources development and management	115	115
Water resources assessment	355	145
Protection of water resources, water quality and aquatic ecosystems	1,000	340
Impacts of climate change on water resources	100	44
Water and sustainable urban development	20,000	4,500
Water for sustainable food production and rural development	13,200	4,500
Drinking water supply and sanitation	20,000	7,400
Total programme	*54,770*	*17,040*

Source: Young et al. (1994)

To empower those in the developing world to embark on sustainable development of river basins will be extremely expensive. Fortunately Agenda 21 hazarded estimates of the costs of each part of its programme related to water (Table 9.9). Transfers of funds to build capacity are seen as preferable to the traditional transfer of technologies suited mainly to the country of origin or of funds for the remote and unpopular 'megaproject'.

9.8   CATCHMENTS, BASINS, CORRIDORS AND VALLEYS: A FINAL NOTE ON SCALE

It is the geographer's privilege to pick out the scale elements of a scientific, political or administrative problem. Gamble and Meentemeyer (1996) have recently reiterated that any remedies for the currently unsustainable use of the Ganges and Brahmaputra must come from the explicit use of scale in both research and applications. The whole system may be considered, as in Figure 9.2, as a complex of resources, including rural, urban, water and human resources, each requiring management and impact assessment at the appropriate scale.

I have frequently claimed in this book that the river basin concept has been promoted mainly by geographers; however, the topic of stream restoration introduces the very important, hitherto latent role of the biologist. Academic biologists have for many years stressed the continuity of the river basin system (e.g. Hynes, 1975; Vannote *et al.*, 1980); issues of conservation and restoration have brought biologists and fluvial geomorphologists together, if not in research programmes on contentious sites. Their contribution is particularly noteworthy, for example, in the journals *Regulated Rivers: Research and Management* and *Aquatic Conservation.* Biologists have achieved many of the senior posts in the reorganised water industry in England and Wales. Biological assessment of water quality standards is becoming formally incorporated in UK monitoring systems. Definitions of land-take for conservation of wetlands are increasingly 'bullish', particularly as doubts are raised by economists and politicians about the validity of production-orientated farming.

Petersen *et al.* (1987) first pointed out to river managers the centrality of the riparian zone; huge costs involved with river purification can be saved by what the authors call a 'holistic ecosystem approach', in which the first goal is not catchment control but riparian control. In many countries of Europe stream restoration is now associated with the achievement of water quality objectives and, where relevant, compliance with EU legislation – a fusion of idealistic conservation and hard-nosed economics. In Denmark, for example, formerly straightened streams are being 're-meandered' and *buffer strips*, akin to those used in US forestry activities (Chapter 4), are legally applied to prevent soil erosion and nutrient leaching from polluting streams. Germany, the Netherlands and Denmark are close behind. Petersen and colleagues set down five principles of 'repeated patterns of stream management':

(a) Watershed management is the goal, riparian control the starting point.
(b) The riparian zone is the important interface between the terrestrial and stream ecosystem.
(c) Short-term events may be far more damaging than average conditions.
(d) Good management requires holistic approaches.
(e) Some stream management problems are the result of global environmental problems.

If a practical agenda of riparian control can be introduced as part of developed-world stream restoration projects it can be introduced as part of international environmental management for those 200 major river basins whose boundaries cross national boundaries, including the 'water crisis' zone.

Petersen *et al.* (1992) have now developed a management sequence for the riparian zone, effectively the valley floor:

(a) Buffer strips.
(b) Revegetation (natural, cf. agricultural species).
(c) Horseshoe wetlands (where productive agriculture discharges drainage waters to the valley floor).
(d) Reduction of channel bank slopes to reduce erosion.
(e) Restoration of meanders.
(f) Restoration of riffle/pool sequences in channels.
(g) Restoration of wetland valley floors and swamp forests.

The future is one of great interest. It represents, as has been stressed, a pattern of the application of knowledge within an increasingly public context. The compromise will exist between two extremes, the traditional one of water power, as quoted by Ingram (1990):

> Water still symbolizes such values as opportunity, security and self-determination. Water represents these values less because the water itself has economic value than because control over it signals social organization and political power... Strong communities are able to hold on to their water and put it to work.
>
> (Ingram, 1990, p. 5)

and the new context of crisis and opportunity provided by water as an element of our natural environment for which we require sustainable management strategies, compliance with which will depend on public acceptability.

Allan (1995) removes issues of scale from the debate by adopting the American term 'problem-shed' within which the key solutions in future will be:

(a) Movement from the management of water supply to management of demand.
(b) Movement from water supply as a 'free good', or uncosted incentive to development, to an economic entity.
(c) Change from inequitable practices (social, temporal and spatial) to equity.
(d) Incorporation of ecological impacts in all aspects of water-related development.

Perhaps less ordered and practical have been the many pleas in the last five years for a 'water ethic' to guide sustainable development. Feldman (1991) advocates the lead role for water in educating those charged with environmental policy-making; the outcome will be 'dignified, orderly survival on this planet'. Falkenmark and Lunqvist (1995) are similarly convinced of the primacy of water management:

> The concern for the multifunctional value of water has so far been neglected. There is a need for a new professionalism which does not lead to another generation of compartmentalization, nor to a lingering disregard for the linkages between resource endowments, environmental conditions and livelihood situations.
>
> (p. 212)

Postel (1992) is yet more profound in the quest for a water ethic:

> at the heart of the matter is modern society's disconnection from water's life-giving qualities. Grasping the connection between our own destiny

and that of the water world around us is integral to the challenge of meeting human needs while protecting the ecological functions that all life depends on. The essence of such an ethic is to make protection of water ecosystems a central goal in what we do.

(pp. 184, 185)

Let's drink to that!

Bibliography

Abate, Z. (1994) *Water Resources Development in Ethiopia*. Ithaca Press, Reading.

Aberley, D. (1993) *Boundaries of Home: Mapping for Local Empowerment*. New Society Publications, Gabriola, BC.

Abracosa, R. and Ortolano, L. (1988) 'Lessons from EIA for Bicol river development in Philippines', *Journal of Water Resource Planning and Management*, Proceedings of the American Society of Civil Engineers, 114(5), 517–29.

Abu-Maila, Y. S. (1991) 'Water resource issues in the Gaza Strip'. *Area*, 23(3), 209–16.

Abu-Zeid, M. A. and Biswas, A. K. (1992) *Climatic Fluctuations and Water Management*. Butterworth-Heinemann, Oxford.

Acreman, M. (1994) 'The role of artificial flooding in the integrated development of river basins in Africa', in C. Kirby and W. R. White (eds) *Integrated River Basin Development*. Wiley, Chichester, 35–44.

Adams, W. M. (1985) 'River basin planning in Nigeria', *Applied Geography*, 5, 297–308.

Adams, W. M. (1992) *Wasting the Rain: Rivers, People and Planning in Africa*. Earthscan, London.

Adams, B. and Foster, S. S. D. (1992) 'Land-surface zoning for groundwater protection', *Journal of the Institution of Water and Environmental Management*, 6, 312–20.

Adams, W. M., Potkanski, T. and Sutton, J. E. G. (1994) 'Indigenous farmer-managed irrigation in Soiyo, Tanzania', *Geographical Journal*, 160(1), 17–32.

Agarwal, A., Kimondo, J., Moreno, G. and Tinker, J. (1980) *Water, Sanitation, Health – For All? Prospects for the International Drinking Water Supply and Sanitation Decade, 1981–90*. Earthscan, London and Washington.

Ahmed, S. (1990) 'Cleaning the River Ganga: rhetoric and reality', *Ambio*, 19(1), 42–5.

Alabaster, J. S. (1972) 'Suspended solids and fisheries', *Proceedings of the Royal Society of London, Series B*, 180, 395–406.

Al-Ibrahim, A. A. (1990) 'Water use in Saudi Arabia: problems and policy implications', *Journal of Water Resource Planning and Management*, 116(3), 375–88.

Al-Ibrahim, A. A. (1991) 'Excessive use of groundwater resources in Saudi Arabia: impacts and policy options', *Ambio*, 20(1), 34–7.

Allan, J. A. (1992) 'Fortunately there are substitutes for water: otherwise our hydropolitical futures would be impossible', in *Proceedings of the Conference on Priorities for Water Resource Allocation and Management*. Overseas Development Administration, London, 13–26.

Allan J.A. (1995) 'Water: a substitutable resource?', in Faggi, P. and Minoia, P. (eds) *Water and Regional Dynamics*. ERASMUS Workshop, Department of Geography, University of Padua, Italy.

Allan, J. A. (1996) 'The political economy of water: reasons for optimism but long term caution', in J. A. Allan (ed.) *Water, Peace and the Middle East: Negotiating Resources in the Jordan Basin*. Tauris, London, 75–117.

Allan, J. A. and Karshenas, M. (1996) 'Managing environmental capital: the case of water in Israel, Jordan, the West Bank and Gaza, 1947 to 1995', in J. A. Allan (ed.) *Water, Peace and the Middle East: Negotiating Resources in the Jordan Basin*. Tauris, London, 121–33.

Allen, R. G., Gichniki, F. N. and Rosenzweig, C. (1991) 'CO_2-enhanced climatic changes and irrigation water requirements', *Journal of Water Resource Planning and Management*, 117(2), 157–78.

Alvares, C. and Billorey, R. (1988) *Damming the Narmada: India's Greatest Planned Environmental Disaster*. Third World Network/Appen, Penang, Malaysia.

American Society of Civil Engineers (1988a) *Evaluation Procedures for Hydrologic Safety of Dams*. New York.

American Society of Civil Engineers (1988b) *Lessons from Dam Incidents USA II*. New York.

Amerman, C. R. (1965) 'The use of unit-source watershed data for runoff prediction', *Water Resources Research*, 1, 499–507.

Amoros, C., Roux, A. L., Raygrobellet, J. L., Brayard, J. P. and Paton, G. (1987) 'A method for applied ecological studies of fluvial hydrosystems', *Regulated Rivers: Research and Management*, 1(1), 17–36.

Amphlett, M. B. (1990) 'A field study to assess the benefits of land husbandry in Malawi', in J. Boardman, I. D. L. Foster and J. A. Dearing (eds) *Soil Erosion on Agricultural Land*, Wiley, Chichester, 575–88.

Antle, L. G. (1983) 'Evaluation of completed projects: why is it necessary?', in G. G. Green and E. E. Eiker (eds) *Accomplishments and Impacts of Reservoirs*. American Society of Civil Engineers, New York, 6–19.

Arnell, N. (1989a) 'The influence of human activities on hydrological characteristics: an introduction', in N. Arnell (ed.) *Human Influences on Hydrological Behaviour: An International Literature Survey*. UNESCO, Paris, 1–18.

Arnell, N. (ed.) (1989b) *Human Influences on Hydrological Behaviour: an International Literature Survey*. UNESCO, Paris.

Arnell, N. W., Jenkins, A. and George, D. G. (1994) *The Implications of Climate Change for the National Rivers Authority*. HMSO, London.

Arntzen, J. (1995) 'Economic instruments for sustainable resource management: the case of Botswana's water resources', *Ambio*, 24(6), 335–42.

Arya, S. L., Agnihotri, Y. and Samra, J. S. (1994) 'Watershed management: changes in animal population structure, income and cattle migration, Shiwaliks, India', *Ambio*, 23(7), 446–50.

Ashby, E. (1978) *Reconciling Man with the Environment*. Oxford University Press, Oxford.

Ashworth, P. J. and Ferguson, R. I. (1986) 'Interrelationships of channel processes, changes and sediments in a proglacial braided river', *Geografiska Annale*, 68A, 361–71.

Atampugre, N. (1993) *Behind the Lines of Stone: The Social Impact of a Soil and Water Conservation Project in the Sahel*. Oxfam, Oxford.

Atkinson, T. C. (1978) 'Techniques for measuring subsurface flow on hillslopes', in M. J. Kirkby (ed.) *Hillslope Hydrology*. Wiley, Chichester, 73–120.

Bagader, A. A., El-Sabbagh, A. T. E., Al-Glayand, M. A. and Samarrai, M. Y. I. (1994) *Environmental Protection in Islam*. IUCN, Gland, Switzerland, 2nd edition (1–36 in English).

Bagnold, R. A. (1977) 'Bedload transport by natural rivers', *Water Resources Research*, 13, 302–12.

Baker, D. R. (1981) *Environmental Crisis in Kenya: Social Crisis or Environmental Crisis?* DEV Discussion Paper 82, School of Development Studies, University of East Anglia, Norwich.

Baldock, D. (1984) *Wetland Drainage in Europe: The Effects of Agricultural Policy in Four EEC Countries*. Institute for European Environmental Policy/International Institute for Environment and Development, London.

Baldwin, A. B. (1978) 'Quality aspects of water in upland Britain', in R. B. Tranter (ed.) *The Future of Upland Britain*. Centre for Agricultural Strategy, University of Reading, 322–7.

Bandyopadhyay, J. (1987) *The Indian Drought 1987–88. The Ecological Causes of Water Crisis: What to Do*. Third World Science Movement, Penang, Malaysia.

Bandyopadhyay, J. and Gyawali, D. (1994) 'Himalayan water resources: ecological and political aspects of management', *Mountain Research and Development*, 14(1), 1–24.

Barbier, E. (1991) 'Environmental degradation in the Third World', in D. Pearce (ed.) *Blueprint 2: Greening the World Economy*. Earthscan, London, 75–108.

Barney, G. O. (Study Director) (1982) *The Global 2000 Report to the President*. Penguin Books, Harmondsworth.

Barrow, C. J. (1987) *Water Resources and Agricultural Development in the Tropics*. Longman, Harlow.

Barrow, C. J. (1995) *Developing the Environment: Problems and Management*. Longman, London.

Batchelor, M. and Brown, K. (eds) (1992) *Buddhism and Ecology*. Cassell, London.

Bates, S. F., Getches, D. H., Macdonnell, L. J. and Wilkinson, C. F. (1993) *Searching out the Headwaters: Change and Rediscovery in Western Water Policy*. Island Press, USA.

Beauclerk, J., Narby, J. and Townsend, J. (1988) *Indigenous Peoples: A Fieldguide for Development*. Oxfam, Oxford.

Beaumont, P. (1978) 'Man's impact on river systems: a world-wide view', *Area*, 10, 38–41.

Beaumont, P. (1983) 'Water resource management in the USA: a case study of large dams', *Applied Geography*, 3, 259–75.

Beaumont, P. (1989) *Environmental Management and Development in Drylands*. Routledge, London.

Beven, K. (1985) 'Distributed models', in M. G. Anderson and T. P. Burt (eds) *Hydrological Forecasting*. Wiley, Chichester, 405–35.

Beven, K. (1993) 'Riverine flooding in a warmer Britain', *Geographical Journal*, 159(2), 157–61.

Binnie, G. M. (1981) *Early Victorian Water Engineers*. Thomas Telford, London.

Binnie, G. M. (1987) *Early Dam Builders in Britain*. Thomas Telford, London.

Binns, A. (1990) 'Is desertification a myth?', *Geography*, 75(2), 106–13.

Bissio, B. (ed.) (1988) *Third World Guide*. Third World Editors, Montevideo.

Biswas, A. K. (1967) 'Hydrologic engineering prior to 600 BC', *Proceedings of the American Society of Civil Engineers Journal, Hydraulics Division*, HY5, 118–31.

Biswas, A. K. (1993) 'Land resources for sustainable agricultural development in Egypt', *Ambio*, 22(8), 556–60.

Biswas, A. K., Jellali, M. and Stout, G. (eds) (1993) *Water for Sustainable Development in the 21st Century*. Oxford University Press, Delhi.

Black, P. E. (1970) 'The watershed in principle', *Water Resources Bulletin*, 6(2), 153–62.

Black, P. E. (1982) *Conservation of Water and Related Land Resources*. Praeger, Westport, Conn.

Blackie, J. R., Edwards, K. A. and Clarke, R. T. (eds) (1979) 'Hydrological research in East Africa', *East African Agriculture and Forestry Journal*, special issue 43.

Blackie, J. R. and Eeles, C. W. O. (1985) 'Lumped catchment models', in M. G. Anderson and T. P. Burt (eds) *Hydrological Forecasting*. Wiley, Chichester, 311–45.

Blaikie, P. (1985) *The Political Economy of Soil Erosion in Developing Countries*. Longman, Harlow.

Blench, T. (1952) 'Regime theory for self-formed sediment-bearing channels', *Transactions of the American Society of Civil Engineers*, 117, 383–400.

Boardman, J. (1988) 'Public policy and soil erosion in Britain', in J. M. Hooke (ed.) *Geomorphology in Environmental Planning*. Wiley, Chichester, 33–50.

Boardman, J. (1990) *Soil Erosion in Britain: Costs, Attitudes and Policies*. Social Audit Paper 1, Education Network for Environment and Development, Brighton.

Boardman, J., Dearing, J. A., and Foster, I. D. L. (1990) 'Soil erosion studies: some assessments', in J. Boardman, I. D. L. Foster and J. A. Dearing (eds) *Soil Erosion on Agricultural Land*. Wiley, Chichester, 659–72.

Bonell, M. and Balek, J. (1993) 'Recent scientific developments and research needs in hydrological processes of the humid tropics', in M. Bonell, M. M. Hufschmidt and J. S. Gladwell (eds) *Hydrology and Water Management in the Humid Tropics*. Cambridge University Press, Cambridge, 167–260.

Bonell, M. Hufschmidt, M. M. and Gladwell, J. S. (1993) *Hydrology and Water Management in the Humid Tropics*. Cambridge University Press, Cambridge.

Boon, P. J. (1987) 'The influence of Kielder Water on trichopteran (caddisfly) populations in the River North Tyne (northern England)', *Regulated Rivers: Research and Management, 1*, 95–109.

Boon, P. J. (1992) 'Essential elements in the case for river conservation', in P. J. Boon, P. Calow and G. E. Petts (eds) *River Conservation and Management*. Wiley, Chichester, 11–33.

Boon, P. J., Calow, P. and Petts, G. E. (eds) (1992) *River Conservation and Management*. Wiley, Chichester.

Bord, J. and Bord, C. (1986) *Sacred Waters: Holy Wells and Water Lore in Britain and Ireland*. Paladin, London.

Bordas, M. P. and Walling, D. E. (1988) *Sediment Budgets*. International Association of Hydrological Sciences, Publication 174.

Bosch, J. M. and Hewlett, J. D. (1982) 'A review of catchment experiments to determine the effect of vegetation changes on water yield and evapotranspiration', *Journal of Hydrology*, 55, 3–23.

Bottrall, A. (1992) 'Institutional aspects of watershed management', in Overseas Development Agency, *Priorities for Water Resources Allocation and Management*. ODA, London, 81–90.

Bowman, J. A. (1990) 'Ground-water-management areas in United States', *Journal of Water Resources Planning and Management*, 116(4), 484–502.

Bowonder, B. and Ramana, K. V. (1987) 'Environmental degradation and economic development: a case study of a marginally productive area', *Applied Geography*, 7, 301–15.

Brammer, H. (1990a) 'Floods in Bangladesh. I Geographical background to the 1987 and 1988 floods', *Geographical Journal*, 156(1), 12–22.

Brammer, H. (1990b) 'Floods in Bangladesh. II Flood mitigation and environmental aspects', *Geographical Journal*, 156(2), 158–65.

Bratt, G. (1995) *The Bisses of Valais: Man-made Watercourses in Switzerland*. Guy Bratt, Gerrards Cross, UK.

Breuilly, E. and Palmer, M. (eds) (1992) *Christianity and Ecology*. Cassell, London.

British Ecological Society (1990) *River Water Quality*. Ecological Issues, 1. London.

Broen, A. G. (1987) 'Long-term sediment storage in the Severn and Wye catchments', in K. J. Gregory, J. Lewin and J. B. Thornes (eds) *Palaeohydrology in Practice*, Wiley, Chichester, 307–32.

Brookes, A. (1988) *Channelized Rivers: Perspectives for Environmental Management*. Wiley, Chichester.

Brookes, A. (1992) 'Recovery and restoration of some engineered British river channels', in P. J. Boon, P. Calow and G. E. Petts (eds) *River Conservation and Management*. Wiley, Chichester, 337–52.

Brookes, A. (1995a) 'Challenges and objectives for geomorphology in UK river management', *Earth Surface Processes and Landforms*, 20, 593–610.

Brookes, A. (1995b) 'River channel restoration: theory and practice', in A. Gurnell and G. E. Petts (eds) *Changing River Channels*. Wiley, Chichester, 369–88.

Brookes, A. and Gregory, K. J. (1988) 'Channelization, river engineering and geomorphology', in J. M. Hooke (ed.) *Geomorphology in Environmental Planning*. Wiley, Chichester, 145–67.

Brookes, A. and Shields, F. F. (1996) *River Channel Restoration: Guiding Principles for Sustainable Projects*. Wiley, Chichester.

Brown, A. G. (1987) 'Long-term sediment storage in the Severn and Wye catchments', in K. J. Gregory, J. Lewin and J. B. Thomas (eds) *Palaeohydrology in Practice*. Wiley, Chichester.

Brown, B. W. and Shelton, R. A. (1983) 'Fifty years of operation of the TVA reservoir system', in G. G. Green and E. E. Eiker (eds) *Accomplishments and Impacts of Reservoirs*. American Society of Civil Engineers, New York, 138–51.

Bruk, S. (Rapporteur) (1985) *Methods of Computing Sedimentation in Lakes and Reservoirs*. UNESCO, Paris.

Brune, G. M. (1953) 'Trap efficiency of reservoirs', *American Geophysical Union Transactions*, 34(3), 407–17.

Bull, K. R. and Hall, J. R. (1989) *Classification and Comparison of River and Lake Catchments*. Institute of Terrestrial Ecology, Monks Wood Research Station, Cambs.

Buller, H. (1996) 'Towards sustainable water management: catchment planning in France and Britain', *Land Use Policy*, 13(4), 289–302.

Bullock, J. and Darwish, A. (1993) *Water Wars: Coming Conflicts in the Middle East*. Gollancz, London.

Bunyard, P. and Goldsmith, E. (1988) *GAIA, the Thesis, the Mechanisms and the Implications*. Wadebridge Ecological Centre, Camelford, Cornwall.

Burt, J. P. and Walling, D. E. (1984) 'Catchment experiments in fluvial geomorphology: a review of objectives and methodology', in T. P. Burt and D. E. Walling (eds) *Catchment Experiments in Fluvial Geomorphology*. Geo Books, Norwich, 3–18.

Burt, T. P. (1986) 'Runoff processes and solute denudation rates on humid temperate hillslopes', in S. T. Trudgill (ed.) *Solute Processes*. Wiley, Chichester, 193–249.

Burton, L. (1991) *American Indian Rights and the Limits of Law*. University Press of Kansas, Lawrence, Kan.

Cairns, J. R. (1995) 'Maintenance and sustainable use of desirable ecosystem services'. *Regulated Rivers: Research and Management*, 11, 313–23.

Calder, I. R. (1990) *Evaporation in the Uplands*. Wiley, Chichester.

Calder, I. R. (1994) *Eucalyptus: Water and Sustainability*. ODA Forestry Series, 6, Overseas Development Agency, London.

Calder, I. R., Hall, R. L. and Adlard, P. G. (1992) *Growth and Water Use of Forest Plantations*. Wiley, Chichester.

Calder, I. R. and Newson, M. D. (1979) 'Land use and upland water resources in Britain: a strategic look', *Water Resources Bulletin*, 15(6), 1628–39.

Caldwell, L. K. and Shrader-Frechette, K. (1993) *Policy for Land: Law and Ethics*. Rowman & Littlefield, Lanham, Md.

Callicott, J. B. (1987) 'The scientific substance of the land ethic', in T. Tanner (ed.) *Aldo Leopold: the Man and his Legacy*. Soil Conservation Society of America, Ankeny, Iowa, 87–104.

Calow, P. and Petts, G. E. (eds) (1992) *The Rivers Handbook*. Vol. I. Blackwell, Oxford.

Calow, P. and Petts, G. E. (eds) (1994) *The Rivers Handbook*. Vol. II. Blackwell, Oxford.

de Camp, L. S. (1990) *The Ancient Engineers*. Dorset Press, New York.

Canadian Council of Resource and Environment Ministers (1987) *Canadian Water Quality Guidelines*. Environment Canada, Montreal.

Canter, L. (1985) *Environmental Impact of Water Resources Projects*. Lewis Publishers, Chelsea, Mich.

Carroll, J. E. (1988) *International Environmental Diplomacy*. Cambridge University Press, Cambridge.

Carruthers, I. D. (1983) *Aid for the Development of Irrigation*. OECD, Paris.

Carson, M. A. (1984) 'The meandering–braided threshold: a reappraisal', *Journal of Hydrology*, 73, 315–34.

Carter, R. (1992) 'Small-scale irrigation in sub-Saharan Africa: a balanced view', in *Priorities for Water Resource Allocation and Management*. Proceedings of the Conference. Overseas Development Administration, London, 103–16.

Carter, R. C. and Howsam, P. (1994) 'Sustainable use of groundwater for small-scale irrigation, with special reference to sub-Saharan Africa', *Land Use Policy*, 11(4), 275–85.

Central Water Planning Unit (1979) *River Regulation Losses in England and Wales*. CWPU, Reading.

Chambers, R. (1988) *Managing Canal Irrigation: Practical Analysis from South Asia*. Cambridge University Press, Cambridge.

Chandler, W. V. (1984) *The Myth of the TVA Conservation and Development in the Tennessee Valley, 1933–1983*. Ballinger, Cambridge, Mass.

Charlton, F. C., Brown, P. M. and Benson, R. W. (1978) 'The hydraulic geometry of some gravel rivers in Britain', *Report IT180*, Hydraulics Research, Wallingford, Oxon.

Chauhan, S. K., Bihua, Z., Gopalakrishnan, K., Lala Rukh, H., Yeboak-Afari, A. and Leal, F. (1983) *Who Puts the Water in the Taps? Community Participation in Third World Drinking Water, Sanitation and Health*. Earthscan, London and Washington.

Chesworth, P. M. (1990) 'The history of water use in Sudan and Egypt', in P. P. Howell and J. A. Allan (eds) *The Nile*. School of Oriental and African Studies/ Royal Geographical Society, London, 41–58.

Chettri, R. and Bowonder, B. (1983) 'Siltation in Nizamsagar reservoir: environmental management issues', *Applied Geography*, 3, 193–204.

Chorley, R. J. (ed.) (1969) *Water, Earth and Man*. Methuen, London.

Church, M. (1984) 'On experimental method in geomorphology', in T. P. Burt and D. E. Walling (eds) *Catchment Experiments in Geomorphology*. Geo Books, Norwich, 563–80.

Colborn, T. E., Davidson, A., Green, S. N., Hodge, R. A., Jackson, C. I. and Liroff, R. A. (1990) *Great Lakes: Great Legacy?* Conservation Foundation, Washington DC; Institute for Research on Public Policy, Ottawa, Ontario.

Coles, B. and Coles, J. (1989) *People of the Wetlands: Bogs, Bodies and Lake-dwellers*. Thames & Hudson, London.

Collingridge, D. and Reeve, C. (1986) *Science Speaks to Power*. Frances Pinter, London.

Collins, R. O. (1990) *The Waters of the Nile: Hydropolitics and the Jonglei Canal 1900–1988*. Clarendon Press, Oxford.

Conacher, A. J., Combes, P. L., Smith, P. A. and McLellan, R. L. (1983a) 'Evaluation of throughflow interceptors for controlling secondary soil and water salinity in dryland agricultural areas of southwestern Australia: I. Questionnaire surveys', *Applied Geography*, 3, 29–44.

Conacher, A. J., Combes, P. L., Smith, P. A. and McLellan, R. L. (1983b) 'Evaluation of throughflow interceptors for controlling secondary soil and water salinity in dryland agricultural areas of southwestern Australia: II. Hydrological study', *Applied Geography*, 3, 115–32.

Conley, A. H. (1989) The value of high-technology for managing intersectoral and international sharing of inadequate water resources in a developing region of Africa', in *Water '89: Water Decade and Beyond*, World Water, Bangkok, Thailand, 1–7.

Conley, A. H. (1995) 'The evolution of measures to manage conflicts between competing users of scarce water resources in South Africa', in *International Conference on Water Resource Management in Arid Countries, Sultanate of Oman*, 1, 329–36.

Conley, A. H. (1996) 'A synoptic view of water resources in Southern Africa'. Southern African Society of Aquatic Scientists, Victoria Falls Conference.

Conroy, C. and Litvinoff, M. (1988) *The Greening of Aid: Sustainable Livelihoods in Practice*. Earthscan, London.

Conway, V. M. and Millar, A. (1960) 'The hydrology of some small peat covered catchments in the north Pennines', *Journal of the Institution of Water Engineers*, 14, 415–24.

Cooke, A. (1973) *America*. BBC Publications, London.

Cooke, R. U., Brunsden, D., Doornkamp, J. C. and Jones, D. K. C. (1982) *Urban Geomorphology in Drylands*. Oxford University Press, Oxford.

Cosgrove, D. (1990) 'An elemental division: water control and engineered landscape', in D. Cosgrove and G. Petts (eds) *Water, Engineering and Landscape*. Belhaven Press, London, 1–11.

da Costa, J. A. and Jacquet, J. (1965) 'Présentation des résultats de l'enquête UNESCO–AIHS sur les bassins représentatifs et expérimentaux dans le monde', *IAHS Bulletin*, x(4), 107–19.

Costa, J. E. (1988) 'Floods from dam failures', in V. R. Baker, R. C. Kochel and P. C. Patton (eds) *Flood Geomorphology*. Wiley, New York, 439–63.

Cox, W. E. (1988) 'Water and development: a complex relationship', *Journal of Water Resource Planning and Management*, 114(1), 91–8.

Crabb, P. (1991) 'Resolving conflicts in the Murray-Darling basin', in J. W. Handmer, A. H. J. Dorcey and D. I. Smith (eds) *Negotiating Water: Conflict Resolution in Australian Water Management*. Centre for Resource and Environmental Studies, Australian National University, Canberra.

Critchley, W. (1991) *Looking after our Land: New Approaches to Soil and Water Conservation in Dryland Africa*. Oxfam, Oxford.

Crow, B. (1995) *Sharing the Ganges: The Politics and Technology of River Development*. Sage, London.

Cummings, B. J. (1990) *Dam the Rivers, Damn the People*. Earthscan, London.

Cummings, R. G., Brajer, V., McFarland, J. W., Trava, J. and El-Ashry, M. T. (1989) *Waterworks: Improving Irrigation Management in Mexican Agriculture*. World Resources Institute, WRI Paper 5, Washington DC.

Custers, P. (1992) 'Banking on a flood-free future? Flood mismanagement in Bangladesh', *Ecologist*, 22(5), 241–7.

Dallas, R. (1990) 'The agricultural collapse of the arid midwest', *Geographical Magazine* (October), 16–20.

Dankelman, I. and Davidson, J. (1988) *Women and Environment in the Third World*. Earthscan, London.

Darby, H. C. (1983) *The Changing Fenland*. Cambridge University Press, Cambridge.

Darian, S. G. (1978) *The Ganges in Myth and History*. University of Hawaii Press, Honolulu.

Davies, H. R. J. (1986) 'The human factor in development: some lessons from rural Sudan?', *Applied Geography*, 6, 107–21.

Davis, J. R., Nanninga, P. M., Biggins, J. and Lant, P. (1991) 'Prototype decision support system for analysing the impact of catchment policies', *Journal of Water Resource Planning and Management*, 117(4), 399–414.

Davis, W. M. (1899) 'The geographical cycle', *Geographical Journal*, 14, 481–504.

Day, J. C. (1985) 'Canadian interbasin diversions: Inquiry on Federal Water Policy', Research Paper 6, Simon Fraser University, BC.

Decamps, H., Fortune, M. and Gazelle, F. (1989) 'Historical changes of the Garonne River, Southern France', in G. E. Petts (ed.) *Historical Change of Large Alluvial Rivers: Western Europe*. Wiley, Chichester, 249–67.

Decamps, H. and Naiman, R. J. (1990) 'Towards an ecotone perspective', in R. J. Naiman and H. Decamps (eds) *The Ecology and Management of Aquatic-Terrestrial Ecotones*. Parthenon, Carnforth, Lancs, 1–5.

Department of the Environment (1988) *Privatisation of the Water Authorities in England and Wales*, Cmnd 9734. HMSO, London.

Department of the Environment/Ministry of Agriculture Fisheries and Food/Welsh Office (1987) *The National Rivers Authority: The Government's Proposals for a Public Regulatory Body in a Privatised Water Industry*. London.

Department of the Environment/Welsh Office (1988) *Integrated Pollution Control, A Consultation Paper*. London.

Department of Water Affairs (1986) *Management of the Water Resources of the Republic of South Africa*. Pretoria.

Department of Water Affairs and Forestry (1994) *Water Supply and Sanitation Policy: An Indivisible National Asset*. White Paper. Cape Town.

Department of Water Affairs and Forestry (1996) *Water Law Principles*. Discussion Document. Pretoria.

Devenay, W. T. (1978) 'Water supply in upland Scotland', in R. B. Tranter (ed.) *The Future of Upland Britain*. Centre for Agricultural Strategy, University of Reading, Reading, 328–35.

Dhruva Narayana, V. V. (1987) 'Downstream impacts of soil conservation in the Himalayan region', *Mountain Research and Development*, 7(3), 287–98.

Dickinson, N. W. T., Rudra, R. P. and Wall, G. J. (1986) 'Identification of soil erosion and fluvial sediment problems', *Hydrological Processes*, 1, 111–24.

Dixon, J. A., Carpenter, R. A., Fallon, L. A., Sherman, P. B., and Manipomoke, S. (1986) *Economic Analysis of the Environmental Impacts of Development Projects*. Earthscan, London.

Dooge, J. C. I. (1974) 'The development of hydrological concepts in Britain and Ireland between 1674 and 1874', *Hydrological Sciences Bulletin*, 19, 279–302.

Doornkamp, J. C., Gregory, K. J., and Burn, A. S. (1980) *Atlas of Drought in Britain 1975–76*. Institute of British Geographers, London.

Douglas, I. (1985) 'Urban sedimentology', *Progress in Physical Geography*, 9(2), 255–80.

Douglass, J. E. (1983) 'The potential for water yield augmentation from forest management in the eastern United States', *Water Resources Bulletin*, 19(3), 351–8.

Downs, P. W., Gregory, K. J. and Brookes, A. (1991) 'How Integrated is River Basin Management?' *Environmental Management*, 15(3), 299–309.

Drought Mitigation Working Group (1993) *Summary of a Workshop on Mitigating Drought in Developing Countries: The Contribution of UK Institutions*. Institute of Hydrology, Wallingford, UK.

Duggan, P. (ed.) *Wetland Conservation: A Review of Current Issues and Required Action*. International Union for the Conservation of Nature, World Conservation Union, Gland, Switzerland.

Dunne, T. and Leopold, L. B. (1978) *Water in Environmental Planning*. W. H. Freeman, San Francisco, Cal.

Earthscan (1984) *Cropland or Wasteland. The Problems and Promises of Irrigation*. Earthscan Press Briefing Document no. 38. London.

Ebisemiju, F. S. (1989) 'The response of headwater stream channels to urbanization in the humid tropics', *Hydrological Processes*, 3, 237–53.

Eckerberg, K. (1990) *Environmental Protection in Swedish Forestry*. Gower, Aldershot.

Eckholm, E. (1976) 'The politics of soil conservation', *The Ecologist*, 6(2), 54–9.

Edwards, R. W. and Brooker, M. P. (1982) 'The ecology of the Wye', *Monographiae Biologicae*, 50, Junk, The Hague.

Eiseltova, M. and Biggs, J. (1995) *Restoration of Stream Ecosystems: An Integrated Catchment Approach*. International Waterfowl and Wetlands Research Bureau, Slimbridge, UK.

El Arifi, S. A. (1988) 'Problems in planning extensive agricultural projects: the case of New Halfa, Sudan', *Applied Geography*, 8, 37–52.

Elliot, C. (1982) *Making Excellence Useful*. Royal Society of Arts Conference on Technical Assistance Overseas and the Environment (16 November), London, 20–5.

Elmendorf, M. (1978) 'Public participation and acceptance', in C. G. Gunnerson and J. M. Kalbermatten (eds) *Environmental Impacts of International Civil Engineering Projects and Practices*. American Society of Civil Engineers, New York, 184–201.

Englebert, G. A. and Scheuring, A. F. (1984) *Water Scarcity: Impacts on Western Agriculture*. University of California Press, Berkeley, CA.

English Nature (1996) *Impact of Water Abstraction on Wetland SSSIs*. English Nature Freshwater Series, 4, Peterborough, UK.

Environment Canada (1989) *Federal Water Policy*. Ottawa.

Environment Canada (1992) *State of Environment for the Lower Fraser River Basin*. SOE Report 92—1, Environment Canada, Ottawa.

Environment Ontario (1988) *Controlling Industrial Discharges to Sewers*. Queens Printer.

Ericksen, N. J. (1986) 'Creating flood disasters?', Water and Soil Miscellaneous Publication 77, National Water and Soil Conservation Authority, Wellington, New Zealand.

Ericksen, N. J. (1990) 'New Zealand water planning and management: evolution or revolution?', in B. Mitchell (ed.) *Integrated Water Management*. Belhaven, London, 45–87.

Ericksen, N. J., Handmer, J. W., and Smith, D. I. (1988) 'ANUFLOOD: Evaluation of a computerised urban flood-loss assessment policy for New Zealand', Water and Soil Miscellaneous Publication 115, National Water and Soil Conservation Authority, Wellington, New Zealand.

Evans, R. (1990) 'Soils at risk of accelerated erosion in England and Wales', *Soil Use and Management*, 6(3), 125–31.

Evans, T. E. (1990) 'History of Nile flows', in P. P. Howell and J. A. Allan (eds) *The Nile*. School of Oriental and African Studies/Royal Geographical Society, London, 5–39.

Eybergen, F. A. and Imeson, A. C. (1989) 'Geomorphological processes and climatic change', *Catena*, 16(4), 307–20.

Evers, Y. D. (1995) 'Supporting local natural resource management institutions: experience gained and guiding principles', in D. Stiles (ed.) *Social Aspects of Sustainable Dryland Management*. Wiley, Chichester, 93–103.

Fahim, H. M. (1981) *Dams, People and Development: The Aswan High Dam Case*. Pergamon, New York.

Fahmy, S. H. (1992) 'Effect of the previous drought event on Nile river yields', in M. A. Abu-Zeid and A. K. Biswas (eds) *Climatic Fluctuations and Water Management*. Butterworth-Heinemann, Oxford, 272–81.

Falkenmark, M. (1986) 'Fresh water: time for a modified approach', *Ambio*, 15(4) 192–200.

Falkenmark, M. (1989) 'The massive water scarcity now threatening Africa: why isn't it being addressed?', *Ambio*, 18(2), 112–18.

Falkenmark, M. (1996) 'Stockholm Water Symposium, 1995: Human Dimensions of the Water Crisis', *Ambio*, 25(3), 216.

Falkenmark, M. and Chapman, T. (1989) *Comparative Hydrology: An Ecological Approach to Land and Water Resources*, UNESCO, Paris.

Falkenmark, M. and Lunqvist, J. (1995) 'Looming water crisis: new approaches are inevitable', in L. Ohlsson (ed.) *Hydropolitics: Conflicts over Water as a Development Constraint*. Zed Books, London, 178–212.

Falkenmark, M. and Rockström, J. (1996) 'Escaping from ongoing land/water mismanagement', *Ambio*, 25(3), 211–12.

Farrimond, M. S. (1980) 'Impact of man in catchments. (iii) Domestic and industrial wastes', in A. M. Gower (ed.) *Water Quality in Catchment Ecosystems*. Wiley, Chichester, 113–44.

Feldman, D. L. (1991) *Water Resources Management: In Search of an Environmental Ethic*. Johns Hopkins Press, Baltimore.

Ferguson, R. I. (1981) 'Channel form and channel changes', in J. Lewin (ed.) *British Rivers*. Allen & Unwin, London, 90–125.

Ferguson, R. (1987) 'Hydraulic and sedimentary controls of channel pattern', in K. S. Richards (ed.) *River Channels: Environment and Process*. Blackwells, Oxford, 129–58.

Finkel, H. J. (ed.) (1977) *Handbook of Irrigation Technology*, Vol. 1. CRC Press, Boca Raton, Fla.

Fisher, R. and Ury, W. (1981) *Getting to Yes*. Houghton Mifflin, Boston, MD.

Fleming, G. (1969) 'Design curves for suspended load estimation', *Proceedings of the Institution of Civil Engineers*, 43.

Flug, M. and Ahmed, J. (1990) 'Prioritizing flow alternatives for social objectives', *Journal of Water Resource Planning and Management*, 116(5), 610–24.

Fontane, D. G. and Tuvel, H. N. (eds) (1994) *Water policy and management: Solving the Problems*. American Society of Civil Engineers, New York.

Food and Agriculture Organisation (1977) *Assessing Soils Degradation*. FAO Soils Bulletin no. 34. Rome.

Food and Agriculture Organisation (1986) *World Agricultural Statistics*. Rome.

Fookes, P. G. and Vaughan, P. R. (1986) *A Handbook of Engineering Geomorphology*. Surrey University Press, London.

Forestry Authority (1993) *Forests and Water Guidelines* (3rd edn). HMSO, London.

Forestry Commission (1979) *The Wood Production Outlook in Britain*. Forestry Commission, Edinburgh.

Forestry Commission (1988) *Forests and Water Guidelines*. Forestry Commission, Edinburgh.

Forestry Commission (1991) *Forests and Water Guidelines* (2nd edn). HMSO, London.

Foster, I. D. L., Dearing, J. A. and Grew, R. (1988) 'Lake-catchments: an evaluation of their contribution to studies of sediment yield and delivery processes', in M. P. Bordas and D. E. Walling (eds) *Sediment Budgets*. International Association of Hydrological Sciences Publication 174, 413–29.

Foster, S. S. D. and Chilton, P. J. (1993) 'Groundwater systems in the humid tropics', in M. Bonell, M. M. Hufschmidt and J. S. Gladwell (eds) *Hydrology and Water Management in the Humid Tropics*. Cambridge University Press, Cambridge, 261–72.

Franks, T. R. (1994) 'The Green River in Bangladesh: the Lower Atrai Basin', in C. Kirby and W. R. White (eds) *Integrated River Basin Development*. Wiley, Chichester, 367–76.

Frost, C. A., Speirs, R. B. and McLean, J. (1990) 'Erosion control for the UK: strategies and short-term costs and benefits', in J. Boardman, I. D. L. Foster and J. A. Dearing (eds) *Soil Erosion on Agricultural Land*. Wiley, Chichester, 559–67.

Furuseth, O. and Cocklin, C. (1995) 'Regional perspectives on resource policy: implementing sustainable management in New Zealand', *Journal Env. Planning & Management*, 38(2), 181–200.

Gadgil, M. and Guha, R. (1992) *This Fissured Land: An Ecological History of India*, Oxford University Press, Delhi.

Gamble, D. W. and Meentemeyer, V. (1996) 'The role of scale in research on the Himalaya – Ganges – Brahmaputra interaction', *Mountain Research and Development*, 16(2), 149–55.

Gangstad, E. O. (1990) *Natural Resource Management of Water and Land*. Van Nostrand Reinhold, New York.

Gardiner, J. (1996) 'The use of EIA in delivering sustainable development through integrated water management'. *European Water Pollution Control*, 6(1), 50–9.

Gardiner, J. L. (1988) 'Environmentally sensitive river engineering: examples from the Thames catchment', in G. Petts (ed.) *Regulated Rivers Research and Management*, 2.

Gardiner, J. L. (ed.) (1991) *River Projects and Conservation: A Manual for Holistic Appraisal*. Wiley, Chichester.

Gardiner, J., Thompson, K. and Newson, M. (1994) 'Integrated watershed/river catchment planning and management: a comparison of selected Canadian and United Kingdom experiences', *Journal of Environmental Planning and Management*, 37(1), 53–67.

Garrod, G. D. and Willis, K. G. (1996) 'Estimating the benefits of environmental enhancement: a case of willingness to pay to improve low flows', *Journal of Environmental Planning and Management*, 39(2), 189–203.

George, R. J. (1990) 'Reclaiming sandplain seeps by intercepting perched groundwater with Eucalyptus', *Land Degradation and Rehabilitation*, 2, 13–25.

Gerrard, J. (1990) *Mountain Environment: An Examination of the Physical Geography of Mountains*. Belhaven, London.

Ghassemi, F., Thomas, G. A. and Jakeman, A. J. (1988) 'Effect of groundwater interception and irrigation on salinity and piezometric levels in an aquifer', *Hydrological Processes*, 2, 369–82.

Gibbs, R. (1970) 'Mechanisms controlling world water chemistry', *Science*, 170, 1088–90.

Gilbertson, D. D. (1986) 'Runoff (floodwater) farming and rural water supply in arid lands', *Applied Geography*, 6, 5–11.

Gilmour, D. A. (1988) 'Not seeing the trees for the forest: a reappraisal of the deforestation crisis in two hill districts of Nepal', *Mountain Research and Development*, 8(4), 343–50.

Gilmour, D. A., Bonell, M. and Cassells, D. S. (1987) 'The effects of forestation on soil hydraulic properties in the Middle Hills of Nepal: a preliminary assessment', *Mountain Research and Development*, 7(3), 239–49.

Glacken, C. J. (1956) 'Changing ideas of the habitable world', in W. L. Thomas (ed.) *Man's Role in Changing the Face of the Earth*. University of Chicago Press, Chicago, Ill., 70–92.

Gladwell, J. S. (1993) 'Urban water management problems in the humid tropics: some technical and non-technical considerations', in M. Bonnell, M. M. Haufschmidt and J. S. Gladwell (eds) *Hydrology and Water Management in the Humid Tropics*. Cambridge University Press, Cambridge, 414–36.

Gleick, P. H. (1987) *Global Climatic Changes and Regional Hydrology: Impacts and Responses*. International Association of Scientific Hydrology Publication 168, 389–402.

Gleick, P. H. (ed.) (1993) *Water in Crisis: A Guide to the World's Fresh Water Resources*. Oxford University Press, Oxford.

Goldsmith, E. and Hildyard, N. (eds) (1984) *The Social and Environmental Effects of Large Dams, Vol. 1, Overview*. Wadebridge Ecological Centre, Wadebridge, Cornwall.

Goldsmith, E. and Hildyard, N. (eds) (1986) *The Social and Environmental Effects of Large Dams, Vol. 2, Case studies*. Wadebridge Ecological Centre, Wadebridge, Cornwall.

Gomez, B. and Church, M. (1989) 'An assessment of bed load sediment transport formulae for gravel bed rivers', *Water Resources Research*, 25(6), 1161–86.

Goodell, L. L. (1988) 'Water management in the Delaware river basin', in S. K. Majumdar, E. W. Miller, and L. E. Sage (eds) *Ecology and Restoration of the Delaware River Basin*. Pennsylvania Academy of Science, Phillipsburg, NJ, 286–94.

Gore, J. A. (1985) *The Restoration of Rivers and Streams: Theories and Experience*. Ann Arbor Science (Butterworth), Stoneham, Mass.

Gore, J. A., Layzer, J. B. and Russell, I. A. (1992) 'Non-traditional applications of instream flow techniques for conserving habitat of biota in the Sabie River of Southern Africa', in P. J. Boon, P. Calow and G. E. Petts (eds) *River Conservation and Management*. Wiley, Chichester, 161–77.

Graf, W. L. (1985) *The Colorado River: Instability and Basin Management*. Association of American Geographers, Washington, DC.

Graf, W. L. (1992) 'Science, public policy, and western American rivers', *Transactions of the Institute of British Geographers*, 17, 5–19.

Grainger, A. (1990) *The Threatening Desert: Controlling Desertification*. Earthscan, London.

Green, G. G. and Eiker, E. E. (eds) (1983) *Accomplishments and Impacts of Reservoirs*. American Society of Civil Engineers, New York.

Greenwell, J. R. (1978) 'Aridity, human evolution and desert primate ecology', *Arid Lands Newsletter*, 8, 10–18.

Gregory, K. J. (ed.) (1977) *River Channel Changes*. Wiley, Chichester.

Gregory, K. J. (1985) *The Nature of Physical Geography*. Edward Arnold, London.

Grewal, S. S., Mital, S. P. and Singh, G. (1990) 'Rehabilitation of degraded lands in the Himalayan foothills – peoples' participation', *Ambio*, 19(1), 45–8.

Guerrieri, F. and Vianello, G. (1990) 'Identification and reclamation of erosion-affected lands in the Emilia-Romagna region, Italy', in J. Boardman, I. D. L. Foster and J. A. Dearing (eds) *Soil Erosion on Agricultural Land*. Wiley, Chichester, 621–5.

Guest, P. (1987) 'Who values our waterways?', *Soil and Water*, 23(3), 8–12.

Guijt, I. and Thompson, J. (1994) 'Landscapes and livelihoods: environmental and socioeconomic dimensions of small-scale irrigation', *Land Use Policy*, 11(4), 294–308.

Guillerme, A. E. (1988) *The Age of Water*. Texas, Texas A&M University Press.

Gunnerson, C. G. and Kalbermatten, J. M. (eds) (1978) *Environmental Impacts of International Civil Engineering Projects and Practices*. American Society of Civil Engineers, New York.

Gustard, A., Cole, G., Marshall, D. and Bayliss, A. (eds) (1987) *A Study of Compensation Flows in the UK*. Report no. 99, Institute of Hydrology, Wallingford, UK.

Guy, S. and Marvin, S. J. (1996) 'Managing water stress: the logic of demand side infrastructure planning', *Journal of Environmental Planning and Management*, 89(1), 123–8.

Haagsma, B. (1995) 'Traditional water management and state intervention: the case of Santo Antao, Cape Verde', *Mountain Research and Development*, 15(1), 39–56.

Haigh, N. (1986) 'Public perceptions and international influences', in G. Conway (ed.) *The Assessment of Environmental Problems*. Imperial College Centre for Environmental Technology, London, 73–83.

Hall, C. (1989) *Running Water*. Robertson McCarta, London.

Hallsworth, E. G. (1987) *Anatomy, Physiology and Psychology of Erosion*. Wiley, Chichester.

Hamilton, L. S. (1988) 'Forestry and watershed management', in J. Ives and D. C. Pitt (eds) *Social Dynamics in Watersheds and Mountain Ecosystems*. Routledge, London, 99–131.

Hamley, W. (1990) 'Hydrotechnology, wilderness and culture in Quebec', in D. Cosgrove and G. E. Petts (eds) *Water, Engineering and Landscape*. Belhaven Press, London, 144–58.

Handmer, J. W. (1987) 'Guidelines for floodplain acquisition', *Applied Geography*, 7, 203–21.

Handmer, J. W., Dorcey, A. H. J. and Smith, D. I. (eds) (1991) *Negotiating Water: Conflict Resolution in Australian Water Management*. Centre for Resource and Environmental Studies, Australian National University, Canberra.

Hare, F. K. (1984) 'The impact of human activities on water in Canada', Trinity College, University of Toronto Inquiry on Federal Water Policy Research, paper 2.

Harper, D. E. (1988) 'Improving the accuracy of the Universal Soil Loss Equation in Thailand', in S. Runwanich (ed.) *Land Conservation for Future Generations*. Department of Land Development, Bangkok, 531–40.

Harper, D. M. and Ferguson, A. J. D. (1995) *The Ecological Basis for River Management*. Wiley, Chichester.

Harriman, R. (1978) 'Nutrient leaching from fertilised forest watersheds in Scotland', *Journal of Applied Ecology*, 15, 933–42.

Harris, T. and Boardman, J. (1990) 'A rule-based expert system approach to predicting waterborne soil erosion', in J. Boardman, I. D. L. Foster and J. A. Dearing (eds) *Soil Erosion on Agricultural Land*. Wiley, Chichester, 401–12.

Hartmann, L. (1990) 'Methodological guidelines for integrated environmental evaluation of water resources development', in UNEP/UNESCO *The Impact of Large Water Projects on the Environment*. UNESCO, Paris, 467–86.

Hasfurther, V. R. (1985) 'The use of meander parameters in restoring hydrological balance to reclaimed stream beds', in J. A. Gore (ed.) *The Restoration of Rivers and Streams*. Butterworth, Boston, Mass., 21–40.

Haslam, S. M. (1991) *The Historic River: Rivers and Culture down the Ages*. Cobden, Cambridge.

Hauck, G. F. W. and Novak, R. A. (1987) 'Interaction of flow and incrustation in the Roman aqueduct of Nimes', *Journal of Hydraulic Engineering*, 113(2), 141–57.

Hausler, S. (1995) 'Listening to the people: the use of indigenous knowledge to curb environmental degradation', in D. Stiles (ed.) *Social Aspects of Sustainable Dryland Management*. Wiley, Chichester, 179–88.

Hawkes, J. (1976) *The Atlas of Early Man*. Macmillan, London.

Heathcote, R. L. (1969) 'Drought in Australia: a problem of perception', *Geographical Review*, 49(2), 175–94.

Heede, B. H. and King, R. M. (1990) 'State-of-the-art timber harvest in an Arizona mixed conifer forest has minimal effect on overland flow and erosion', *Hydrological Sciences Journal*, 35(6), 623–35.

Hellawell, J. M. (1986) *Biological Indicators of Freshwater Pollution and Environmental Management*. Elsevier, London.

Hellen, J. A. and Bonn, P. (1981) 'Demographic change and public policy in Egypt and Nepal: some long-term implications for development planning', *Science and Public Policy*, 308–36.

Helley, E. J. and Smith, W. (1971) 'Development and calibration of a pressure difference bedload sampler', US *Geological Survey Open File Report*, 8037–01, Menlo Park, CA.

Hellier, C. (1990) 'Running the rivers dry', *Geographical Magazine*, July, 32–5.

Hewlett, J. D. and Nutter, W. L. (1970) 'The varying source area of streamflow from upland basins: Interdisciplinary aspects of watershed management', American Society of Civil Engineers, New York, 65–83.

Hey, R. D. (1994) 'Environmentally sensitive river engineering', in P. Calow and G. E. Petts (eds) *The Rivers Handbook*. Vol. II. Blackwell, Oxford, 337–62.

Hey, R. D. (1995) 'River processes and management', in T. O'Riordan (ed.) *Environmental Science for Environmental Management*. Longman, London, 131–50.

Hibbert, A. R. (1967) 'Forest treatment effects on water yield', in W. E. Soppen and H. W. Lull (eds) *International Symposium on Forest Hydrology*. Pergamon, New York, 527–43.

Hickin, E. J. (1983) 'River channel changes: retrospect and prospect', Special Publications International Association of Sedimentologists, 6, 61–83.

Higgs, G. and Petts, G. E. (1988) 'Hydrological changes and river regulation in the UK', *Regulated Rivers: Research and Management*, 2, 349–68.

Higgins, G. M., Dielman, P. J. and Abernethy, C. L. (1988) 'Trends in irrigation development and their implications for hydrologists and water resource engineers', *Hydrological Sciences Journal*, 33(112), 43–59.

Hill, A. R. (1990) 'Groundwater cation concentrations in the riparian zone of a forested headwater stream', *Hydrological Processes*, 4, 121–30.

Hillel, D. (1992) *Out of the Earth: Civilization and the Life of the Soil*. Aurum Press, London.

Hillel, D. (1994) *Rivers of Eden: The struggle for water and the quest for peace in the Middle East*. Oxford University Press, Oxford.

Hjulstrom, F. (1935) 'Studies of the morphological activity of rivers as illustrated by the River Fyris', *Bulletin of the Geological Institute*, University of Uppsala, 25, 221–527.

Hodges, R. D. and Arden-Clarke, C. (1986) *Soil Erosion in Britain: Levels of Soil Damage and their Relationship to Farming Practices*. Soil Association, Bristol.

Hofer, T. (1993) 'Himalayan deforestation, changing river discharge and increasing floods: myth or reality?', *Mountain Research and Development*, 13(3), 213–33.

Holeman, J. N. (1968) 'The sediment yield of major rivers of the world', *Water Resources Research*, 4(4), 737–47.

Hollis, G. E. (1988) 'Rain, roads, roofs and runoff: hydrology in cities', *Geography*, 73(1), 9–18.

Hollis, G. E. (1990) 'Environmental impacts of development on wetlands in arid and semi-arid lands', *Hydrological Sciences Journal*, 35(4), 411–28.

Hooke, J. M. and Kain, R. J. P. (1982) *Historical Change in the Physical Environment*. Butterworth, London.

Horberry, J. (1983) *Environmental Guidelines Survey: An Analysis of Environmental Procedures and Guidelines Governing Development Aid*. International Institute for Environment and Development, Washington and London.

Horta, K. (1995) 'The mountain kingdom's white oil: the Lesotho Highlands Water Project', *Ecologist*, 25(6), 227–31.

Horton, R. E. (1933) 'The role of infiltration in the hydrologic cycle', *American Geophysical Union Transactions*, 14, 446–60.

Horton, R. E. (1945) 'Erosional development of streams: quantitative physiographic factors', *Bulletin of the Geological Society of America*, 56, 275–370.

Houghton, J. T., Callander, B. A. and Varney, S. K. (1992) *Climate Change 1992*. Cambridge University Press, Cambridge.

Howard, P. J. A., Thompson, T. R. E., Hornung, M. and Beard, G. R. (eds) (1989) *An Assessment of the Principles of Soil Protection in the UK* (3 vols). Institute of Terrestrial Ecology/HMSO, London.

Howarth, W. (1988) *Water Pollution Law*. Shaw & Sons, London.

Howell, P. P. and Allan, J. A. (eds) (1990) *The Nile: Resource Evaluation, Resource Management, Hydropolitics and Legal Issues*. Conference Proceedings, Royal Geographical Society, School of Oriental and African Studies, London.

Howell, P. P. and Allan, J. A. (eds) (1994) *The Nile: Sharing a Scarce Resource. An Historical and Technical Review of Water Management and of Economic and Legal Issues*. Cambridge University Press, Cambridge.

Howell, P., Lock, M., and Cobb, S. (1988) *The Jonglei Canal: Impact and Opportunity*. Cambridge University Press, Cambridge.

Hudson, J. A. and Gilman, K. (1993) 'Long-term variability in the water balances of the Plynlimon catchments', *Journal of Hydrology*, 143, 355–80.

Huff, D. D. and Swank, W. T. (1985) 'Modelling changes in forest evapotranspiration', in M. G. Anderson and T. P. Burt (eds) *Hydrological Forecasting*. Wiley, Chichester, 125–51.

Hughes, F. M. R. (1990) 'The influence of flooding regimes on forest distribution and composition in the Tana River floodplain, Kenya', *Journal of Applied Ecology*, 27, 475–91.

Hulme, M. (1986) 'The adaptability of a rural water supply system to extreme rainfall anomalies in central Sudan', *Applied Geography*, 6, 89–105.

Hulme, M. (1990) 'Global climate change and the Nile Basin', in P. P. Howell and J. A. Allan (eds) *The Nile*. School of Oriental and African Studies/Royal Geographical Society, London, 59–82.

Hume, M. (1992) *The Run of the River: Portraits of Eleven British Columbia Rivers*. New Star Books, Vancouver, BC.

Hunt, R. L. (1993) *Trout Stream Therapy*. The University of Wisconsin Press, Madison.

Johnson, A. K. L., Shrubsole, D. and Merrin, M. (1996) 'Integrated catchment management in northern Australia, from concept to implementation', *Land Use Policy*, 13(4), 303–16.

Johnson, S. P. (1993) *The Earth Summit*. Graham & Trotman, London.

Hurni, H. (1983) 'Soil erosion and soil formation in agricultural ecosystems: Ethiopia and Northern Thailand', *Mountain Research and Development*, 3(2), 131–42.

Huxley, J. (1943) *TVA: Adventure in Planning*. Architectural Press, London.

Hynes, H. B. N. (1975) 'The stream and its valley', *Verh Internat Verein Limnol*, 19, 1–15.

International Association of Hydrological Sciences (IAHS) (1970) *Proceedings of the Wellington Symposium*, 2 vols. Wallingford, Oxon.

International Association of Hydrological Sciences (1980) 'The influence of man on the hydrological regime with special reference to representative and experimental basins', *Proceedings of the Helsinki Symposium*. Wallingford, Oxon.

IAHS–UNESCO (1970) *Results of Research on Representative and Experimental Basins*. UNESCO, Paris.

Independent Commission on International Development Issues (1980) *North–South: A Programme for Survival*. Pan Books, London.

Ingram, H. (1990) *Water Politics: Continuity and Change*. University of New Mexico Press, Albuquerque, N. Mex.

Institute of Hydrology (1980) *Low Flow Studies*. Wallingford, Oxon.

International Union for the Conservation of Nature (1980) *World Conservation Strategy*. IUCN–UNEP–WWF, Gland, Switzerland.

Ives, J. D. (1988) 'Development in the face of uncertainty', in J. Ives and D. C. Pitt (eds) *Deforestation: Social Dynamics in Watersheds and Mountain Ecosystems*. Routledge, London, 54–74.

Ives, J. and Pitt, D. C. (eds) (1988) *Deforestation: Social Dynamics in Watersheds and Mountain Ecosystems*. Routledge, London.

Ives, J., Messerli, B. and Thompson, M. (1987) 'Research strategy for the Himalayan region', *Mountain Research and Development*, 7(3), 332–44.

Jansen, J. M. L. and Painter, R. B. (1974) 'Predicting sediment yield from climate and topography', *Journal of Hydrology*, 21, 371–80.

Johnson, P. (1988) 'River regulation: a regional perspective – Northumbrian Water Authority', *Regulated Rivers: Research and Management*, 2, 233–55.

Johnston, W. B. (1985) 'Sector and place: the place of environment in government administration', *Proceedings of the 13th New Zealand Geography Conference*, Hamilton, 96–8.

Kalbermatten, J. M. and Gunnerson, C. A. (1978) 'Environmental impacts of international engineering practice', in C. G. Gunnerson and J. M. Kalbermatten (eds) *Environmental Impacts of International Civil Engineering Projects and Practices*. American Society of Civil Engineers, New York, 232–54.

Kally, E. (1993) *Water and Peace: Water Resources and the Arab-Israeli Peace Process*. Praeger, London.

Kalpavriksh and The Hindu College Nature Club (1986) 'The Narmada Valley Project: development or destruction?', in E. Goldsmith and N. Hildyard (eds) *The Social and Environmental Effects of Large Dams*. Cambridge University Press, Cambridge, 224–44.

Karpiscak, M. M., Foster, K. E. and Rawles, R. L. (1984) 'Water harvesting and evaporation suppression', *Arid Lands Newsletter*, 21, 11–17.

Keller, E. A. (1978) 'Pools, riffles and channelization', *Environmental Geology*, 2, 119–27.

Kellerhals, R. (1967) 'Stable channels with gravel-paved beds', *Journal of the Waterways and Harbours Division*, Proceedings of the American Society of Civil Engineers, 93, 63–84.

Kern, K. (1992) 'Rehabilitation of streams in South-west Germany', in P. J. Boon, P. Calow and G. E. Petts (eds) *River Conservation and Management*. Wiley, Chichester, 332–5.

Khalid, F. and O'Brien, J. (eds) (1992) *Islam and Ecology*. Cassell, London.

King, J. M., de Moor, F. C. and Chutter, F. M. (1992) 'Alternative ways of classifying rivers in Southern Africa', in P. J. Boon, P. Calow and G. E. Petts (eds) *River Conservation and Management*, Wiley, Chichester, 213–28.

Kinnersley, D. (1988) *Troubled Water: Rivers, Politics and Pollution*. Hilary Shipman, London.

Kirby, C., Newson, M. D. and Gilman, K. (1992) *Plynlimon Research: The First Two Decades*. Institute of Hydrology, report 109, Wallingford, Oxon.

Kirby, C. and White, W. R. (eds) (1994) *Integrated River Basin Development*. Wiley, Chichester.

Kirpich, P. Z. (1990) 'Technology, society and water management – discussion' (of a paper by W. Viessman), *Journal of Water Resource Planning and Management*, 116(6), 846–7.

Kittredge, J. 1948 (1973) *Forest Influences*. Dover, New York.

Kliot, N. (1986) 'Man's impact on river basins: an Israeli case study', *Applied Geography*, 6, 163–78.

Kliot, N. (1994) *Water Resources and Conflict in the Middle East*. Routledge, London.

Knight, D. W. (1987) 'Dissemination of information to practising engineers and researchers in the water industry', *Journal of the Institution of Water and Environmental Management*, 1(3), 315–24.

Knight, M. S. and Tuckwell, S. B. (1988) 'Controlling nitrate leaching in water supply catchments', *Journal of the Institution of Water and Environmental Management*, 2(3), 248–52.

Knighton, A. D. (1984) *Fluvial Forms and Processes*. Edward Arnold, London.

Knox, J. C. (1989) *Long- and Short-term Episodic Storage and Removal of Sediment in Watersheds of Southwestern Wisconsin and Northwestern Illinois*. IAHS Publication 184, 157–64.

Kogan, M. and Henkel, M. (1983) *Government and Research*. Heinemann, London.

Konrad, J. A., Baumann, J. S. and Ott, J. A. (1986) 'Non-point source planning and implementation in Wisconsin', in J. F. Solb (ed.) *Effects of Land Use on Fresh Waters*. Ellis Horwood, Chichester, 283–95.

Kovacs, G. (1990) 'Decision support systems for managing large international rivers', in UNEP/UNESCO *The Impact of Large Water Projects on the Environment*. UNESCO, Paris, 435–48.

Kruse, E. G., Burdick, C. R., and Yousef, Y. A. (eds) (1982) *Environmentally Sound Water and Soil Management*. American Society of Civil Engineers, New York.

Krutilla, J. V. and Eckstein, O. (1958) *Multiple Purpose River Development Studies in Applied Economic Analysis*. John Hopkins University Press, Baltimore, Md.

Laki, S. L. (1994) 'The impact of the Jonglei Canal on the economy of the local people', *International Journal of Sustainable Development and World Economy*, 1, 89–96.

Lal, R. (1993) 'Challenges in agriculture and forest hydrology in the humid tropics', in M. Bonell, M. M. Hufschmidt and J. S. Gladwell (eds) *Hydrology and Water Management in the Humid Tropics*. Cambridge University Press, Cambridge, 273–300.

Lambert, A. (1988) 'Regulation of the River Dee', *Regulated Rivers: Research and Management* 2, 293–308.

Lambert, C. P. and Walling, D. E. (1988) 'Measurement of channel storage of suspended sediment in a gravel-bed river', *Catena*, 15(1), 65–80.

Langford-Smith, T. and Rutherford, J. (1966) *Water and Land: Two Case Studies in Irrigation*. Australian National University, Canberra.

Law, F. (1956) 'The effect of afforestation upon the yield of water catchment areas', *Journal of the British Waterworks Association*, 38, 484–94.

Leeks, G. J. L. and Newson, M. D. (1989) 'Responses of the sediment system of a regulated river to a scour valve release: Llyn Clywedog, Mid-Wales, UK', *Regulated Rivers: Research and Management*, 3, 93–106.

Lemon, M., Seaton, R. and Park, J. (1994) 'Social enquiry and the measurement of natural phenomena: the degradation of irrigation water in the Argolid Plain, Greece', *International Journal of Sustainable Development and World Ecology*, 1, 206–20.

Leopold, L. B. (1968) 'Hydrology for urban land planning', US Geological Survey, Circular 554.

Leopold, L. B. (1974) *Water: A Primer*. W. H. Freeman, San Francisco, Cal.

Leopold, L. B. (1991) 'Closing remarks', in GCES Committee, *Colorado River Ecology and Dam Management*. National Academy Press, Washington DC, 254–7.

Leopold, L. B. and Langbein, W. B. (1962) 'The concept of entropy in landscape evolution', US Geological Survey Professional Paper 500-A, Menlo Park, Cal.

Leopold, L. B. and Maddock, T. (1954) *The Flood Control Controversy*. Donald, New York.

Leopold, L. B. and Wolman, M. G. (1957) 'River channel patterns: braided, meandering and straight', US Geological Survey Professional Paper 282-B, Menlo Park, Cal.

Leopold, L. B., Wolman, M. G. and Miller, J. P. (1964) *Fluvial Processes in Geomorphology*. W. H. Freeman, San Francisco, Cal.

Levi, E. (1995) *The Science of Water: The Foundation of Modern Hydraulics*. American Society of Civil Engineers Press, New York.

Lewin, J., Macklin, M. G. and Newson, M. D. (1988) 'Regime theory and environmental change: irreconcilable concepts?', in W. R. White (ed.) *International Conference on River Regime*. Wiley, Chichester, UK, 431–45.

Lewis, G. and Williams, G. (1984) *Rivers and Wildlife Handbook: A Guide to Practices which Further the Conservation of Wildlife on Rivers*. RSPB/Royal Society for Nature Conservation, Sandy, UK.

Liebscher, H. J. (1987) *Palaeohydrologic Studies Using Proxy Data and Observations*. International Association of Hydrological Sciences Publication 168, Wallingford, Oxon, 111–21.

Likens, G. E., Bormann, F. H., Pierce, R. S. and Reiners, W. A. (1978) 'Recovery of a deforested ecosystem', *Science, 199*, 192–6.

Livingstone (1963) 'Chemical composition of rivers and lakes: data of geochemistry', US Geological Survey, Professional Paper, Menlo Park, Cal., 440–9.

Lockwood, J. G. (1994) 'Impact of global warming on evapotranspiration', *Weather*, 49(9), 318–21.

Logan, B. I. (1987) 'A micro-level approach to rural development planning: the case of piped water in Sierra Leone', *Applied Geography*, 7, 29–40.

Lowi, M. (1993) *Water and Power: The Politics of a Scarce Resource in the Jordan River Basin*. Cambridge University Press, Cambridge.

Lvovitch, M. I. (1973) 'The global water balance', US *International Hydrological Decade Bulletin*, 23, 28–42.

Lynch-Stewart, P., Wiken, E. B. and Ironside, G. R. (1986) *Acid Deposition on Prime Resource Lands in Eastern Canada*. Canada Land Inventory, Report 18. Environment Canada, Ottowa.

McClean, C. J., Watson, P. M., Wadsworth, R. A., Blaiklock, J. and O'Callaghan, J. R. (1995) 'Land use planning: a decision support system', *Journal of Environmental Planning and Management*, 38(1), 77–92.

McClimans, J. (1980) 'Best management practices for forestry activities', *Watershed Management 1980*. American Society of Civil Engineers, New York, 694–705.

McColl, R. H. S. and Gibson, A. R. (1979) 'Downslope movement of nutrients in hill pasture, Taita, New Zealand: III Amounts involved and management implication', *Journal of Agricultural Research*, 22, 279–86.

Macklin, M. G. and Dowsett, R. B. (1989) 'The chemical and physical speciation of trace metals in fine grained overbank flood sediments in the Tyne basin, North East England', *Catena*, 16(2), 135–51.

Madej, M. A. (1984) *Recent changes in channel-store sediment, Redwood Creek, California, Technical Report*. Redwood National Park, Cal.

Magalhaes, A. R. (1994) 'Sustainable development planning and semi-arid regions', *Global Environment Change*, 4(4), 275–9.

Maheshwari, B. L., Walker, K. F. and McMahon, T. A. (1995) 'Effects of regulation on the flow regime of the River Murray, Australia', *Regulated Rivers, Research and Management*, 10, 15–38.

Mahoo, H. (1989) 'Deforestation of a tropical humid rainforest and resulting effects on soil properties, surface and subsurface flow, water quality and crop evapotranspiration', unpublished PhD thesis, University of Sokoine, Tanzania.

Majumdar, S. K., Miller, E. W. and Sage, L. E. (eds) (1988) *Ecology and Restoration of the Delaware River Basin*. Pennsylvania Academy of Science, Phillipsburg, NJ.

Makhoalibe, S. (1984) *Suspended Sediment Transport Measurement in Lesotho*. International Association of Hydrological Sciences Publication 144, 313–21.

Malanson, G. P. (ed.) (1993) *Riparian Landscapes*. Cambridge University Press, Cambridge.

Maltby, E. (1986) *Waterlogged Wealth: Why Waste the World's Wet Places?* Earthscan, London.

McMahon, T. A., Finlayson, B. L., Haines, A. and Srikanthan, R. (1987) *Runoff Variability: A Global Perspective*. International Association of Scientific Hydrology Publication 168, Wallingford, Oxon, 3–11.

Marchand, M. and Toornstra, F. H. (1986) *Ecological Guidelines for River Basin Development*. Centrum voor Milienkunde, Dept 28, Rijksuniversiteit, Leiden.

Marsh, G. P. 1864 (1965) *Man and Nature, or Physical Geography as Modified by Human Action* (reprinted). Belknap Press of Harvard University Press, Cambridge, Mass.

Marsh, T. J. and Turton, P. S. (1996) 'The 1995 drought: a water resource perspective', *Weather*, 51(2), 46–53.

Mascarenhas, O. and Veit, P. G. (1994) *Indigenous Knowledge in Resource Management: Irrigation in Msanzi, Tanzania*, World Resources Institute, Washington DC.

Mas'ud, A. F. (1987) 'Land use and physical hydrology of selected mesoscale catchments in Wales', unpublished PhD thesis, University College North Wales, Bangor.

Mathur, K. and Jayal, N. G. (1993) *Drought, Policy and Politics: The Need for a Long-term Perspective*. Sage, New Delhi.

Meade, R. H. (1982) 'Sources, sinks and storage of river sediment in the Atlantic drainage of the United States', *Journal of Geology*, 90(3), 235, 252.

Mearns, L. (1993) 'Implications of global warming for climate variability and the occurrence of extreme events', in D. A. Wilhite (ed.) *Drought Assessment, Management and Planning*. Theory and Case Studies. Kluwer, Boston, Mass., 109–30.

Mehta, G. (1993) *A River Sutra*. Minerva, London.

Melton, M. A. (1957) *An Analysis of the Relations among Elements of Climate, Surface Properties and Geomorphology*. US Office of Naval Research, Project NR389—042, Columbia University, New York.

Mensching, H. (1986) 'Is the desert spreading? Desertification in the Sahel zone of Africa', *Applied Geography and Development*, 27, 7–18.

Mersey Basin Campaign (1988) *Reviving the Region's Rivers*. Department of the Environment, Manchester (information pack).

Metalie, J. P. (1987) 'The degradation of the Pyrenees in the nineteenth century', in V. Gardiner (ed.) *International Geomorphology*. Wiley, London, 533–44.

Meybeck, M. (1979) 'Concentrations des eaux fluviales éléments majeurs, et apports en solution aux océans', *Revue de Géologie Dynamique et de Géographie Physique*, 21, 215–46.

Meybeck, M. (1983) *Atmospheric Inputs and River Transport of Dissolved Substances*. International Association of Hydrological Sciences Publication 141, 173–92.

Michener, J. A. (1975) *Centennial*. Corgi Books, London.

Milliman, J. D., Broadus, J. M. and Gable, F. (1987) 'Environmental and economic implication of rising sea level and subsiding deltas: the Nile and Bengal examples', *Ambio*, 18(6), 340–5.

Ministry for the Environment New Zealand (1989) *Update on the Resource Management Law Reform*. Wellington, New Zealand.

Ministry of Health (1948) *Gathering Grounds: Public Access to Gathering Grounds, Afforestation and Agriculture on Gathering Grounds*. HMSO, London.

Ministry of Natural Resources, Ontario (1986) *Conservation Areas Guide*. Government Bookstore, Toronto.

Mitchell, B. (ed.) (1990) *Integrated Water Management*. Belhaven, London.

Mitchell, B. and Pigram, J. J. (1989) 'Integrated resource management and the Hunter Valley Conservation Trust, NSW, Australia', *Applied Geography*, 9, 196–211.

Mitchell, J. K. and Bubenzer, G. D. (1980) 'Soil loss estimation', in M. J. Kirkby and R. P. C. Morgan (eds) *Soil Erosion*. Wiley, Chichester, 17–62.

Montz, B. and Gruntfest, E. C. (1986) 'Changes in American urban floodplain occupancy since 1958: the experiences of nine cities', *Applied Geography*, 6, 325–38.

Moore, D. J. (ed.) (1982) 'Catchment management for optimum use of land and water resources: documents from an ESCAP seminar, Part 2: New Zealand contributions', Water and Soil Miscellaneous Publication 45, National Water and Soil Conservation Authority, Wellington.

Moorehead, A. (1973) *The White Nile*. Penguin Books, Harmondsworth.

Moorehead, A. (1983) *The Blue Nile*. Penguin Books, Harmondsworth.

Morgan, R. P. C. (1979) *Soil Erosion*. Longman, London.

Morgan, R. P. C. (1980) 'Implications', in M. J. Kirkby and R. P. C. Morgan (eds) *Soil Erosion*. Wiley, Chichester, 253–301.

Morgan, R. P. C. (1992) *Soil Erosion and Conservation*, 2nd edn. Longman, Harlow.

Morgan, R. P. C., Morgan, D. D. V. and Finney, H. J. (1984) 'A predictive model for the assessment of soil erosion risk', *Journal of Agricultural Engineering Research*, 30, 245–53.

Mormont, M. (1996) 'Towards concerted river management in Belgium', *Journal of Environmental Planning and Management*, 39(1), 131–41.

Mortimore, M. (1989) *Adapting to Drought: Farmers, Famines and Desertification in West Africa*. Cambridge University Press, Cambridge.

Muckleston, K. W. (1990) 'Integrated water management in the United States', in B. Mitchell (ed.) *Integrated Water Management*. Belhaven, London, 22–44.

Nace, R. (1974) *General Evolution of the Concept of the Hydrological Cycle*. UNESCO/WMO/IAHS, Paris, 40–51.

Naiman, R. J. and Decamps, H. (1990) *The Ecology and Management of Aquatic-terrestrial Ecotones*. Parthenon, Carnforth, Lancs.

Naiman, R. J., Magnuson, J. J., McKnight, D. M. and Stanford, J. A. (1995) *The Freshwater Imperative: A Research Agenda*. Island Press, Washington DC.

Napier, T. L. (1990) 'The evolution of US soil-conservation policy: from voluntary adoption to coercion', in J. Boardman, I. D. L. Foster and J. A. Deaning (eds), *Soil Erosion on Agricultural Land*. Wiley, Chichester.

National Research Council (1992) *Restoration of Aquatic Ecosystems: Science, Technology and Public Policy*, National Academy Press, Washington DC.

National Research Council (1995) *Wetlands. Characteristics and Boundaries*. National Academy Press, Washington DC.

National Rivers Authority (1992) *Policy and Practice for the Protection of Groundwater*, National Rivers Authority, Bristol.

National Rivers Authority (1994) 'The implications of climate change for the National Rivers Authority', Research and Development Report 12, HMSO, London.

National Rivers Authority (1996) *River Habitats in England and Wales: A National Overview*, National Rivers Authority, Bristol.

National Water Commission (1973) *Water Policies for the Future*. Water Information Center, Port Washington, New York.

National Water Commission (1976) *We Didn't Wait for the Rain*. National Water Council, London.

National Water and Soil Conservation Authority (NWASCA) (NZ) (1987) *Farming the Hills: Mining or Sustaining the Resource?* Streamland 62, Wellington.

National Environment Research Council (NERC) (1975) *Flood Studies Report* (5 vols). Institute of Hydrology, Wallingford, Oxon.

Nature Conservancy Council (1986) *Nature Conservation and Afforestation in Britain*. Peterborough.

Nearing, M. A., Lane, L. J. and Lopes, V. L. (1994) 'Modelling soil erosion', in R. Lal (ed.) *Soil Erosion Research Methods*. Soil and Water Conservation Society, Ankeny, Iowa.

Nelson, D. (1979) 'A national watershed inventory', *Journal of Nepal Research Centre*, 2/3, 81–96.

Newbold, C., Honnor, J. and Buckley, K. (1989) *Nature Conservation and the Management of Drainage Channels*. Nature Conservancy Council, Peterborough.

Newson, M. D. (1979) *Hydrology: Measurement and Application*. Macmillan, Basingstoke.

Newson, M. D. (1980) 'The geomorphological effectiveness of floods: a contribution stimulated by two recent events in mid-Wales', *Earth Surface Processes, 5*, 1–16.

Newson, M. D. (1986) 'River basin engineering: fluvial geomorphology', *Journal of the Institution of Water Engineers and Scientists*, 40(4), 307–24.

Newson, M. D. (1988) 'Upland land use and land management: policy and research aspects of the effects on water', in J. M. Hooke (ed.) *Geomorphology in Environmental Planning*. Wiley, Chichester, 19–32.

Newson, M. D. (1989) 'Flood effectiveness in river basins: progress in Britain in a decade of drought', in K. Beven and P. Carling (eds) *Floods: Hydrological, Sedimentological and Geomorphological Implications*. Wiley, Chichester, UK, 151–69.

Newson, M. D. (1990) 'Forestry and water, "good practice" and UK catchment policy', *Land Use Policy*, 7(1), 53–8.

Newson, M. D. (1991) 'Catchment control and planning: emerging patterns of definition, policy and legislation in UK water management', *Land Use Policy*, 9(1), 9–15.

Newson, M. D. (1992a) *Geomorphic Thresholds in Gravel-bed Rivers: Refinements for an Era of Environmental Change. Gravel-bed Rivers*. Wiley, Chichester, UK, 3–20.

Newson, M. D. (1992b) 'River conservation and catchment management: UK perspective', in P. Boon, P. Calow and G. Petts (eds), *River Conservation and Management*. Wiley, Chichester, 385–96.

Newson, M. D. (1992c) *Land, Water and Development*. Routledge, London.

Newson, M. D. (1992d) 'Land and water: convergence, divergence and progress in the UK policy', *Land Use Policy*, 9(2), 111–21.

Newson, M. D. (1994a) 'Sustainable integrated development and the basin sediment system: guidance from fluvial geomorphology', in C. Kirby and W. R. White (eds) *Integrated River Basin Development*. Wiley, Chichester, 1–10.

Newson, M. D. (1994b) 'The roles and potential of development planning and catchment management planning in bringing about sustainable use of freshwater capacity', in N. Ward and G. Garrod (eds) *Water Quality: Understanding the Benefits and Meeting the Demands*. Centre for Rural Economy, Department of Agricultural Economics and Food Marketing, University of Newcastle upon Tyne, 85–104.

Newson, M. D. (1995a) 'Catchment-scale solute modelling in a management context', in S. T. Trudgill (ed.) *Solute Modelling in Catchment Ecosystems*. Wiley, Chichester, 445–60.

Newson, M. D. (1995b) 'Fluvial geomorphology and environmental design', in A. Gurnell and G. Petts (eds) *Changing River Channels*, Wiley, 413–32.

Newson, M. D. (1997a) (in press) 'Land, water and development: key themes driving international policy on catchment management', in *Proceedings of International Conference on Multiple Land Use and Catchment Management*. Macaulay Land Use Research Institute, Aberdeen.

Newson, M. D. (1997b) (in press) *Forests and Water: Land-use and Management Guidance in an Era of Catchment Planning*. Institute of Chartered Foresters, York.

Newson, M. D. and Calder, I. R. (1989) 'Forests and water resources: problems of prediction on a regional scale', *Philosophical Transactions Royal Society of London*, B324, 283–98.

Newson, M. D. and Ghazi, I. (1995) 'River basin management and planning in the Zayandeh Rud basin, Iran', University of Isfahan, Isfahan.

Newson, M. D. and Leeks, G. J. (1987) 'Transport processes at the catchment scale', in C. R. Thorne, J. C. Bathurst and R. D. Hey (eds) *Sediment Transport in Gravel-bed Rivers*. Wiley, Chichester, UK, 187–223.

Newson, M. D. and Lewin, J. (1991) 'Climatic change, river flow extremes and fluvial erosion: scenarios for England and Wales', *Progress in Physical Geography*, 15(1), 1–17.

Newson, M. D. and Robinson, M. (1983) 'Effects of agricultural drainage on upland streamflow: case studies in mid-Wales', *Journal of Environmental Management*, 17, 333–48.

Newson, M. D. and Sear, D. A. (1994) 'Sediment and gravel transportation in rivers including the use of gravel traps', Research and Development Note C5/384/2. National Rivers Authority, Bristol.

Newson, M. D., Marvin, S. and Slater, S. (1996) *Pooling our Resources: A Campaigner's Guide to Catchment Management Planning*. Council for the Protection of Rural England, London.

Obeng, L. E. (1978) 'Environmental impacts of four African impoundments', in C. G. Gunnerson and J. M. Kalbermatten (eds) *Environmental Impacts of International Civil Engineering Projects and Practices*. American Society of Civil Engineers, New York, 29–43.

O'Callaghan, J. R. (1995) 'NELUP: an introduction', *Journal of Environmental Planning and Management*, 38(1), 5–20.

Odemerho, F. O. and Sada, P. O. (1984) 'The role of urban surface characteristics on the extent of gullying in Auchi, Bendel State, Nigeria', *Applied Geography*, 4, 333–44.

Ohlsson, L. (ed.) (1995) *Hydropolitics: Conflict over Water as a Development Constraint*, Zed Books, London.

Okidi, C. O. (1990) 'History of the Nile and Lake Victoria Basins through treaties', in P. P. Howell and J. A. Allan (eds) *The Nile*. School of Oriental and African Studies/Royal Geographical Society, London, 193–224.

Olofin, E. A. (1984) 'Some effects of the Tiga Dam on valleyside erosion in downstream reaches of the River Kano', *Applied Geography*, 4, 321–32.

Organisation for Economic Co-operation and Development (1989) *Water Resource Management: Integrated Policies*. Paris.

O'Riordan, J. (1986) 'Some examples of land and water use planning in British Columbia, Canada', in F. T. Last, M. C. B. Hotz and B. G. Bell (eds) *Land and Its Uses – Actual and Potential*. Plenum, New York, 193–211.

O'Riordan, T. (1976) 'Policy making and environmental management: some thoughts on purposes and research issues', *Natural Resources Journal*, 16, 55–72.

O'Riordan, T. (1977) 'Environmental ideologies', *Environment and Planning*, Series A, 9, 3–14.

O'Riordan, T. (ed.) (1995) *Environmental Science for Environmental Management*, Longman, London.

O'Riordan, T. and More, R. J. (1969) 'Choice in water use', in R. J. Chorley (ed.) *Water, Earth and Man*. Methuen, London, 547–73.

O'Riordan, T. and Rayner, S. (1991) 'Risk management for global environmental change', *Global Environmental Change*, 1, 91–108.

Otago Catchment Board and Regional Water Board (1986) *Clutha, Kawarau and Hawea Rivers. Management Plan*. Dunedin, New Zealand.

Palmer, T. (1986) *Endangered Rivers and the Conservation Movement*. University of California Press, Berkeley, Cal.

Palutikov, J. P. (1987) *Some Possible Impacts of Greenhouse Gas Induced Climatic Change on Water Resources of England and Wales.* International Association of Scientific Hydrology Publication 168. Wallingford, Oxon, 585–96.

Park, C. C. (1977) 'World-wide variations in hydraulic geometry exponents of stream channels: an analysis and some observations', *Journal of Hydrology*, 35, 133–46.

Parker, R. (1976) *The Common Stream.* Paladin, London.

Patrick, R. (1992) *Surface Water Quality: Have the Laws been Successful?* Princeton University Press, Princeton, NJ.

Patterson, A. (1987) *Water and the State.* Geographical Paper no. 98. Department of Geography, University of Reading, Reading.

Paylore, P. and Greenwell, J. R. (1979) 'Fools rush in: pinpointing the arid zone', *Arid Lands Newsletter*, 10, 17–18.

Pearce, D. (ed.) (1993) *Blueprint 3: Measuring Sustainable Development.* Earthscan, London.

Pearce, F. (1994) 'High and dry in Aswan', *New Scientist*, 142, 28–32 (7 May).

Pearse, P. H., Bertrand, F. and Maclaren, J. W. (1985) *Currents of Change.* Environment Canada, Ottowa.

Pease, R. (1982) *The River Keeper: Caring for the Angler's Trout Stream.* David & Charles, Newton Abbot.

Peck, A. J. (1983) *Response of Groundwaters to Clearing in Western Australia.* Papers of the International Conference on Groundwater and Man, Australian Water Research Council conference series no. 8, Canberra.

Pereira, H. C. (1973) *Land Use and Water Resources.* Cambridge University Press, Cambridge.

Pereira, H. C. (1989) *Policy and Practice in the Management of Tropical Watersheds.* Belhaven, London.

Peters, E. (1992) 'Protecting the land under modern land claims agreements: the effectiveness of the environmental regime negotiated by the James Bay Cree in the James Bay and Northern Quebec Agreement', *Applied Geography*, 12, 133–45.

Petersen, R. C., Peterson, B.-M. and Lacoursire, J. (1992) 'A building-block model for stream restoration', in P. Boon, G. Petts and P. Calow (eds) *River Conservation and Management.* Wiley, Chichester, 293–309.

Petersen, R. C., Madsen, B. L., Wilzbach, M. A., Magadza, C. H. D., Paarlberg, A., Kullberg, A. and Cummins, K. W. (1987) 'Stream management: emerging global similarities', *Ambio*, 16(4), 166–79.

Peterson, D. H., Cayan, D. R., Dileo-Stephens, J. and Ross, T. G. (1987) *Some Effects of Climate Variability on Hydrology in Western North America.* International Association of Scientific Hydrology Publication 168, Wallingford, Oxon, 45–62.

Petts, G. E. (1979) 'Complex response of river channel morphology subsequent to reservoir construction', *Progress in Physical Geography*, 3(3), 329–62.

Petts, G. E. (1984) *Impounded Rivers: Perspectives for Ecological Management.* Wiley, Chichester.

Petts, G. E. (1987) 'Timescales for ecological change in regulated rivers', in J. Craig and J. B. Kemper (eds) *Regulated Streams: Advances in Ecology.* Plenum, New York, 257–66.

Petts, G. (1990) 'Water, engineering and landscape: development, protection and restoration', in D. Cosgrove and G. Petts (eds) *Water, Engineering and Landscape.* Belhaven, London.

Petts, G. E., Maddock, I., Bickerton, M. and Ferguson, A. J. D. (1995) 'Linking hydrology and ecology: the scientific basis for river management', in D. M.

Harper and A. J. D. Ferguson (eds) *The Ecological Basis for River Management.* Wiley, Chichester, 1–16.

Petts, G. E. and Thoms, M. C. (1987) 'Morphology and sedimentology of a tributary confluence bar in a regulated river', *Earth Surface Processes*, 12(4), 433–40.

Petts, G. E., Foulger, T. R., Gilvear, D. J., Pratts, J. D. and Thoms, M. C. (1985) 'Wave-movement and water quality variations during a controlled release from Kielder Reservoir, North Tyne River, UK', *Journal of Hydrology*, 80, 371–89.

Pezzey, J. (1989) *Definitions of Sustainability.* Discussion Paper 9, Centre for Economic and Environmental Development, London.

Pinay, A. and Decamps, H. (1988) 'The role of riparian woods in regulating nitrogen fluxes between the alluvial aquifer and surface water: a conceptual model', *Regulated Rivers: Research and Management*, 2, 507–16.

Pinay, G., Decamps, H., Chauvet, E. and Fustec, E. (1990) 'Functions of ecotones in fluvial systems', in R. J. Naiman and H. Decamps (eds) *The Ecology and Management of Aquatic-terrestrial Ecotones*, Parthenon, Carnforth, Lancs, 141–69.

Platt, R. H., Macinko, G. and Hammond, K. (1983) 'Federal environmental management: some land-use legacies of the 1970s', in J. W. House (ed.) *United States Public Policy: A Geographical Review.* Clarendon Press, Oxford, 125–66.

Playfair, J. (1802) *Illustrations of the Huttonian Theory of the Earth.* William Creech, Edinburgh.

Polhemus, Van Dyke (1988) 'Delaware River Basin Commission's river management role: the interface between water users and managing entities', in S. K. Majunder, E. W. Miller and L. E. Sage (eds) *Ecology and Restoration of the Delaware River Basin.* Pennsylvania Academy of Science, Easton, Penn., 312–22.

Polls, I. and Lanyon, R. (1980) 'Pollutant concentrations from homogenous land uses', *Journal of Environmental Engineering Division*, Proceedings of the American Society of Civil Engineers, 106, 69–80.

Poole, A. L. (1983) 'Catchment control in New Zealand', Water and Soil Miscellaneous Publication 48. Ministry of Works and Development, Wellington.

Pope, W. (1980) 'Impact of man in catchments (II) road and urbanisation', in A. M. Gower (ed.) *Water Quality in Catchment Ecosystems.* Wiley, Chichester, 73–112.

Popham, A. E. (1946) *The Drawings of Leonardo da Vinci.* Jonathan Cape, London.

Postel, S. (1992) *The Last Oasis: Facing Water Scarcity.* Earthscan, London.

Postle, M. (1993) *Development of Environmental Economics for the NRA.* National Rivers Authority, Research and Development Report 6. National Rivers Authority, Bristol.

Prime, R. (1992) *Hinduism and Ecology.* Cassell, London.

Prince, H. (1995) 'Floods in the upper Mississippi River basin, 1993: newspapers, official views and forgotten farmlands', *Area*, 27(2), 118–26.

Priscoli, J. D. (1989) 'Public involvement, conflict management: means to EQ and social objectives', *Journal of Water Resource Planning and Management*, Proceedings of the American Society of Civil Engineers, 115(1), 31–42.

Pryor, F. (1991) *Flag Fen: Prehistoric Fenland Centre.* Batsford, London.

Purseglove, J. (1988) *Taming the Flood.* Oxford University Press, Oxford.

Quinn, J. M. and Hickey, W. (1987) 'How well are we protecting the life in our rivers?', *Soil and Water*, 23(4), 7–12.

Raikes, R. (1967) *Water, Weather and Prehistory.* John Baker, London.

Ramsay, W. J. H. (1987) *Deforestation and Erosion in the Nepalese Himalaya: Is the Link Myth or Reality?* International Association of Hydrological Sciences Publication 167, Wallingford, Oxon, 239–50.

Rees, J. (1989) *Water Privatisation*. Research Papers, Department of Geography. London School of Economics, London.

Rees, J. A. (1969) *Industrial Demand for Water: A Study of South East England*. London School of Economics/Weidenfeld and Nicolson, London.

Reid, I. and Frostick, L. E. (1986) 'Dynamics of bedload transport in Turkey Brook', *Earth Surface Processes and Landforms*, 11, 143–55.

Reid, I. and Parkinson, R. J. (1984) 'The nature of the tile-drain outfall hydrograph in heavy clay soils', *Journal of Hydrology*, 72, 289–305.

Reisner, M. (1990) *Cadillac Desert: The American West and its Disappearing Water*. Secker & Warburg, London.

Renard, K. G., Laflen, J. M., Foster, G. R. and McCool, D. K. (1994) 'The revised Universal Soil Loss Equation', in R. Lal (ed.) *Soil Erosion Research Methods*. Soil and Water Conservation Society, Ankeny, Iowa, 105–24.

Rennison, R. W. (1979) *Water to Tyneside*. Newcastle and Gateshead Water Co., Newcastle upon Tyne.

Reynolds, E. R. C. and Leyton, L. (1967) 'Research data for forest policy: the purpose, methods and progress of forest hydrology', *Proceedings of the 9th British Commonwealth Forestry Conference*, University of Oxford.

Richards, K. S. (1982) 'Channel adjustment to sediment pollution by the china clay industry in Cornwall, England', in D. D. Rhodes and G. P. Williams (eds) *Adjustments of the Fluvial System*. Allen & Unwin, London, 309–31.

Richardson, J. J., Jordan, A. G. and Kimber, R. H. (1978) 'Lobbying, administrative reform and policy style: the case for land drainage', *Political Studies*, 26(1), 47–64.

Riebsame, W. E. (1992) 'Social constraints on adjusting water resources management to anthropogenic climate change', in M. A. Abu-Zeid and A. K. Biswas (eds) *Climatic Fluctuations and Water Management*. Butterworth-Heinemann, Oxford, 210–26.

RIZA (National Institute for Inland Water Management and Wastewater Treatment) (1991) *Sustainable Use of Groundwater: Problems and Threats in the European Communities*. Lelystad, Netherlands.

Rizzo, B. (1988) *The Sensitivity of Canada's Ecosystems to Climatic Change*. Newsletter 17, Canada Committee on Ecological Land Clarification, Environment Canada, Ottawa, 10–15.

Roberts, C. R. (1989) 'Flood frequency and urban-induced channel change: some British examples', in K. Beven and P. Carling (eds) *Floods: Hydrological, Sedimentological and Geomorphological Implications*. Wiley, Chichester, UK, 57–82.

Roberts, G. and Marsh, T. (1987) *The Effects of Agricultural Practices on the Nitrate Concentrations in the Surface Water Domestic Supply Sources of Western Europe*. International Association of Hydrological Sciences, Wallingford, Oxon, Publication 164.

Roberts, J. (1983) 'Forest transpiration: a conservative hydrological process?', *Journal of Hydrology*, 66, 133–41.

Robinson, M. and Beven. K. J. (1983) 'The effect of mole drainage on the hydrological response of a swelling clay soil', *Journal of Hydrology, 64*, 205–23.

Robinson, M., Ryder, E. L. and Ward, R. C. (1985) 'Influence on streamflow of field drainage in a small agricultural catchment', *Agricultural Water Management*, 10, 145–58.

Robinson, N. A. (1987) 'Marshalling environmental law to resolve the Himalaya–Ganges problem', *Mountain Research and Development*, 7(3), 305–15.

Rogers, P. (1992) 'Integrated urban resources management', in *International Conference on Water and the Environment, Dublin, Ireland*. Geneva, World Meteorological Organization, 7.7–7.39.

Rogers, P. (1993) *America's Water: Federal Roles and Responsibilities*. MIT Press, Harvard, Mass.

Rolt, L. T. C. (1985) *Navigable Waters*. Penguin Books, Harmondsworth.

Roome, N. (1984) 'A better future for the uplands: a planning critique', *Planning Outlook*, 27(1), 12–17.

Rose, A. (ed.) (1992) *Judaism and Ecology*. Cassell, London, 142pp.

Rosgen, D. L. (1994) 'A classification of natural rivers', *Catena*, 22, 169–99.

Ross, S. M., Thornes, J. B. and Nortcliff, S. (1990) 'Soil hydrology nutrient and erosional response to the clearance of terra firme forest, Maraca Island, Roraima, Northern Brazil', *Geographical Journal*, 156(3), 267–82.

Ross, T. A. (1989) 'Drought in the US 1987–8', US *Geological Survey Yearbook 1988*. US Government Printing Office, Washington DC, 24–7.

Rowley, G. (1990) 'The West Bank: native water-resource systems and competition', *Political Geography Quarterly*, 9(1), 39–52.

Rowntree, K. (1990) 'Political and administrative constraints on integrated river basin development: an evaluation of the Jana and Athi Rivers Development Authority, Kenya', *Applied Geography*, 10, 21–41.

Rowntree, P. R., Murphy, J. M. and Mitchell, J. F. B. (1993) 'Climate change and future rainfall predictions', *Journal of the Institution of Water and Environmental Management*, 7, 464–70.

Royal Commission on Environmental Pollution (1988) *12th Report, Best Practicable Environmental Option*. HMSO, London.

Royal Society for the Protection of Birds (1995) *Water Wise: The RSPB's Proposals for Using Water Wisely*. RSPB, Sandy, Beds.

Ryckborst, H. (1980) 'Geomorphological changes after river meander surgery', *Geologie in Mijnbouw*, 59(2), 121–8.

Rydzewski, J. R. (199) 'Irrigation: a viable development strategy?', *Geographical Journal*, 156(2), 175–80.

Sagoff, M. (1989) *The Economy of the Earth*, Cambridge University Press, Cambridge.

Saha, S. K. and Barrow, C. J. (eds) (1981) *River Basin Planning, Theory and Practice*. Wiley, Chichester.

Said, R. (1993) *The River Nile: Geology, Hydrology and Utilization*. Pergamon, Oxford.

Salau, A. T. (1990) 'Integrated water management: the Nigerian experience', in B. Mitchell (ed.) *Integrated Water Management*. Belhaven, London, 188–202.

Samir, A. (1990) 'Principles and precedents in international law governing the sharing of Nile waters', in P. P. Howell and J. A. Allan (eds) *The Nile*. School of Oriental and African Studies/Royal Geographical Society, London, 225–38.

Satterlund, D. R. and Adams, P. W. (1992) *Wildland watershed management*. Wiley, New York.

Saunders, I. and Young, A. (1983) 'Rates of surface processes on slopes, slope retreat and denudation', *Earth Surface Processes and Landforms*, 8, 473–501.

Saunders, P. (1985) 'The forgotten dimension of central-local relations: theorising the "regional state"', *Environment and Planning*, C, 3, 149–62.

Schick, A. P. (1995) 'Fluvial processes on an urbanizing alluvial fan: Eilat, Israel', in J. E. Costa, A. J. Miller, K. W. Potter and P. R. Wilcock (eds) *Natural and Anthropogenic Influences in Fluvial Geomorphology*. American Geophysical Union, Geophysical Monographs 89. Washington DC, 209–18.

Schramm, G. (1980) 'Integrated river basin planning in a holistic universe', *Natural Resources Journal*, 20(4), 787–806.

Schumacher, E. F. (1973) *Small is Beautiful*. Abacus, London.

Schumm, S. A. (1963) 'A tentative classification of alluvial river channels', US *Geological Survey*, Circular 477, Menlo Park, Cal.

Schumm, S. A. (1969) 'River metamorphosis', *Journal of Hydraulics Division*, American Society of Civil Engineers, 95, 255–73.

Schumm, S. A. (1977) *The Fluvial System*. Wiley, New York.

Schumm, S. A. (1985) 'Patterns of alluvial rivers', *Annual Review of Earth/Planet Sciences*, 13, 5–27.

Schumm, S. A. and Lichty, R. W. (1965) 'Time, space and causality in geomorphology', *American Journal of Science*, 263, 110–19.

Schumm, S. A., Harvey, M. D. and Watson, C. C. (1984) *Incised Channels: Morphology, Dynamics and Control*. Water Resources Publications, Littleton, Col.

Scott, D. (1993) 'New Zealand's Resource Management Act and fresh water', *Aquatic Conservation: Marine and Freshwater Ecosystems*, 3, 53–65.

Scottish Development Department (1990) *First Policy Review of the River Purification Boards*. Edinburgh.

Sear, D. A. (1992) 'Impact of hydroelectric power releases on sediment transport processes in pool-riffle sequences', in P. Billi, R. D. Hey, C. R. Thorne and P. Tacconi (eds) *Dynamics of Gravel-bed Rivers*. Wiley, Chichester, 630–50.

Sear, D. A. (1994) 'River restoration and geomorphology', *Aquatic Conservation: Marine and Freshwater Ecosystems*, 4, 169–77.

Sear, D. A. and Newson, M. D. (1994) 'Sediment and gravel transportation in rivers: a geomorphological approach to river maintenance'. Research and Development Note 315. National Rivers Authority, Bristol.

Sear, D. A., Newson, M. D. and Brookes, A. (1995) 'Sediment-related river maintenance: the role of fluvial geomorphology', *Earth Surface Processes and Landforms*, 20, 629–47.

Sewell, W. R. D., Handmer, J. W. and Smith, D. I. (eds) (1985) *Water Planning in Australia. From Myths to Reality*. CRES, Australian National University, Canberra.

Sewell, W. R. D., Smith, D. I. and Handmer, J. W. (1985) 'From myths to reality: the evolution of Australian water planning', in W. R. D. Sewell, J. W. Handmer and D. I. Smith (eds) (1985a) *Water Planning in Australia. From Myths to Reality*. CRES, Australian National University, Canberra, 227–63.

Seymour, J. and Girardet, H. (1986) *Far from Paradise*. BBC Books, London.

Shah, T. (1993) *Groundwater markets and irrigation development. Political economy and practical policy*. Oxford University Press, Bombay.

Sharma, C. K. (1987) 'The problem of sediment load in the development of water resources in Nepal', *Mountain Research and Development*, 7(3), 316–18.

Shaxon, T. F., Hudson, N. W., Sanders, D. W., Roose, E. and Moldehauer, W. C. (1989) *Land Husbandry: A Framework for Soil and Water Conservation*. Soil and Water Conservation Society, Ankeny, Iowa.

Sheail, J. (1988) 'River regulation in the United Kingdom: an historical perspective', *Regulated Rivers: Research and Management*, 2, 221–32.

Shearer, D. M. (1978) 'Water resource development in upland Britain', in R. B. Tranter (ed.) *The Future of Upland Britain*. Centre for Agricultural Strategy, University of Reading, Reading, 294–306.

Shiklomanov, J. A. (1989) 'Climate and water resources', *Hydrological Sciences Journal*, 34(5), 495–529.

Shuman, J. R. (1995) 'Environmental considerations for assessing dam removal alternatives for river restoration', *Regulated Rivers: Research and Management*, 11, 249–61.

Shuval, H. I. (1987) 'The development of water re-use in Israel', *Ambio*, 16(4), 186–90.

Simpson, R. W. (1991) 'The international scene: the involvement of British hydrogeologists', in R. A. Downing and W. B. Wilkinson (eds) *Applied Groundwater Hydrology: A British Perspective*. Clarendon Press, Oxford, 314–31.

Sivanappan, R. K. (1995) 'Soil and water management in the dry lands of India', *Land Use Policy*, 12(2), 165–75.

Slater, S., Newson, M. D. and Marvin, S. J. (1995) 'Land use planning and the water sector: a review of development plans and catchment management plans', *Town Planning Review*, 65(4), 375–97.

Slaymaker, O. (1982) 'Land use effects on sediment yield and quality', *Hydrobiologia*, 91–2, 93–109.

Smith, C. T. (1969) 'The drainage, basin as an historical basis for human activity', in R. J. Chorley (ed.), *Water, Earth and Man*. Methuen, London, 101–10.

Smith, D. A. (1987) 'Water quality indices for use in New Zealand's rivers and streams', Water Quality Centre Publication 12, Ministry of Works and Development, Hamilton.

Smith, D. I. (1981) 'Actual and potential flood damage: a case study for urban Lismore, New South Wales', *Applied Geography*, 1, 31–9.

Smith, D. I. (1990) 'The worthwhileness of dam failure mitigation: an Australian example', *Applied Geography*, 10, 5–19.

Smith, D. I. and Finlayson, B. (1988) 'Water in Australia: its role in environmental degradation', in R. L. Heathcote and J. A. Mabbutt (eds) *Land, Water and People: Geographical Essays in Australian Resource Management*. Allen & Unwin, Sydney.

Smith, D. I. and Handmer, J. (eds) (1989) *Flood Insurance and Relief in Australia*. CRES, Australian National University, Canberra.

Smith, N. (1972) *A History of Dams*. Citadel Press, Secaucus, NJ.

Smith, T. R. (1974) 'A derivation of the hydraulic geometry of steady-state channels from conservation principles and sediment transport laws', *Journal of Geology*, 82, 98–104.

Smith, W. (1993) *River God*. Macmillan, London.

Snyder, G. (1980) 'Evaluating silvicultural impacts on water resources', *Watershed Management 1980*. American Society of Civil Engineers, New York, 682–93.

Solbé, J. F. de L. G. (ed.) (1986) *Effects of Land Use on Fresh Waters: Agriculture, Forestry, Mineral Exploitation, Urbanisation*. Ellis Horwood, Chichester.

Solley, W. B. (1989) 'Reflections on water use in the United States', US *Geological Survey 1988 Yearbook*, Denver, Col., 28–30.

Solomon, S. I., Beran, M. and Hogg, W. (eds) (1987) *The Influence of Climate Change and Climatic Variability on the Hydrologic Regime and Water Resources*. International Association of Scientific Hydrology Publication 168, Wallingford, Oxon.

Sommers, L. W. and Lounsbury, J. W. (1988) 'Boomtown growth issues along the Colorado River border', *Land Use Policy*, 2(4), 385–93.

Soons, J. M. (1986) 'Erosion rates in a superhumid environment', in V. Gardiner (ed.) *International Geomorphology*, Vol. 1. Wiley, Chichester, 885–96.

Steinberg, T. (1991) *Nature Incorporated: Industrialization and the Waters of New England*. Cambridge Univ. Press, Cambridge.

Stephens, H. G. and Shoemaker, E. M. (1987) *In the Footsteps of John Wesley Powell*. Johnson Books, Boulder, Col.

Stern, P. (1979) *Small-scale Irrigation*. Intermediate Technology Publications, London.

Stiles, D. (1995) *Social Aspects of Sustainable Dryland Management*. Wiley, Chichester.

Stoffberg, F. A., van Zyl, F. C. and Middleton, B. J. (1994) 'The role of integrated catchment studies in the management of water resources in South Africa', in W. R. White and C. Kirby (eds) *Integrated River Basin Development*. Wiley, Chichester, 455–62.

Stocking, M. (1987) *Environmental Crises in Developing Countries: How Much, for Whom and by Whom?* Developing Areas Research Group, Institute of British Geographers, Swansea.

Stokes, E. (1992) 'The Treaty of Waitangi and the Waitangi Tribunal: Maori claims in New Zealand', *Applied Geography*, 12, 176–91.

Stone, P. B. (1992) *The State of the World's Mountains. A Global Report*. Zed Books, London.

Stoner, R. F. (1990) 'Further irrigation planning in Egypt', in P. P. Howell and J. A. Allan (eds) *The Nile*. School of Oriental and African Studies/Royal Geographical Society, London, 83–105.

Strahler, A. N. (1957) 'Quantitative analysis of watershed geomorphology', *Transactions of the American Geophysical Union*, 38, 913–20.

Sutcliffe, J. V. (1974) 'A hydrological study of the Southern Sudd region of the Upper Nile', *Hydrological Sciences Bulletin*, 19, 237–55.

Sutcliffe, J. V. and Knott, D. G. (1987) *Historical Variations in African Water Resources*. International Association of Scientific Hydrology Publication 168, Wallingford, Oxon, 463–75.

Sutcliffe, J. V. and Parks, Y. P. (1987) 'Hydrological modelling of the Sudd and Jonglei Canal', *Hydrological Sciences Journal*, 32(2), 143–59.

Swanson, R. H., Bernier, P. Y. and Woodard, P. D. (eds) (1987) *Forest Hydrology and Watershed Management*. International Association of Hydrological Sciences Publication 167, Proceedings of the Vancouver Symposium. Wallingford, Oxon.

Synnott, M. (1991) 'Market mechanisms and the resolution of conflict in water management', in J. W. Handmer, A. H. J. Dorcey and D. I. Smith (eds) *Negotiating Water: Conflict Resolution in Australian Water Management*. Canberra: Centre for Resource and Environmental Studies, Australian National University, 92–114.

Tanner, T. (ed.) (1987) *Aldo Leopold: The Man and His Legacy*. Soil Conservation Society of America, Ankeny, Iowa.

Taylor, A. and Patrick, M. (1987) 'Looking at water through different eyes: the Maori perspective', *Soil and Water*, 23(4), 22–4.

Tejwani, K. G. (1993) 'Water management issues: population, agriculture and forests – a focus on watershed management', in M. Bonnell, M. M. Hufschmidt and J. S. Gladwell (eds) *Hydrology and Water Management in the Humid Tropics*. Cambridge University Press, Cambridge, 496–525.

Tewari, A. K. (1988) 'Revival of water harvesting methods in the Indian Desert', *Arid Lands Newsletter*, 26, 3–8.

Thames Region, National Rivers Authority (1995) *Thames 21: A Planning Perspective and a Sustainable Strategy for the Thames Region*. Thames Region, NRA, Reading.

Thomas, D. S. G. and Middleton, N. J. (1995) *Desertification: Exploding the Myth*. Wiley, Chichester, 194pp.

Thompson, M. and Warburton, M. (1985) 'Uncertainty on a Himalayan scale', *Mountain Research and Development*, 5(2), 115–35.

Thompson, P. M. and Sultana, P. (1996) 'Distributional and social impact of flood control in Bangladesh', *Geographical Journal*, 162(1), 1–13.

Thorne, C. R. and Lewin, J. (1982) 'Bank process, bed material movement and planform development in a meandering river', in D. D. Rhodes and G. P. Williams (eds) *Adjustments of the Fluvial System*. Allen & Unwin, London, 117–37.

Tollan, A. (1992) 'The ecosystem approach to water management', *World Meteorological Organisation Bulletin*, 41(1), 28–34.

Tomkins, S. C. (1986) *The Theft of the Hills: Afforestation in Scotland*. Ramblers' Association, London.

Toynbee, A. (1976) *Mankind and Mother Earth*. Oxford University Press, Oxford.

Tranter, R. B. (ed.) (1978) *The Future of Upland Britain* (2 vols). Centre for Agricultural Strategy, University of Reading, Reading.

Trudgill, S. T. (1986) 'Introduction', in S. T. Trudgill (ed.) *Solute Processes*. Wiley, Chichester, 1–14.

Turner, B. (1994) 'Small-scale irrigation in developing countries', *Land Use Policy*, 11(4), 251–61.

Turner, R. K. (1993) *Sustainable Environmental Economies and Management: Principles and Practice*. Belhaven, London.

Tyler, S. (1987) 'River birds and acid water', *RSPB Conservation Review*, 1, 68–70.

UN Commission on Desertification (1977) Desertification: Its Causes and Consequences. Pergamon, Nairobi.

Ungate, C. D. (1996) 'Tennessee Valley Authority's Clean Water Initiative: building partnerships for watershed improvement', *Journal of Environmental Planning and Management*, 39(10), 113–22.

United Nations (1990) *Global Outlook 2000: Economic, Social, Environmental*. New York.

United Nations Economic Commission for Europe (1995) *Protection and Sustainable Use of Waters: Recommendations for ECE governments*. United Nations, New York.

United Nations Environment Programme/UNESCO (1990) *The Impact of Large Water Projects on the Environment*. UNESCO, Paris.

United States Environmental Protection Agency (1986) *Water Quality Program Highlights: Arkansas Ecoregion Program*. Washington DC.

United States Environmental Protection Agency (1990) *Ground Water. Contamination and Methodology*. Technomic, Lancaster, Penn.

US National Water Commission (1973) *Water Policies for the Future*. Water Information Center, Port Washington, New York.

Vallentine, H. R. (1967) *Water in the Service of Man*. Penguin Books, Harmondsworth.

Van Dijk, G. M., Marteija, E. C. L. and Schulte-Wülwer-Leidig, A. (1995) 'Ecological rehabilitation of the River Rhine: plans, progress and perspective', *Regulated Rivers: Research and Management*, 11, 377–88.

van Dyke, J. A. (1995) *Taking the Waters: Soil and Water Conservation among Settling Beja Nomads in Eastern Sudan*. Avebury/African Studies Centre, London.

van Niekerk, A. W. and Heritage, G. L. (1994) 'The use of GIS techniques to evaluate sedimentation patterns in a bedrock controlled channel in a semi-arid region', in W. R. White and C. Kirby (eds) *Integrated River Basin Development*. Wiley Chichester, 257–69.

Vannote, R. L., Minshall, G. W., Cummins, K. W., Sedell, J. R. and Cushing, C. E. (1980) 'The river continuum concept', *Canadian Journal of Fish and Aquatic Sciences*, 37, 130–7.

Veit, P. G., Mascarenhas, A. and Ampadu-Agyei, O. (1995) *Lessons from the Ground Up: African Development that Works*, World Resources Institute, Washington DC.

Vischer, D. (1989) 'Impact of 18th and 19th century river training works: three case studies from Switzerland', in G. E. Petts (ed.) *Historical Change of Large Alluvial Rivers: Western Europe*. Wiley, Chichester, 19–40.

de Voto, B. (1953) *The Journals of Lewis and Clark*. Houghton Mifflin, Boston, Mass.

Wadeson, R. A. (1994) 'A geomorphological approach to the identification and classification of instream flow environments', *South African Journal of Aquatic Sciences*, 20(1), 1–24.

Waggoners, P. E. (ed.) (1990) *Climate Change and US Water Resources*. New York, Wiley.

Walling, D. E. (1977) 'Assessing the accuracy of suspended sediment rating curves for a small basin', *Water Resources Research*, 13(3), 531–8.

Walling, D. E. (1979) 'The hydrological impact of building activity: a study near Exeter', in G. E. Hollis (ed.) *Man's Impact on the Hydrological Cycle in the UK*. Geo Books, Norwich, 135–51.

Walling, D. E. (1983) 'The sediment delivery problem', *Journal of Hydrology*, 65, 209–37.

Walling, D. E. (1990) 'Linking the field to the river: sediment delivery from agricultural land', in J. Boardman, I. D. L. Foster and J. A. Dearing (eds) *Soil Erosion on Agricultural Land*. Wiley, Chichester, 129–52.

Walling, D. E. (1994) 'Measuring sediment yield from river basins', in R. Lal (ed.) *Soil Erosion Research Methods*. Ankeny, Iowa, Soil and Water Conservation Society.

Walling, D. E. and Webb, B. (1986) 'Solutes in river systems', in S. T. Trudgill (ed.) *Solute Processes*. Wiley, Chichester, 281–327.

Walling, D. E., Foster, S. D. D. and Wurzel, P. (eds) (1984) *Challenges in African Hydrology and Water Resources*. International Association of Hydrological Sciences Publication 144, Proceedings of the Harare Symposium.

Ward, D., Holmes, N. and Jose, P. (eds) (1994) *The New Rivers and Wildlife Handbook*. RSPB, Sandy, UK.

Ward, J. V. and Stanford, J. A. (1983) 'The serial discontinuity concept of lotic ecosystems', in T. D. Fountain and S. M. Bartell (eds) *Dynamics of Lotic Ecosystems*. Ann Arbor Science Publishers, Ann Arbor, Mich., 29–42.

Ward, R. C. (1967) *Principles of Hydrology*. McGraw Hill, New York.

Ward, R. C. (1971) *Small Watershed Experiments: An Appraisal of Concepts and Research Developments*. Occasional Papers in Geography 18, University of Hull.

Ward, R. C. (1982) *The Fountains of the Deep and the Windows of Heaven*. University of Hull, Hull.

Ward, T. (ed) (1995) *Watershed Management: Planning for the 21st Century*. American Society of Civil Engineers, New York.

Warren, D. M. and Rajasekaran, B. (1995) 'Using indigenous knowledge for sustainable dryland management: a global perspective', in D. Stiles (ed.) *Social Aspects of Sustainable Dryland Management*. Wiley, Chichester, 193–209.

Watson, C. (1980) 'Watershed management: the California experience', *Watershed Management 1980*. American Society of Civil Engineers, New York, 1048–59.

Welbank, M. (1978) 'Irrigation and people', in C. G. Gunnerson and J. M. Kalbermatten (eds) *Environmental Impacts of International Civil Engineering Projects and Practices*. American Society of Civil Engineers, New York, 44–70.

Wertz, A. (1982) 'Integration of land and water management: political, administrative and planning problems', in P. Laconte and Y. Y. Haimes (eds) *Water Resources and Land-Use Planning: A Systems approach*. Nijhoff, The Hague, 283–95.

Westcoat, J. L. (1991) 'Managing the Indus River basin in the light of climate change', *Global Environmental Change*, 1(5), 381–95.

Weyman, D. R. (1975) *Runoff Processes and Streamflow Modelling*. Oxford University Press, Oxford.

Wheeler, B. D., Shaw, S. C., Foyt, W. J. and Robertson, R. A. (1995) *Restoration of Temperate Wetlands*. Wiley, Chichester.

Whipple, W. Jr and Van Abs, D. J. (1990) 'Principles of a ground-water strategy', *Journal of Water Resources Planning and Management*, American Society of Civil Engineers, 116(4), 503–16.

White, W. R. (1982) *Sedimentation Problems in River Basins: Studies and Reports in Hydrology*. UNESCO, Paris.

Whitlow, J. R. and Gregory, K. J. (1989) 'Changes in urban stream channels in Zimbabwe', *Regulated Rivers: Research and Management*, 4, 27–42.

Whittenmore, C. (1981) *Land for People: Land Tenure and the Very Poor*. Oxfam, Oxford.

Wiersum, K. F. (1984) 'Surface erosion under various tropical agroforestry systems', in C. O'Loughlin and A. Pearce (eds) *Effects of Forest Land Use on Erosion and Slope Stability*. East–West Center, Honolulu, 231–9.

Wilhite, D. A. (1993) *Drought Assessment, Management and Planning: Theory and Case Studies*. Kluwer, Boston, Mass.

Wilhite, D. A. (1993) 'Planning for drought: a methodology', in D. A. Wilhite (ed.) *Drought Assessment, Management and Planning: Theory and Case Studies*. Kluwer, Boston, Mass., 87–108.

Winid, B. (1981) 'Comments on the development of the Awash Valley, Ethiopia', in S. K. Saha and C. J. Barrow (eds) *River Basin Planning, Theory and Practice*. Wiley, Chichester, 147–65.

Winpenny, J. T. (1991) *Values for the Environment: A Guide to Economic Appraisal*. HMSO, London.

Winpenny, J. T. (1994) *Managing Water as an Economic Resource*. Routledge, London.

Wischmeier, W. H. and Smith, D. D. (1965) *Predicting Rainfall Erosion from Cropland East of the Rocky Mountains*. Agriculture Handbook no. 282. United States Department of Agriculture, Washington DC.

Wisdom, A. S. (1979) *The Law of Rivers and Watercourses*, Shaw & Sons, London.

Witter, S. G. and Carrasco, D. A. (1996) 'Water quality: a development bomb waiting to explode', *Ambio*, 25(3), 199–204.

Wittfogel, K. A. (1956) 'The hydraulic civilisations', in W. L. Thomas (ed.) *Man's Role in Changing the Face of the Earth*. University of Chicago Press, Chicago, Ill., 152–64.

Wittfogel, K. A. (1957) *Oriental Despotism*. Yale University Press, New Haven, Conn. and London.

Wolman, M. G. (1967) 'A cycle of sedimentation and erosion in urban river channels', *Geografiska Annale*, 49A, 385–95.

Woolhouse, C. H. (1989) *Managing the Effects of Urbanisation in the Upper Lee*. 2nd National Hydrology Symposium, British Hydrological Society, Wallingford, Oxon, 2.9–2.17.

World Bank (1993) *Water resources management*, policy paper. Washington, DC.

World Commission on Environment and Development (1987) *Our Common Future*. Oxford University Press, Oxford.

World Health Organization (1991) *Surface Water Drainage for Low-income Communities*. Geneva.

World Resources Institute and United Nations Development Programme (1992) *World Resources 1992–93*. Oxford University Press, New York.

Worster, D. (1992) *Rivers of Empire: Water, Aridity and the Growth of the American West*. Oxford University Press, Oxford.

Young, G. J., Dooge, J. C. I. and Rodda, J. C. (1994) *Global Water Resource Issues*. Cambridge University Press, Cambridge.

Youngman, R. E. and Lack, J. (1981) 'New problems with upland waters', *Water Services*. 85, 13–14.

Zazueta, A. (1995) *Policy Hits the Ground: Participation and Equity in Environmental Policy-making*. World Resources Institute, Washington, DC.

Ziman, J. R. (1984) *An Introduction to Science Studies*. Cambridge University Press, Cambridge.

Zon, R. (1912) *Forests and Water in the Light of Scientific Investigation*. US Senate, Document 469.

Name index

References to figures and tables are shown in italic.

Subject index

References to figures and tables are shown in italic.